THIS IS AN OPEN LETTER ON GRAVITY PART 2

Asking the Questions That Science Never Asks Because Science can never Answer such Questions when Asked

ISBN-13: 978-1505661071

ISBN-10: 1505661072

WRITTEN BY PEET SCHUTTE

© KOSMOLOGIESE EN ASTRONOMIESE TEGNIKA

All rights and entitlement to copyrights in this book are reserved. No part, parts or the entirety of this book may be reproduced by publishing, electronically copied, duplicated by whatever means that form reproduction or duplication, without the prior written consent of the copyright owner.

THIS LETTER ANNOUNCES FOUR MORE BOOKS, WHICH IS A CULMINATION OF LETTERS I WROTE IN SIX YEARS TRYING TO ALERT PHYSICS ACADEMICS, INFORMING THOSE CONCERNED ABOUT A MISTAKE IN NEWTON'S APPROACH TO SCIENCE CONCERING GRAVITY, WHICH THEN BECAME MY CONCERN. *An open letter ON GRAVITY* IS ALSO THE NOTIFYING OF ANOTHER BOOK ENTITLED *MATTER'S TIME IN SPACE: THE THESIS* vol. 1 to 7 I.S.B.N. 0-9584410-8-1.

Before attempting any further reading, please give heed to the following exceptionally serious and urgent warning about the extensive deliberateness and the unacceptable behaviour that may amount to displaying possible petty provocation on the part of the author concerning the accepted norms that form the foundation of civilisation.

To whom it may concern:
My introduction as well as introducing the readers to general cosmology in a very brief and compressed manner but first, I have to give the emphatic warning to all prospective contemplating readers.

Please take note of a conscientious warning about the gravity of the misgiving there is on the part of the most respected Academics in physics about a much concerning matter. Where the stating declares the possibility that the content in this book has been (written by…) then don't take the announcing Written By Peet Schutte (Petrus S. J. Schutte) very seriously for there are grievous doubts leaving considerable dispute about the possibility, which underwrites the authenticity of Peet Schutte achieving the (written by…Peet Schutte) status. Please take note of the following dehortation. In the light of the reference to me serving in the capacity as being responsible for authoring, (written by…) in line of keeping fairness and justice to members of society, where all civil beings should carry reputed honesty, then: Please be warned before any reader starts reading about the following extremely serious admonition: I am bound by my conscience to warn all intended readers that I am placed under caution by the Academics in Physics. Those most esteemed members responsible for the guardianship and maintaining the ethos in physics are of the opinion that I, Peet Schutte am unable to write any book on the science of Physics as well as Astrophysics. Therefore I, Peet Schutte, must declare that I should be considered as not very able to write anything, because I am incapable thereof. I suppose, I merely generate new information, which I establish as thoughts and then gather as concepts. I further collect the result as words, which I put on paper using alphabetic symbols. I then compile that in a format that others may confuse with a book, but a book it cannot be, since the Masters in science found me unable to write a book. Please do not allow me to fool you, for this then cannot be, or represent a book. Now I have done my duty in warning everyone and in that, I denounce further participating with any purposive intention to wilfully bring down the crux of civilization by acting unacceptable and irresponsible.

I didn't write a book since I am not schooled to do so. It is my guess that I merely generated uninformed thoughts, which I collected as alphabetic symbols and plotted that in ink on paper. This effort I achieved from harbouring my delusional ideas spawned by a dehumanised brain. It only proves my weak and under developed mentality, due to my lack of an informed insight that is a typical symptom that all those have that is suffering from a disadvantaged past that one can only have when the person obviously lacks formal education. While you are reading the letter deciding to regard or dismiss my work, then also please keep in mind when reading my language used and also please give credit where it belongs…if you do find linguistically improper use of words or misspelling, then remember that I am a feeble minded motor mechanic and not a literal giant. I do find much pride in my status as being Afrikaner and would like to have my names used by pronouncing it in the manner Afrikaans dictates…therefore I would sincerely appreciate the courtesy when readers will take note that my name and last name are pronounced in Afrikaans, which is originally from Dutch and must be pronounced that way. **Peet** one would pronounce "here" which is the closest English to the pronouncing of the "ee". The "Sch" in Schutte is pronounced exactly as school is where both actually are pronounced Skutte or "skool". By pronouncing my name in Afrikaans you do me the utmost courtesy any one can. Being an Afrikaner and a Boer is what I am most proud of.

An Open Letter On Gravity Part 2

To Whom It May Concern,

I am Petrus Stephanus Jacobus Schutte going by the name of Peet and who is the author of the above-mentioned book(s). I hope you find your reading of this book presented as an open letter a most fruitful experience. I feel I need to warn you whom are reading this letter that this work contained in this letter strays widely from mainstream science and for that there is a very good reason but I should add that in the least it is thought provoking. I researched the work of a man that is most exceptional and even more prominent in the history of mankind. His role in the gathering of information furthering knowledge accumulating of the human species' efforts stands second to none while most of everyone is not even aware of the full implication of his work. While recognising his work Mainstream science bluntly ignores his work and in that they miss the full vastness of the wide influencing of his work. It is therefore almost absolutely realistic to say that what you are about to read in this open letter sent to you for your attention was never yet printed in the near or the far past although the work has been with us for about four hundred years during which time it went unnoticed. It seems to me that any research predating Newton never came into use or in practise. My investigation of Kepler's work brought about a conclusion that no one yet arrived at concerning the findings of Kepler because no one scrutinised Kepler's formula. Kepler found planets rotating around a centre but Newton saw a circle and added what is mathematically required to indicate such a circle. Newton added a mathematical $4\Pi^2$ to the formula of Kepler and removed the distance symbolising measure that Kepler introduced using k. On the other side Newton changed the symbol of k by using the symbols $G (m + m_p)$. This is just a longer and probably a more detailed manner of indicating k and better defining of k but it symbolises precisely to the point what k stands for nonetheless.

Has any one reading this ever knocked on a door to find no one home...quite frustrating experience that turns out to be. Has any one reading this called on another person and knew the person was home, but for some unknown reason the person would not answer the knocking? I have to ask there questions as to determine how informed are you on feeling frustrations. Have you driven for hours to go visit someone just to find no one is answering your ringing their front door bell and the only reply to your ringing the front door bell is the sound of vacant no replying? All you hear in return to all the ringing you do is the hollow emptiness of silence...? You have been on the road for hours and now that you have arrived, being tired, exhausted but also being relieved to be at your destination and for all the good that does to you is just to find that all those inside pretend not to be home, embarrassing wouldn't you say? This is what I have been through for the largest part of seven years.

Here I am busy complaining while you are reading but that is just about what I have been doing for seven years non stop...complaining about not finding an audience to listen to my plight. I hope you find your reading of this book presented as an open letter a most fruitful experience. I feel I need to warn you, the person reading this letter, that the work contained herein strays widely from mainstream science and for that there is a very good reason. However, in the least, the content is thought provoking.

An Open Letter On Gravity Part 2

I am bringing my case to you to find honest judgement on evidence I wish to introduce as evidence with the introduction I am about to do.

I researched the work of a man that is most exceptional and therefore should be placed much more prominent in the allocated position his work has in the history of mankind. His contribution in the gathering of information that furthered the entire human species in their accumulation of knowledge as well as the human understanding in cosmic affairs stands second to none in comparison to most others whilst most people are not even aware of the full implication of his work. Whilst recognising the work of Johannes Kepler, Mainstream science bluntly ignores the impact of his work, and in that, they miss the full vastness of the wide influence of his work. Newton shrouded Kepler under a blanket and every one since kept Kepler there. It is therefore almost absolutely realistic to say that what you are about to read in this open letter sent to you for your attention was never yet printed in the near or the far past although the work has been with us for about four hundred years during which time it went unnoticed. It seems to me that any research predating Newton never came into use or into practise. My investigation of Kepler's work brought about a conclusion that no one yet arrived at concerning the findings of Kepler because no one scrutinised Kepler's formula. Kepler found planets rotating around a centre but Newton saw a circle and added what is mathematically required to indicate such a circle. Newton added a mathematical $4\Pi^2$ to the formula of Kepler and removed the distance symbolising measure that Kepler introduced using k. On the other side, Newton changed the symbol of k by using the symbols G $(m + m_p)$. This is just a longer and probably a more detailed manner of indicating k and better defining of k but it symbolises precisely to the point what k stands for nonetheless. I wish to draw your attention to the matter of Johannes Kepler's findings that Mainstream science considers as resolved and closed for many a century while it is not. My investigating Kepler helped me too resolve other unresolved matters but it was only possible by using Kepler's work.

I am all too well aware that at first glance you will immediately arrive at the opinion that the theme of the letter has to be considerably below the standard of an intellectual Master such as you must be, due to the position you hold, and because of that, the normal research work you do. Nevertheless, I hope that this writing may spark interest even at such a low academic level and grade in scientific sophistication and development because I am about to prove that I discovered. It is the formula that Kepler contributed to science. This might seem a little modest but this changes the entirety as how we do look at science verses the way we should look at science.

$a^3 = kT^2$ where ($k > T^2$) or ($k > T^2$).

I uncovered a mistake in science. From the onset, the mistake seems as insignificant as it is small. Because the rest of the book is about the mistake, I do not intend on elaborating about the mistake itself. The mistake came about with the culture of education and the mistake in itself seems harmless. When admitting that, one must also admit that any pilgrim that got lost and died of starvation through an incorrect travelling direction, made the very first part of his ultimate mistake by looking in the wrong direction. How harmful does looking in a specific direction seem, and yet such a mistake leads to his ultimate mortality. The traveller could when taking his first directional flaw with that the first incorrect step, only put his foot skew in avoiding a rock. Or he could have turned his face to avoid a branch and that move pointed him in a direction that leads to his fatality.

It is not the mistake that becomes the penalty and it is not the origin of such a mistake that leads to the penalising, but the ignoring of accepting signs telling the wonderer of an impending error and his stubborn ignoring of such telling sign that makes the lost party pay the ultimate price. By ignoring the mistake, for whatever reason, the ignoring of such a mistake is his undoing because the price due comes from the inability in recognising the sign indicating the presence of the mistake forming the reason for his final demise. The sooner such a person sees and admits the wrong, the less will be the consequences of his final price to pay.

I am most aware that there is this perception that Newton has never made a mistake. It is more than a perception...it is religiosity. Newton can do no wrong because Newton never did wrong. This entire concept is wrong making Newton's entire concept incorrect.

In the book I, the author, explain gravity. This achievement is possible because I saw a way to break away from invalid concepts Mainstream physics hold. I recognised the impossible double standards Mainstream physics apply to promote their much shady explaining. The inconsistencies brought them double vision and to compensate their incredible theories they simplify issues to a level where what they embark on to understand is becoming meaningless. What they say can't be supported and authenticated any investigation even in simple terms. It is as if they never read with interest that they explain and they never scrutinise that which they advocate. They give values that are senseless and make that which they say meaningless.

In this article I am going to investigate how much truth there are in mass pulling by the force of gravity. To most if not to all of the persons reading this and just the thought about me embarking on the investigating of the issue is totally senseless to investigate. It is senseless because the concept it carries became accepted as household practise and life science.

Do you think of astrophysics as the department that is run by the wise and the level minded, the sober thinking, and the absolute trustworthy? If you are a student there is no other choice you have. If you think those in charge of astrophysics are the pillars of trust, then get wise and read the following. What you are about to read is simply mystifyingly simple and yet to this day I have not challenged one academic any where that had the honesty to admit to the fact of Newton being wrong.

After you have considered the following you might agree with me that even small Children can reach a higher level of clear-minded logic and find more sensibility than what those scientists promoting astrophysics have because science lives in a make believe fool's paradise. If you are a student, then ask your Educated Masters to please explain to you the following abnormalities and inconsistencies they promote, which I present in this letter and then get wise instead of brainwashed. By teaching you to believe untruths in a method of forcing you to repeat what they declare as being the truth without proving it as the truth is brainwashing and mind control.

I say brainwash again because they force-feed you fabrications, as you will come to see. They can't explain the facts as the facts but hide the fact that the facts are in fact untruths. Tell them to prove that planets have mass. Tell them to prove that it is mass that generates gravity that pulls the planets. Ask them to explain gravity in detail.

Science has the opinion that according to their ethos I am not allowed to write about physics because they don't like what I write. It is not my writing about what they say I don't understand that they dislike. What they detest is my saying about what they have, being wrong. They claim I am not schooled to have an opinion but on the contrary, if they listen to what I say, they can't defend their opinion. For the first time in almost half a century there is a challenge on the dogma that science uses and I dispute the very basics they use to prove science. It is a question of questioning the opinion of those so opinionated, no one ever this far dared question their opinion…until now that is. Those Brilliant Minds guarding the ethics and the ethos that form the pillar of physics has a Universe expanding in a Big Bang in the face of Newton's contracting Universe by the power of all the mass found through the entire Universe. The Big Bang shows explicit expanding of the entirety, while Newton says there is only a contracting Universe.

The Masters hold evidence of a massive shift of all material growing from all centre points outwards. This shift is expanding space by creating more space between materials. The Universe is growing or is expanding according to E.P Hubble. The Universe is expanding at a precise given rate. The Hubble Constant is what proves this massive expanding that is so accurate that Physics used it to date the age of the Universe. Then we have Newton claiming that the Universe is contracting at a rate that the total measured value of all mass filling the entirety of the Universe produces. They have the Universe expanding by Hubble's rate, which is so well documented it is indisputable while Newton is claiming the Universe is contracting by mass. To cover this contradiction, they launched the second biggest fraud ever brought to mind. Physics has this contradiction covered by a theory called the Critical Density Theory. This theory is researching the problem, but the reasons they are looking for is not WHY the Universe is not contracting at present, because that would disrepute Sir Isaac Newton's physics. No, the question is WHEN will the contraction start for the contraction were never in doubt. There just has to be sufficient mass for the contraction, leaving the only outstanding question being when would the mass start producing the gravity to bring about the contraction. When the mass was found to be wanting in quantity, the next mandate was to look for mass hiding in darkness because such mass backing up Sir Isaac Newton's has to be there. They spend billions searching for concealed matter hiding from human view but never offer any explanation why is the hiding mass dormant by not already sustaining gravity? No one ever asks why the hidden mass never started applying the contraction as Sir Isaac Newton predicted. The question in need of answering is if the mass is there now, why isn't the mass active? Why is the activating of gravity suspended until such time when the mass will start to contract? If the mass is there, it must produce gravity or be absent by not producing gravity! It is either working or it is not there at all. It can't linger while waiting for starting orders. This is a cover up that hides the biggest fraud ever committed by any group of persons.

Science teaches that a feather and a hammer have different mass while they fall equal in time through an equal distance travelled. All things fall equal in time and distance when subject to the same environment. If gravity was mass related, then this was not possible, because then objects must fall according to mass. Falling objects bears no evidence of mass playing any part in falling. Any two objects holding different mass fall equal in time and in distance when sharing similar conditions, which suspends mass altogether as an influencing factor. Galileo proved different sized objects fall equally under similar conditions. That fact about Galileo, science does embrace although this strongly contradicts Newton's impressions about mass inflicting gravity. Acknowledging Galileo must make the work of Newton incorrect and also corrupt but science holds them agreeing. On TV we see everyday how all objects, such as cars, humans and bags fall at the same pace, and this is totally contradicting any view about mass being responsible for the falling. Mass can't provide gravity at all and the equal falling undermines Newton principles. The formula $F = \frac{M_1 M_2}{r^2} G$ would suggest that mass takes all the responsibility for such falling taking place. Newtonians declare gravity as the force F, that is = equal to a gravitational constant G, that is multiplied by the mass M_1 and the mass M_2 after which the product of the three factors influencing gravity then is divided by the radius square r^2 forming the distance between mass pulling the mass that destroys the radius square r^2. If mass pulls mass as Newton said, the Big Bang is not possible, but the Universe is, notwithstanding Newton's claims, expanding (growing apart). If the mass destroys the radius separating the objects, then every comet has to crash into the Sun, but they don't. If mass forms gravity, every planet must orbit according to the velocity provided by mass, which they don't. All planets orbit the Sun equal in maintaining an even pace. Planets disregard mass as they ignore Newton's suggested cosmic laws. The truth is that mass only is the resistance of any independent material to deform. To acquire mass, the individual object relinquishes independent motion. When a falling body is captured by a larger structure, mass forms as the captured body shows reluctance to deform by resisting integrating into the larger structure because of density. Mass retains shape and remaining in form produces the mass. Mass stops the falling by blocking movement, preventing further falling, it does not sustain further falling. Gravity is the moving of the object to the centre of the Earth while falling. I say gravity is movement while mass is obstructing independent movement, which is what gravity is. Mass is not forming the factor responsible for gravity or movement, but prevents further movement. A body falls by gravity. Mass obstructs further falling, while gravity remains present as a factor that brings the tendency or inclination to move or the attempt to continue moving. Mass hinders movement and therefore mass can't enhance or produce movement or gravity. Mass prevents or blocks gravity. Gravity is the motion that defines the individual identity of any object's structural form by rendering motion while reserving independence in granting free space from other manipulating objects. When having gravity, the object has freedom, not mass. Mass and gravity doesn't culminate, but opposes each other. My saying this awards me the cloak of death by Academics that deliberately ignore my correspondence as if I never addressed their mailbox.

Mass is relenting motion by retaining independence in form. Gravity is relenting mass by retaining independence in motion. Academics know my arguments are solid, but they refuse to read or recognize any possibility that my work is founded. Mass is a fact by human interpretation, creating a usable human standard or a quantifiable norm and therefore my work does not affect general physics. However, in outer space there is no remote trace of mass. The proof I bring of mass and gravity conflicting, trashes other material written on astrophysics, including Steven Hawkins' book that sold tens of millions. There is no remote evidence of mass in astrophysics, except in a despicable manner where science has to bludgeon the truth to present facts. Again, I state that gravity is motion and that motion places all falling objects equal in movement. That is not the way mass influence objects because by implication thereof does mass bring differentiation and falling objects all descend equally. Notwithstanding large or small, objects fall at equal speed.

There are four phenomena in the Universe that Newtonian science could never explain. I took the phenomena and trashed mass as the product responsible for gravity. I dissected the inner working of the four phenomena science would never admit existing. I found that when the four phenomena combine, it produces material movement becoming gravity. I prove this mathematically. This suspends existing dogma forming the insult and therefore they ignore me, for this is what no academic would admit to. Moreover, the injury is that I have no formal education. While I prove Newton fallible am just an ordinary motor mechanic by trade. This is more than what the academics can stomach. I prove mathematically that gravity forms when the Bode law is interfaced with the Lagrangian system and the Roche limit and from this joint motion derives the Coanda effect that creates space-time.

The Bode law is allocating positions that show how the nine planets arrange their orbital positions they have in accordance with the Sun.

The Lagrangian system is the manner in which natural moons or satellites arrange their orbital positions according to the centre planet.

The Roche limit is the closest positions stars can come to each other before the superior star dissolves the inferior star in the partnership. However, although nature shows clearly the phenomena, it is science that disputes even the official existing thereof in cosmology, just because these phenomena clearly disputes the fact of Sir Isaac Newton mass.

I say when an object falls, the mass factor becomes 1 with all that fall being equal and when the object is in mass the gravity factor becomes 1, with the object's movement being equal to the Earth moving By saying this and me never saying anything is the same since science says that no sober mind will ever contradict Sir Isaac Newton To them Sir Isaac Newton is utmost religiosity. Therefore, whatever I said in this writing up to now, I just as well as never said. Academics in science are unable to read past any criticism targeted at Sir Isaac Newton. Sir Isaac Newton is correct leaving all else mistaken and that includes the Universe. By contradicting Newton's genius in the Big Bang expanding, the cosmos is explicitly deliberately and stupidly incorrect. Sir Isaac Newton said the cosmos has to contract to comply with Newton and the cosmos is out of order in not doing that. To Physicists in regarded Academic circles, the entirety of this book is not about whether Sir Isaac Newton made a mistake, for that is a total impossibility. Sir Isaac Newton's never made a mistake. Science holds the opinion that Sir Isaac Newton never made a mistake since they believe no one could ever prove that Sir Isaac Newton was mistaken. Thus, any mistake is never about Sir Isaac Newton making a mistake, for Sir Isaac Newton never could make any mistake. The fact that the Universe expands, implicates the incorrectness there is on the part of the cosmos and not on the part of Sir Isaac Newton. This Universe's expanding is diverting from the Sir Isaac Newton's law and with the cosmos going against Sir Isaac Newton, cosmos forced the establishment in physics to launch an investigation about detecting the underlining irregularities on the part of the cosmos and therefore on the part of God. Such a case file is well documented and named as the Critical Density Theory and this theory had the most brilliant mind on Earth set out to calculate (and count) all the mass available that could inflict gravity that will provide all the contracting to pull the Universe together again, just as Sir Isaac Newton's law insists.

It is the Universe that went out of line by expanding and not Sir Isaac Newton that is at fault by predicting contraction instead of expanding. It was the task of Albert Einstein to find out when the Universe will abide by Sir Isaac Newton and the laws of Sir Isaac Newton. It is not Sir Isaac Newton that is wrong about the Universe not coming together, it is the Universe that is wrong by not doing what Sir Isaac Newton said it has to do in order to be a good complying Universe and adhere to Sir Isaac Newton's laws. It is the cosmos going out of order by expanding and by that disobedience the cosmos outraged the Paternity in physics to the extent that they have spent billions of dollars to find out what is wrong with the Universe not complying with Newton. Why would the Universe hide matter in dark places just to spite Sir Isaac Newton and not launch the mass to start providing the gravity so that the cosmos will start doing what Sir Isaac Newton ordered it to do? This act of the cosmos just cannot be tolerated and so the Establishment must soon find a way to correct the Universes' spiteful expanding as they ignore the Big Bang since they presume that no one will notice that this theory will never match Sir Isaac Newton's contracting by mass theory.

It is the comet's mistake that the comet goes astray and loses focus of Sir Isaac Newton's laws by not colliding with the centre of the Sun. It is the comet being spiteful and vengeful to escape the gravity of the Sun pulling by mass, using gravity to pull the comet close after which the comet escapes the gravity effort and speed into the darkness of outer space. When it doesn't hit the centre of the Sun, it speeds off into the distance and darkness of outer space contradicting Sir Isaac Newton's formula. In ignoring the comet's despicable ignominious behaviour, one saves the comet of facing the shame of the obvious inappropriateness, as the blame can't be pinned on Sir Isaac Newton.

It must be Galileo's mistake that all things fall equal and not Sir Isaac Newton's mistake because Sir Isaac Newton says mass has things falling while Galileo said al things fall as if even in mass. Only the most ardent Newtonians find Newton and Galileo agreed while in fact Galileo contradicts Newton. Only Newtonians can see that by falling with mass implicating gravity, all objects falling will fall equal and that puts mass playing a centre part by never intervening. One may never dispute the reputation of Sir Isaac Newton as Sir Isaac Newton is always correct! Any mistake that can connect to Sir Isaac Newton or put any connection between Sir Isaac Newton and such a possible mistake becomes the mistake. The mistake becomes a mistake simply while thinking in terms of Sir Isaac Newton and about a mistake at the same time. Therefore, the mistake is in terms of the view the person has when connecting a mistake to Sir Isaac Newton and not in terms of Sir Isaac Newton's ability to make mistakes. Any person that has an opinion about Sir Isaac Newton being able to make a mistake is making a mistake in terms of the idea the person has and not the fact of Sir Isaac Newton committing any possible mistake. Any opinion of accusing Sir Isaac Newton of having made a mistake is then creating a mistake. I don't share the opinion about Sir Isaac Newton's ability not to make a mistake, because I did detect serious mistakes on his part. Since no one this far proved Sir Isaac Newton wrong, this book and my undertaking in proving Sir Isaac Newton wrong, renders me automatically as the mistaken party. By attempting to prove Sir Isaac Newton wrong, in my attempt, I prove myself wrong. If this is senseless to you, then more confusion awaits you in this book, because there is a lot about Newtonian science that is not making sense and the biggest thing about these things not making sense, is the all out effort every person that holds any connection with Science is having in their attempt to hide these facts that doesn't make sense and their actions in covering up the mistakes is that which makes the least sense of all.

However, it is most urgent to note that the enveloping mistake uncovering any mistake about Sir Isaac Newton, is the possibility that the mistake was detected by a person with the most underachieving degree of not achieving any degree and the lacking of such formal education is that part which holds the entirety of the mistake possibly being present or having no possibility of being present whatsoever. Seeing that the uneducated never had the brains to accumulate knowledge even as far as an education degree would present the man, the thinking capability is very suspicious indeed. Academics with the utmost infinite wisdom will never allow any presumptuous unschooled mindless barbaric halfwit grease covered spanner pushing labourer the opportunity to teach those with such brilliant minds and with that endless lucidity, anything about the splendour of Sir Isaac Newton's peterhuman infallibility!

In my final letter that I ever intend to direct to academics I end the letter as follows and this book is what promise I made that I fore fill that final promise that I made when I sent nine letters to nine South African Universities promising I shall never contact them again on the matter of the mistakes I saw that Newton (as well as they) made but I also promised them that a fight is on. This is precisely the words I used when I ended the final letter to the nine academics of Universities being the head of the institutions in cosmology.

Sir, Madam, when you address your students about the wonders of Newton, then tell your students tomorrow how much the Earth drew closer to the Sun since the days of Kepler, and Madam, announce to the world how much the Moon came closer to the Earth since the days of Tycho Brahe while you then remain convinced that Newton is still flawless after three hundred and fifty years. Notwithstanding all the correctness you do attach to Newton, please take note. Do not take this, which I am about to say and what I am about to say as a threat but rather as my promise.

A threat to you and your address it can't be because you are so high and mighty I have no ability to ever harm you so a threat it can't be and neither is there a possibility where one can see what I have to say to you in conclusion as any threat. From where you stand, you will not even notice me down where I am, let alone take note of any threat I might make. Rather see this following remark as a promise. I promise you that I am coming after all Newtonians with everything I dare to use, and that is a promise I do make. This remark means nothing to any person holding the position that you hold since from where you are, all Academics in your position, including you, being in your position, is beyond approach by any one with the likes of me and I am quite unable to reach any person being where you now are.

Therefore you correctly see yourself being where you are also as being outside my reach. You are all-powerful and I am powerless in every sense. That might be very true. Nobody, and that includes me, has the ability touch any academic with your standing. I admit to that very readily. However, there is judgment waiting for all and some comes sooner and some comes later. That which I wished to share with you I am going to try and share with the public at large.

The thing about all Newtonians is that they are boastful and arrogant and self-centred. I have not met one Academic in physics that is not fitting these criteria. These qualities are remarkably the essence of what any person should have that does comply with the demand there is required in being a Newtonian. From my position I have, in relation to the position you have, you may observe me as not worth noticing. Please hear this. All people on Earth think they know everything because what that person knows is everything in that person's Universe. Whatever that person knows fills that person's entire Universe. On the other hand whatever the person does not know, cannot and therefore does not exist in terms of the knowledge that person regard as worth having and therefore to that person, everything that the person does not know in terms of everything the person do know, then the part that the person does not know comes down to being nothing. All persons are in a position where such a person does not know what the person is unaware of. What I know you don't know and for that reason what I know is so inferior to you that it is not worth your effort of taking note of what I know. If it were worth your appreciation, you would have known what I know because it would have been worth your effort to know what there is that I do know. What there is that I do know is to a man with your field of knowledge non-existing insignificant. Since that, which I think I know, has never been to your mind, therefore it was never to you worth knowing, it puts what I know in the bracket of nothing worth you're while to know. What there is that I do know has no place in your Universe you fill. What I know is of such unimportance that it has never been worth finding out. Then in your

An Open Letter On Gravity Part 2

Universe, what you know accumulates to everything there is that is worth knowing because to you that amounts to everything that has any worth for you to know. What I know is nothing to you because that is what you don't know and that which you don't know doesn't exist to you. What you do know is an entire Universe filled with everything there is to know in terms of what you know and only that which fills your Universe is of significance and only that do exist to your knowledge. Nevertheless, please take note of my promise I make.

You might find the level of my education being much below your high standards and you are of the opinion that anyone with such a low qualification such as what I have, can never offer anything of value in the form of information which can further and advance your knowledge you have in your field. Sir, Madam that might be true and then, that might be a very incorrect assessment on your part because and it might just be extremely costly as no person ever holds a position where such a person has in his or her position all there is to know and has no more to gain from others. But knowing what you know and not knowing what you know is not a quantifiable measure but more a degree of fallibility and that makes it much more a state of mind concealing Newtonian utter arrogance. You may know everything there is to know that is in your Universe and you may judge my Universe so empty of substance that you might think in my Universe there is nothing worth knowing but in that there is a surprise install and waiting on you. There might be some things in my Universe that you are not aware of since it is in my Universe and to you my Universe is not worth your regarding. This might turn out to be expensive to you, who would know beforehand.

That what you don't know, to you don't exist because it falls outside the Universe you do know. On the other hand that which you know is everything there is to know in the Universe because to you there is one Universe and that is your Universe. Please do not be surprised if you do miscalculate your Universe and underestimate my Universe. This letter is one of ten letters that I address to academics indicating my concerns and after this letter there will be no further communication by me to any academic in any way. If there are still no response coming from any academic, after I sent off the letters mentioned then this is the last time that any academic, which includes you in person, as well as the institution that you represent, will hear from my address. That you will hear from me again, that is the promise I leave you with, but also that it will be along another avenue when you hear from me again, is also part of my promise.

That, what I know, you think does not exist, because you do not know what I know and if you don't know it then that item is not worth being in you Universe. This attitude is not only connected to the Newtonian mentality but is typical human behaviour.

That what you think you know, you accept I don't know, but that is because my Universe is much smaller since you are the intellectual and others with lesser degrees are lesser intellectually inclined. What I know and the information in my Universe to you in your Universe does not exist, and with you having the more intellectual Universe, what I have in my Universe can't be as important as what you have in your Universe. I am about to prove that concepts you think you are sure about such as Newton, I am going to dismiss and that I can do only by changing interpretation of information and not changing information as fact. When I do that I am planning to address the general public and try to show that audience how there can be another side of perception about science no one ever realised. I am going to open the closet and reveal all the Newtonian corruption hidden under a dark cloud of falsified facts and also reveal all the rotten bones you in science conceal. However, in your Universe, you are not even aware there are corrupted, distorted and unreliable facts because to you your Newtonian Universe is perfect. That is how I ended my last intentional communication with any and all Newtonian High priests anywhere. From then on I tried to reach out to the public.

After that I went on my merry way and I did write this book where I now aim to reach the general public with information in spite of the Newtonian view about my scholastic restraint. Again as not to confuse any reader, when I say I wrote this book then be clear about the Newtonian opinion about my inabilities and with that I am glad to say it still is my guess that they will hold the opinion that I merely generated uninformed thoughts, which I collected as alphabetic symbols and plotted that in ink on paper. That is the information they have to their disposal in their Universe and that I can never change. This effort that I achieved I did so from harbouring my delusional ideas spawned by a dehumanised brain. Criticizing Newton only proves my weak and under developed mentality, due to my lack of an informed insight that is a typical symptom that all those have that is suffering from a disadvantaged past as I so clearly have, that which my views represent one can only have when the person obviously lacks formal education. While you are reading the letter and in reading you obviously decided to regard and not to dismiss my work, then also please keep in mind when reading my language used and also please give credit where it belongs…if you do find linguistically improper use of words or misspelling, then remember that I am a feeble minded motor mechanic and not a literal giant and because of financial restraining I went the course on my own.

In this book you hold and of which I am the author believe it or not and which I hope you are about to read, which is I explains gravity. This achievement is possible because I saw a way to break away from invalid concepts Mainstream physics hold. I recognised the impossible double standards Mainstream physics apply to promote their much shady explaining. The inconsistencies brought them double vision and to compensate their incredible theories they simplify issues to a level where what they embark on to understand is becoming meaningless. What they say can't be supported and authenticated any investigation even in simple terms. It is as if they never read with interest that they explain and they never scrutinise that which they advocate. They give values that are senseless and make that which they say meaningless.

Up till now **Sir Isaac Newton** was revered as a God and as it is in the case of all religions in science it is no different for also in the religiosity of **Sir Isaac Newton** science followed this motto of not asking their God **Sir Isaac Newton** questions for it is not for us to ask why but to accept. It is not for us to doubt but to accept. If god **Sir Isaac Newton** said then what **Sir Isaac Newton** becomes law and no doubting is tolerated.

An Open Letter On Gravity Part 2

Then I came along after almost four hundred years and expressed doubt. This was what never was tolerated by Newtonians. You accept **Sir Isaac Newton** or you are guilty of blasphemous conduct.

If you are a student in physics then this information you are about to read is most important. In the classes you attend in physics has any one confirmed a location where one might find the centre of the Universe? Have you been told precisely what causes gravity to pull? Have you as a student in terms of the fact that you are being a student been informed how mass confirms gravity? What evokes the force that establishes the pulling that confirms the mass that produces the gravity? If no one went to the trouble to tell you this is it not about time that someone exerts himself and do the honours? On the other hand have you been asking what evokes the force that establishes the pulling that confirms the mass that produces the gravity? If not why have you not gone to the trouble and just ask this simple question.

It would be most interesting to hear the answers those lecturers will come up with since these questions have not found answers, up to now that is. I wrote a book in which I found a means to define gravity. I did accomplish this for the first time ever since the time Newton introduced gravity. This is more than what Newton achieved and it is more than what the whole lot of Newtonians achieved ion three hundred and fifty years. I could do that by accomplishing one thing all others thought not to be possible! Before I achieved finding what gravity is I first had to find the centre of the Universe because it is there that anyone and I could locate gravity. I can now show how gravity forms because I have detected the centre of the Universe.

Is there any Newtonian applauding my effort and congratulating me in my achievement? If there is one such a Newtonian that Newtonian still awaits birth. I couldn't find one Newtonian even being prepared to read what I have to say about what they have nothing to say about. I could therefore not locate one publisher that was prepared to publish my work because before publishing they first have to read my work and no one was prepared to even glance at my work let alone read it intensively with publishing in mind. But I need to get the information out to everyone to get anyone to read my work. In achieving that I had to resort to private publishing because from the nature of my work I take Mainstream science head on and am confrontational on most aspects of astronomy including astrophysics and the founding principle guarding the authenticity of physics.

To have a publisher backing me in order to publish my book the publisher had to find an academic prepared to back up my statements that Newton is a criminal that committed extensive scientific and mathematical fraud! In that sense there does not seem to be any publisher that wants to go head bashing with the Physics Custodian establishment of science on official science principles, which I have to do to convey my message in no uncertain language.

I argue that if it is the correct practise to use to calculate gravity then the radius holding the gravitational constant must lead one to the centre of the Universe. This fact

Newton nullified by using the argument that the rotation nullifies work done because the rotation is in repeat of the process and through that the radius between the centre and the point rotating is nullified. If you don't believe me then explain what he says in this statement.

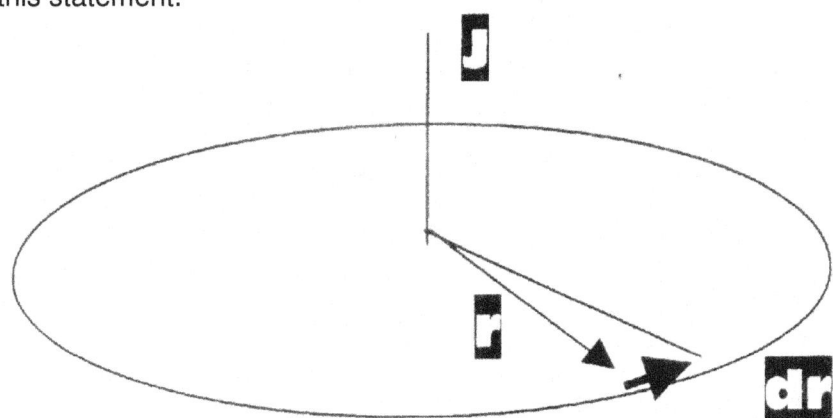

This picture in its entirety condemns Newton's statement of the rotating distance nullifying the motion totally. $\frac{dJ}{dt} = 0$ Newton, and science, made one enormous blunder, from taking this stance. It is as if they took the idea that when a wheel spins the radius of a wheel has not to any influence on the wheel. In doing that, they removed the very fact that keeps the wheel at a radius and size and cosmologically the universal attachment together. They put two objects in an attaching relevancy and then announced that there just is no relevancy applying between the two. When one divides into another there is an irremovable ratio in place. Removing that ratio is breaking the most fundamental mathematical principle.

$\frac{dJ}{0} = dt$ or $\frac{0}{dt} = dJ$ This disputes mathematics. DJ / DT can have any number except the only number not possible to have is zero. This is mathematical and physics fraud and there is no other way to put it but to put it into its correct context. I challenge any academic to show how the radius does vanish by the circle running a full rotation or how the radius removes the circle after it ran a full circle…and make no further judgement error when mathematically expressing that $\frac{dJ}{dt} = 0$ then the statement must directly translate to also saying that $\frac{dJ}{0} = dt$ or $\frac{0}{dt} = dJ$ placing the following is true $\frac{dJ}{0} = dt = 0$ or $\frac{0}{dt} = dJ = 0$ $0 = dt$ or $0 = dJ$. Now match this statement with the sketch on the previous page and reconcile that with the disappearing of either the radius (becoming nil as in disappearing altogether because that is what nil is expressing) or the circle (also becoming nil as in disappearing altogether because that is what nil is expressing in the case of becoming 0) while the other is still remaining and see how mathematically corrupt that is. Then tell you Professor to explain how the supposedly best mathematical brain ever could nully willy come to this conclusion without corrupting mathematics and physics. Tell the professor to convince you why he as the academic is backing up such corruption in mathematical law while he remains as innocent as a newborn baby in the matter of promoting corruption and deceit. What the statement expresses is that the radius between the Earth and the Sun disappears and the Sun and the Earth combines as a unit. There is a distance between the Sun and the Earth notwithstanding Newton's fraud. But let's return firstly to my quest in avoiding the consequences of Newton's corruption by locating the true gravity that is out there waiting since the arrival of the birth of the cosmos to be discovered. First let's go in search of the centre of the Universe where the birth took place.

If you think scientists know what gravity is do not be duped that easily because no one in science remotely knows what gravity is…not even Newton knew what gravity is except Kepler… and because of what Kepler introduced now I know I can prove what gravity is. Gravity is precisely what Kepler said gravity is and only Kepler new where to find the centre of the Universe because only Kepler knew what gravity is all about. I used what Kepler brought into the world to locate the centre of the Universe and therefore the point where gravity starts. But I did not use Kepler in the way that Newton corrupted and raped Kepler because one cannot commence when corruption destroyed the truth.

Try to get an answer from any physics academic about where the centre of the Universe is, is like trying to touch the moon. Science can't see past Newton and Newton couldn't see past mass and mass does not exist so to Newtonians it is an endless cycle of getting no where as far as detecting the truth while it is taking them an eternity not to get there very fast. So Newtonians put it on an elusive undetectable and unexplainable force that can conceal their stupidity and Newton's fraud.

By merely putting gravity in the Universe by telling everyone that gravity is acting as a mysterious FORCE that is pulling towards a common point in an allocated general centre is rather avoiding the question with simplicity because the question about how and why remains unanswered. Not knowing the answer to where the centre of the Universe is, will leave you feeing empty and unfulfilled because of being a student and not knowing is the same as suicide on a mental level. That is why you are primarily a student. Being a student is being in search of information and knowing you might never achieve the prime information in physics must be devastating to eager minds such as yours. Ask yourself the following: If gravity pulls towards a centre and gravity holds the Universe attached the question arising from that simplistic answer is then … where is the centre of the Universe? Newton was unable to find the answer. Newtonians took all of three hundred and fifty years not to find the answer. Do you wish to spend a lifetime searching and never find the answer? Then become the next generation of Newtonian Masters. However if you discard the falsifying of facts that I charge Newton with, physics will present you with an answer as we follow Kepler's lead.

Should you decide to purchase and read this book, it will bring along a new perception about Kepler. Science sees to it that Kepler stays the least appreciated Cosmologist where as in truth Kepler proved gravity, proved singularity, proved space-time, proved the Big Bang, proved every dynamic most of the wise persons afterwards thought about. Yet no one gave Kepler any recognition up to now because science denies Kepler his limelight. All they can see is the way Newton raped Kepler by falsifying everything Kepler introduced.

I wish to return to my previous statement that I made in **An Open Letter on Gravity Part 1** about light because in this one may find the purpose of the Universe. Where I did mention this in the previous part, the purpose was to introduce the concept. Now that the introduction should have sunk in, I wish to elaborate on both the significance of the dark night concept and the manner in which we see the context of outer space. If light came as individual streams of photon flurries, then our visage would translate that as such shown in the fragmented picture above. It does not. The flow of light is precise and coherent which renders the Newtonian concept of light to be rather sheepish. If light were individual photons it would leave a picture that is completely unconnected.

It is Kepler that introduced the formula $a^3 = kT^2$. Since I am aiming to mostly reach students as a target for my audience I have to bring a very small perspective on mathematics and equations concerning mathematics as evidence. It is known in mathematical terms that when I express any symbol in terms of the third power I then am referring to a cube, a volume of defined space having at least six sides with three opposing thee other sides.

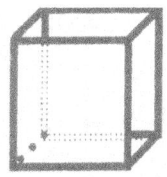
First we attend to Kepler's first symbol a^3 that must be a cube. A cube has six sides where three sides opposes three other sides hence the three in a^3 and therefore the picture represent a cube or something to that likeness that holds a space in terms of defining a space that is apart from a containing space in which the defined space is. What this says is that in the event where I may refer to a box holding a measured volume value of a^3 the a^3 can only be practical and defined if this box being a^3 **was** within but also still part of another bigger container also being to the third power. What this says is that a^3_1 has to be either bigger or smaller than a^3_2. If the two boxes were equal in size the two had to be the same box because there has to be a distinct difference in size putting differentiation between the two boxes holding one while the other is within the first one that is containing the second one. If the two were two different boxes then either $a^3_1 < a^3_2$ **or** $a^3_1 > a^3_2$ but never can they be in a position where they are $a^3_1 = a^3_2$. Being a cube means you can put something into it as it has a top, a bottom and sides and as a result from that it can contain something you put into it. The cube uses a square as a base to hold the third dimension.

Then we have another factor forming part of the Kepler equation. This to which I refer is T^2. In this case where we are investigating the square we find four sides where tow sides opposes two other sides bringing about a square. This factor refers to a square where the previous factor was a cube.

The one is not the same as the other because the cube has three dimensions or three sides and the square has two sides, hence being a square. The square can be a part of the cube but it then needs one more factor but on its own T^2 only refers to a flat two sided surface.

Being a square means you can put something onto it but not into it because it can only form the base of a container but it can never contain something you put into it. The square can act as a base but only represents at best two of the three sides which forms a sphere.

In the final part of Kepler's equation that man uses a symbol he named **k**. The factor **k** refers to a single line holding one dimension. One can't remove this line from any equation because this line has value as it forms part of the integral value of Kepler's equation. In the mentioned equation $a^3 = kT^2$ the factors each has a specific place forming a specific duty when they are equated in placed in relevancy by forming a factor such as they do in $a^3 = kT^2$. This is primal mathematics and a principle that is taught to very small and less developed mathematics students. If one does not know not to change this as it will change the outcome of the result that the equation should deliver, then such a person still has a long road to travel on the path of being schooled and educated in the science of Mathematics.

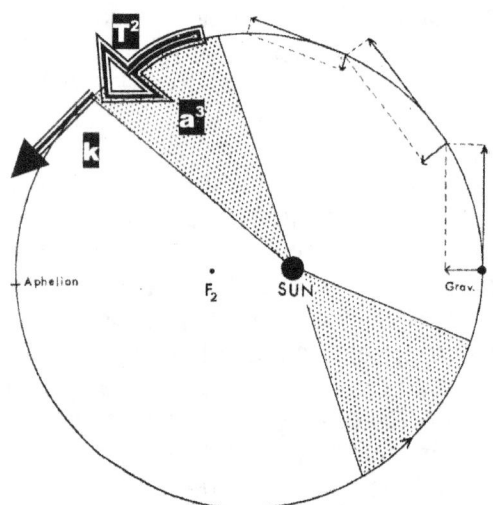

Kepler said $a^3 = kT^2$ and that says that everything that is going in a circle has to follow a straight line. All space a^3 is equal to = the circle T^2 it will follow starting as a straight line **k** that diverse in a specific ratio. Every time a circle forms it is a straight line that is rerouted.

By the way it is noteworthy to realise that the formula $a^3 = kT^2$ Kepler received from the cosmos written in mathematics and it is rather presumptions of any person including Sir Isaac Newton to consider him such an authority that the person can change what the Universe gave as information. If you remove any part of any ratio, then the ratio is gone to hell and that is just what Sir Isaac Newton finally achieved.

Kepler said $a^3 = kT^2$. This means that every line that goes straight redirects the direction tat it follows. ...And also it says so much more.

It means that space a^3 that forms is as space a^3 and that space a^3 is formed by going straight **k**, while also bending by route of T^2. Most of all it says that the formula forms a unit $a^3 = kT^2$ with three separate and individual factors a^3/ **k**/ T^2 that eventually unite in relation to one another to combine as something representative of everything there is throughout and inside what we think of as the all inclusive Universe.

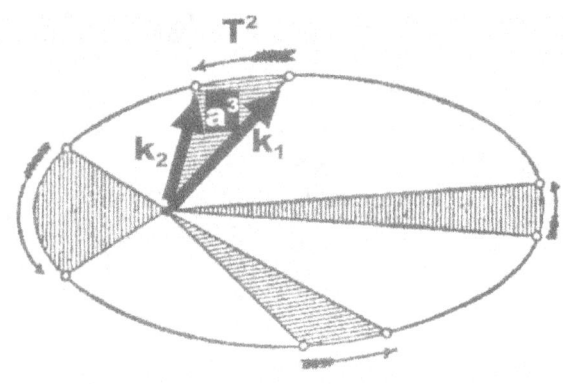

Kepler said that $a^3 = kT^2$ which is that the circle is forming in relation to the distance that there is measured from the centre of the circle forming and the centre distance that is also a radius. The space there is forming in relation to the circle forming according to the distance that is measured from the centre to where the circle starts and the circle ends will be equal to what the radius brings in relation to the circle limiting the area. That is what Kepler said when Kepler said $a^3 = kT^2$.

Kepler definitely places a ration between the movement at a distance and the space such movement limits from the rest of the space containing the movement and more than the movement. The ratio that Kepler introduced hold the space being in the third dimension to a particular ratio there is between the movement and the movement at a specific distance.

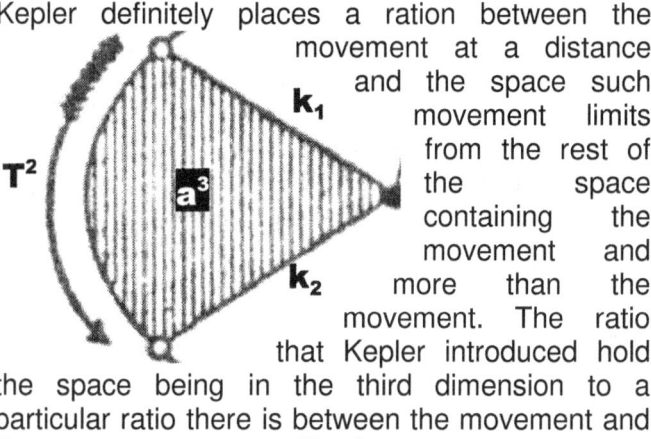
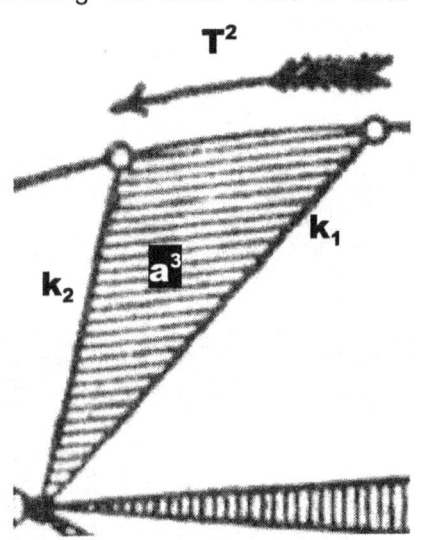

Kepler said that $a^3 = kT^2$ which is what he said when viewing the picture to the below.

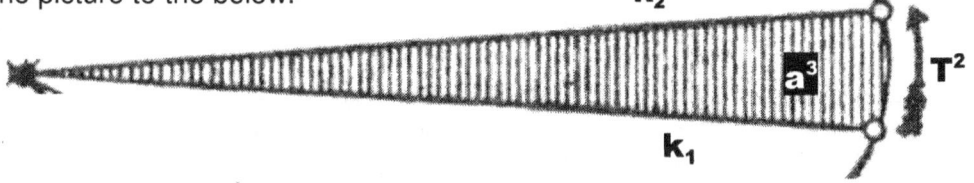

Sir Isaac Newton insisted that Kepler said $a^3 = T^2$, which is something Kepler never said because it ism on the record and part of the history that Kepler said $a^3 = kT^2$ where Kepler

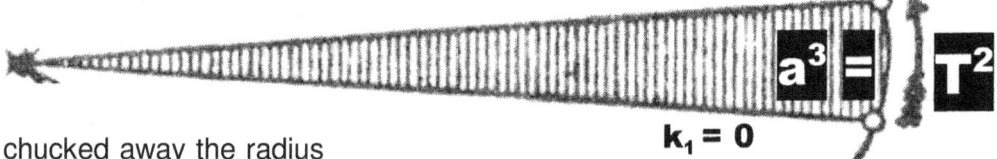

never chucked away the radius facto being **k**. This reckless mathematical corrupting of basic principles is only attributed to the arrogance we find in the attitude and the behaviour of **Sir Isaac Newton**.

At the same venue $a^3{}_1$ can't be = to T^2. But in the event where Sir Isaac Newton's says that $a^3 = T^2$ I have to sit back quietly and believe him, because they say he is Sir Isaac Newton. No other reason is forthcoming except that Sir Isaac Newton $a^3 = T^2$ and without proving that statement I have to accept it as true. Remember Kepler never said $a^3 = T^2$. Kepler said $a^3 = kT^2$ which places three dimensions on one side holding three dimensions equal on the other side of the equation. There is a^3 on the one side of = and then there is kT^2, which is $k^1 \times T^2$ which is $k \times T^2 (^{1+2=3})$ and that makes $a^3 = kT^2$ having three dimensions on the one side being equal to three dimensions on the other side. There is no way in heaven or hell that one can have the third power being equal to the second power or have a cube that is equal to a square, even if you are Sir Isaac Newton. There is no one on Earth that will tell me that $10^3 = 10^2$. There is a case that $10^3 = 10^2 \times 10$ or that $2^3 = 2^2 \times 2$ but never can it be that $2^3 = 2^2$. Not even when Sir Isaac Newton is doing the saying so. If one says that in the event where $a^3 = kT^2$ one may assume that $a^3 = a \times a^2$ or $k^3 = k \times k^2$ or even that using $T^3 = T \times T^2$ will also bring equality but never can $a^3 = T^2$...and then there are academics that try to convince me that $a^3 = T^2$ because Sir Isaac Newton was of the opinion that $a^3 = T^2$ and furthermore it also is true that Sir Isaac Newton has never been wrong and because no one could ever find Sir Isaac Newton to be wrong I have to accept that $a^3 = T^2$ and take it without questioning this abnormality!

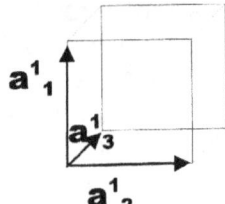

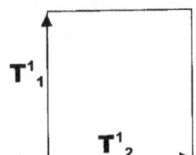

The one image is a cube. The other totally different image is a square. Sir Isaac Newton said the two are equal while they can never be equal since they are one dimension apart. Sir Isaac Newton convinced so many generations of idiots to the point where these stooges are willing to believe they are wise enough to believe that a cube is equal to a square and only on the ground that new Sir Isaac Newton

Sir Isaac Newton proposed and moreover convinced the world of science, and this includes every one and all members that should be the most intellectual bunch living on Earth in human form that they and the entire world should accept that the inexplicable is correct $a^3 = T^2$ and that the biggest trick in fraud can be played on a bunch of fools all willing to be stupid enough to pretend they are clever enough to see that $a^3 = T^2$ and they are so stupid they pretend to be so clever that they will accept that $a^3 = T^2$ which when translated in words means that two dimensions are equal to three dimensions. This is the same as stating that a person in the mirror is the same as the person standing and looking at his image in the mirror. In this group hosting the most advanced minds man can produce there are a big enough bunch of zombies pretending to be mentally superior while being big enough idiots that are foolish enough not to think and not to ask questions but be small minded to the point that they will accept that a cube is equal to a sphere just simply going on the say so of Sir Isaac Newton.

This does not prove that $a^3 = T^2$. It only proves Sir Isaac Newton was the worlds biggest and best silver tongue devil and cheated an entire Earth load of scientists for almost four hundred years. He fooled the wisest there can be to pretend to be wise so that they can hide their stupidity while they only focus on their stupidity by not questioning the validity of $a^3 = T^2$. You bring me one other con artist and fraudster that can manage that. It takes some doing to fool so many people for so long and leave all those fooled feeling good about themselves in that they are fooled. Sir Isaac Newton was the biggest con artist ever to live and never again will the world experience a joker equal to Sir Isaac Newton

Trying to convince the fools filling the physics departments in Universities throughout the world that they are fooled takes some doing and that I found to be more true than that the statement of Sir Isaac Newton being $a^3 = T^2$ is a lie. It seems to be the general idea that we have consensus on the matter that Sir Isaac Newton finds him to be so special that he Sir Isaac Newton can alter mathematical principles on his say so and by only using his command…but thinking about this I still beg to differ. There is no mathematical principle allowing Sir Isaac Newton to say that $a^3 = T^2$. Sir Isaac Newton can't just throw away the **k** in Kepler's formula $a^3 = kT^2$ just berceuse he fancies himself being the most superior god of physics as Sir Isaac Newton and therefore just on the grounds of being Sir Isaac Newton he Sir Isaac Newton can change mathematical principles as much as he sees himself equipped to change the work of Kepler and when all the changes he does by only using the grounds that he i**s** Sir Isaac Newton the one and the only Sir Isaac Newton. But I found that no one is allowed to question Sir Isaac Newton because doing so will lead to immediate castration…Or is the word used for such penalty incarceration, I am nor sure which one it is.

PLANET	PERIOD (Years) (T)	MOVEMENT (T^2)	DISTANCE k	SPACE (a^3)	RATIO
Mercury	0.241	0.058	0.39	0.059	0.983
Venus	0.615	0.378	0.728	0.381	0.992
Earth	1.000	1.000	1.000	1.000	1.000
Mars	1.881	3.54	1.524	3.54	1.000
Jupiter	11.86	140.66	5.20	140.6	1.000
Saturn	29.46	867.9	9.54	868.25	0.999
Uranus	84.008	7069	19.19	7067	1.000
Neptune	164.8	27159	30.07	27189	0.999
Pluto	248.4	61703	39.46	61443	1.004

In the above table that Kepler configured as $a^3 = T^2k$ we have three distinct factors

The way to read this table Kepler gave the world is to see the numbers that Kepler derived from all the studies Tycho Brae as well as he concluded. Every number is a column and the column is named either a^3 or T^2 or **k** or whatever. But the column is named as the symbol; and there it ends. The column consists of numbers and the numbers can't be ignored by simply diminishing it to zero. The numbers hold a ratio and in the ratio the numbers form a result that forms a cosmic principle, which I prove in my books. The one column multiplies with two other numbers in the other columns to give a final result and from that result one may form an opinion. But one just can't be as opinionated as to remove one column altogether by giving all the values in that column a measure of zero. When multiplying zero with the other numbers the result will be zero and Sir Isaac Newton should have been aware of this seeing that he was the great physicist / mathematician. The mistakes are so obvious even I can recognise the mistake, and then there are those with PhD to fill a hall that lacks the ability to see. They never saw this travesty of mathematical principle! To top this lot I am suppose to believe those that so ardently believe in Sir Isaac Newton on the basis that they believe in Sir Isaac Newton and therefore I am not suppose to argue the issue!

Trying to get Newtonians to see this deliberate mathematical fraud just seems to be impossible because I have tried as much as I could and all my trying got me nowhere. If only once there was one academic that could show me ware in the cosmos other than in the mind of Sir Isaac Newton could I find proof that this of $a^3 = T^2$ or going from a cube to a square or that the statement Sir Isaac Newton made is true anywhere, I would stop pestering them and accept I was wrong. But until then I am going for the fight I promised them because I will never accept defeat merely because of my social standings in society.

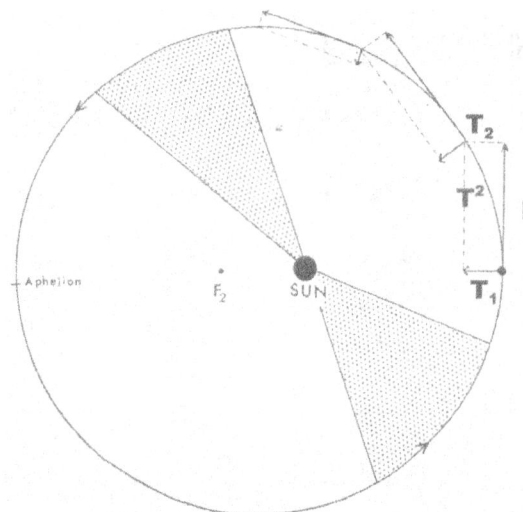

The formula $a^3 = T^2k$ is how the Universe revealed its constructed structure to Tycho Brahe and Johannes Kepler. In that revealed formula coming from no less an authority than the Universe not Kepler or Brahe but the Universe said that we find that it is also true that $k^0 = a^3 / T^2k$. Mathematically it says that the space a^3 forms = when a line k runs from T_1 to T_2 in order to become T^2. This statement is mathematical fact!

The formula $a^3 = T^2k$ we find that it is also true that $k^0 = a^3 / T^2k$. Mathematically it must be that when the space a^3 that incorporates the line k running as T^2 then a centre has to be in place representing singularity k^0 The space a^3 will only form if singularity k^0 is validated = by forming both aspects of a line k as well as T^2. The connection of a circle a^3 in progress of forming T_1 to $T_2 = T^2$ has to be in relation to

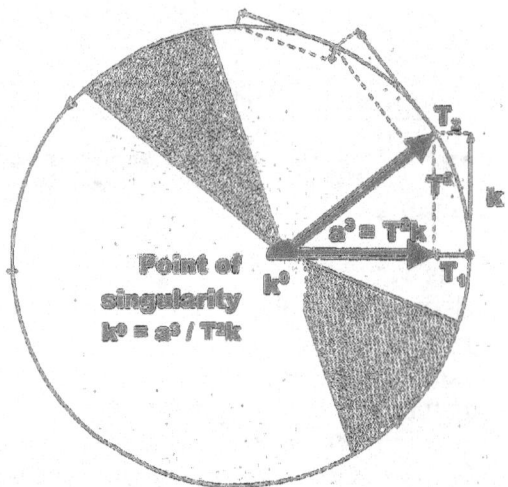

a centre k^0 but also in terms of k. I challenge any mathematician including Sir Isaac Newton and all Newtonians to disprove this statement!

However it is the prerogative of Sir Isaac Newton to change any and all mathematical principles at will and on the command of Sir Isaac Newton as it may please or displease the greatness of Sir Isaac Newton

This is my declaring to go to battle as it is my intention to show everyone willing to read who the true fools are in this world as well as introducing the readers to general cosmology in a very brief and compressed manner but first, I have to give the emphatic warning to all prospective contemplating readers.

Please take note of a conscientious warning about the gravity of the misgiving there is on the part of the most respected Academics in physics about a much concerning matter. As I have indicated, I am fowl of the sin that I don't go by the mere say so of Sir Isaac Newton and for a sin on such a scale I am banished from this Universe and from life! In other words I am officially declared dead because I challenge the greatness of Sir Isaac Newton I have no friends amongst academics especially those in physics. I have no enemies amongst academics especially those in physics because to them I am non- existing and invisible. You see with my attitude I could not get any degree in physics. To be awarded a degree in physics one has to accept the word of Sir Isaac Newton just because it is the word of Sir Isaac Newton **and** being the word that a person such as Sir Isaac Newton utters, no other insisting on more reasons are required or even thinking that more testing is required to find acceptance for what a person such as Sir Isaac Newton declared to be correct. By going so far as to not accept Sir Isaac Newton and therefore not accepting that it is true that $a^3 = T^2$ which is but one point amongst so many other incorrectness never placed me in a position where I could receive the recognition and be awarded a degree. Then not having a degree is equal to not having the ability to write a book on physics because only having a degree allows one the mental ability to be able to write a book on the subject of physics.

In the book where I act as the author, I wish to leave you with concepts science present as being more correct as it is much more truth than what the Holy Bible may claim but when reading my presentation you are about to see how invalid those concepts are and while doing that I go about to explain the truth there is about gravity. This achievement is possible because I saw a way to break away from invalid concepts Mainstream physics hold. I rejected the claim made by Sir Isaac Newton in terms of believing that $a^3 = T^2$ as Newtonians believe. I recognised the impossible double standards Mainstream physics apply to promote their much shady explaining by accepting that when Sir Isaac Newton says $a^3 = T^2$, then you believe anything goes just because Sir Isaac Newton said you have to believe it. The inconsistencies brought them double vision and to compensate their incredible theories they simplify issues to a level where what they embark on to understand is becoming meaningless. What they say can't be supported and authenticated any investigation even in simple terms. It is as if they never read with interest that they explain and they never scrutinise that which they advocate. They give values that are senseless and make that which they say meaningless.

Sir Isaac Newton says a cube is equal to a square and that you are the same as your image reflected in the mirror…well I bet you never thought about the statement that the cube in $a^3 = T^2$ is equal to the square.

In this article I am going to investigate how much truth there are in mass pulling by the force of gravity. To most if not to all of the persons reading this and just the thought about me embarking on the investigating of the issue is totally senseless to investigate. There is a vision that it is senseless to investigate Kepler because the concept such an investigation would carry became synonymous to human culture and part of what is an accepted way of thinking It became the way we accept how the Universe are stitched together and this idea became the household practise of science controlling life.

Do you think of astrophysics as the department that is run by the wise and the level minded, the sober thinking and the absolute trustworthy? If you are a student there is no other choice you have. If you think those in charge of astrophysics are the pillars of trust, then get wise and read the following. What you are about to read is simply mystifyingly simple and yet to this day I have not challenged one academic any where that had the honesty to admit to the fact of Newton being wrong.

After you have considered the following you might agree with me that even small Children can reach a higher level of clear-minded logic and find more sensibility than what those scientists promoting astrophysics have because science lives in a make believe fool's paradise. If you are a student, then ask your Educated Masters to please explain to you the following abnormalities and inconsistencies they promote, which I present in this letter and then get wise instead of brainwashed. By teaching you to believe untruths in a method of forcing you to repeat what they declare as being the truth without proving it as the truth is brainwashing and mind control.

I say brainwash again because they force-feed you fabrications and distortions of truth, as you will come to see. They can't explain these facts I am going to present as they can't prove the validity of these facts but they hide the necessary proof of the facts that will secure that the facts but these facts stands as untruths when not presented with required proof other than Sir Isaac Newton's say so. Tell them to prove that planets have mass by showing how planets line up from large to small in accordance with their planetary mass. Tell them to prove that it is mass that generates gravity that pulls the planets as the planets are therefore heading in the direction of the Sun. Ask them to explain gravity in detail. Ask them to explain $a^3 = T^2$ in mathematical principle by using not the say so of the person in Sir Isaac Newton but where they can use other detailed cases and bring other examples in cases where this abnormality also presents a change in factor valuating just as Sir Isaac Newton suggested.

Science has the opinion that according to their ethos I am not allowed to write about physics because they don't like what I write about physics in example the madness that Sir Isaac Newton propagated namely that $a^3 = T^2$. To them it is fact that we either except that it is true that $a^3 = T^2$ or we face the choice of being branded as inconceivably stupid because if it is Sir Isaac Newton that said so then it must be true and if I don't agree then with the esteemed opinion of the likes of the person such as what a person such as Sir Isaac Newton represents then in that case it is I that know far to little about the complexity there is behind what Sir Isaac Newton is saying when Sir Isaac Newton says that $a^3 = T^2$. It is not my writing about what they say that I don't understand that they hate but it is that which I point out as being folly that they dislike. What they detest is my saying about what they have, being wrong. They claim I am not schooled to have an opinion but on the contrary, if they listen to what I say, they can't defend their opinion. For the first time in almost half a millennium there is a challenge on the dogma that science uses and I dispute the very basics they use to prove science. It is a question of questioning the opinion of those so opinionated, no one ever this far dared question their opinion…until now that is. Those Brilliant Minds guarding the ethics and the ethos that form the pillar of physics has a Universe expanding in a Big Bang in the face of Newton's contracting Universe by the power of all the mass found through the entire Universe. The Big Bang shows explicit expanding of the entirety, while the likes of Sir Isaac Newton say there is only a contracting Universe.

The Masters hold evidence of a massive shift of all material growing from all centre points outwards. This shift is expanding space by creating more space between materials. The Universe is growing or is expanding according to E.P Hubble. The Universe is expanding at a precise given rate according to E.P Hubble. The Hubble Constant is what proves that this massive expanding is in progress and the facts proving the expanding procedure according to E.P Hubble is so accurate that Physics used it to date the age of the Universe. Then we have Sir Isaac Newton claiming that the Universe is contracting at a rate that the total measured value of all mass filling the entirety of the Universe produces. They have the Universe expanding by Hubble's rate, which is so well documented it is indisputable while Sir Isaac Newton is claiming the Universe is contracting by mass. To cover this contradiction, they launched the second biggest fraud ever brought to mind. Physics has this contradiction covered by a theory called the Critical Density Theory. This theory is researching the problem about when the Universe is going to start following the guidelines instated by Sir Isaac Newton, but the reasons they are looking for is not WHY the Universe is not contracting at present, because that would disrepute Sir Isaac Newton's physics. No, the question is WHEN will the contraction start for proving that contraction is a viable certainty, well that were never in doubt. There just has to be sufficient mass for the contraction, leaving the only outstanding question being when would the mass start producing the gravity to bring about the contraction just because Sir Isaac Newton said so. After a search that was second to no other search before the mass wasn't sufficient to bring about contraction and to finally vindicate Sir Isaac Newton once and for always. When the mass was found to be wanting in quantity, the next mandate was to look for mass hiding in darkness because such mass backing up Sir Isaac Newton's has to be there. Why there will be mass hiding in darkness is only part of the question and just because the dark matter mass is hiding in darkness the dark matter mass lost the ability to generate gravity is still an unknown factor. Why is it that just because we can't see the dark matter mass the dark matter mass lost all potency in creating gravity…well that part too was never yet explained at all. What the generating of gravity has to do with the dark matter mass being visible is lost in the detail that never surfaced. They spend billions searching for concealed matter hiding from human view but never offer any explanation why is the hiding mass dormant by not already sustaining gravity? What has the visibility of the dark matter got to do with the dark matter's ability in charging gravity? No one ever asks why the hidden mass never started applying the contraction as Sir Isaac Newton predicted. The question in need of answering is if the dark matter mass is there now, why isn't the dark matter mass active in its ability to generate the required gravity at present as it should? Why is the activating of gravity suspended until such time when the dark matter mass will decide to end its apathy and start to contract? If the mass is there, it must produce gravity or be absent by not producing gravity! It is either working or it is not there at all. It can't linger while waiting for starting orders. This is a cover up that hides the biggest fraud ever committed by any group of persons and on a scale of one to ten as far as hiding fraud it must be close to a hundred. Could anyone just once explain why the luminosity or then rather the lack thereof would suspend the potential gravity and place the dark matter mass in a state of dormancy? What has the luminosity got to do with keeping the gravity pulling by mass dormant? Students, this is your chance! Ask those well informed academic Brainy Bunch with supercharged brain power to explain to you how it is possible that just because no one can see mass, as the dark matter keeps the luminosity suspended for whatever reason, then therefore the mass would disregard normal applying conditions and also disallow the charging of the gravity and therefore deny us the courtesy of the pulling power it should produce! Then the biggest question as far as I am concerned is when the dormancy will end. What will trigger and activate the dormant gravity. We should know this in order to braze ourselves for the enormous jerking that is going to come. Think the lashing in a train that suddenly changes direction in movement. Then place this action across the entire Universe and see what we should be expecting when a sudden and unexpected pulling direction suddenly flows through the entirety of the Universe and jerks every planet and star according to the mass they have in relation to the mass activated by the dark matter. This is the utter most lame thought out cover up for the biggest fraud ever devised to fool the largest group of persons but there is already an equal folly in place. It is keeping people guessing about Newton's mass influencing objects taking part in the falling process by putting a hammer and a feather in a contest to see which will descend first. Galileo never said anything about a feather and a hammer or about that objects must resort to being in a vacuum controlled environment but the Newtonian cheating fraudsters do!

Science teaches that a feather and a hammer have different mass while they fall equal in time through an equal distance travelled. Remember they always use the example of a feather and a hammer because then they can include the part of using a vacuum atmosphere. This explanation in structure presents so much eye blinding that all question fly out of any closed window. A feather will float except if it is in a vacuum atmosphere because if the feather is not in a vacuum atmosphere the hammer will pass the falling feather as if the falling feather never fell. The fact that the hammer passes the feather does not give mass or Newton credence but shows that a feather does not fall naturally. Putting emphasis on the vacuum filled atmosphere puts the atmosphere in the realms of being in a foreign environment that is very unfamiliar to us and very different to what we have. Therefore this will only apply that the hammer and the feather will fall equally when we are allowing such falling to take place on a foreign planet but it will not be valid in our atmosphere because we have an atmosphere filled with air that leaves feathers floating. This gives credit to Newton's claim of mass applying on our planet where feathers float and don't fall. All things fall equal in time and distance when subject to the same environment. By bringing in the vacuum atmosphere the normal person places such conditions applying in an alien environment that is very unusual for humans to encounter. It places the equal falling in situations humans will normally never encounter and by mentioning a feather in a vacuum they bullshit themselves right out of the corner where these detesting lowlife's painted them into by bringing mass into the falling equation. Never do they mention that humans and vehicles fall equally and humans and clothes fall equally and humans and coffee mugs fall equally and humans and…everything falls equally notwithstanding mass of any description playing a part in such a fall. That is what Galileo proved even before Newton was conceived! If gravity was mass related, then this was not possible, because then objects must fall according to mass. Falling objects bears no evidence of mass playing any part in falling. Any two objects holding different mass fall equal in time and in distance when sharing similar conditions, which suspends mass altogether as an influencing factor in the case of any falling taking place. Galileo proved different sized objects fall equally under similar conditions. That fact about Galileo, science does embrace because it is so well proven they can't escape the reality accompanying the facts although this strongly contradicts Newton's impressions about mass inflicting gravity, yet to your garden variety, housebroken Newtonian it is the same thing that people only spell a little different. It is the same meaning to different expressions…and how hey get that together in one unit where all is equal is as much bullshit applying as believing in the principle that $a^3 = T^2$. Acknowledging Galileo must make the work of Newton incorrect and also corrupt but science holds them as if they are agreeing. On TV we see everyday how all objects, such as cars, humans and bags fall at the same pace, and this is totally contradicting any view about mass being responsible for the falling. Mass can't provide gravity at all and the equal falling undermines Newton principles giving mass any credit in the process of falling. The formula $F = \frac{M_1 M}{r^2} G$ would suggest that mass takes all the responsibility for such falling taking place. Newtonians declare gravity as the force F, that is = equal to a gravitational constant G, that is multiplied by the mass M_1 and the mass M_2 after which the product of the three factors influencing gravity then is divided by the radius square r^2 forming the distance between mass pulling the mass that destroys the radius square r^2. If mass pulls mass as Newton said, the Big Bang is not possible, but the Universe is, notwithstanding Newton's claims, expanding (growing apart). If the mass destroys the radius separating the objects, then every comet has to crash into the Sun, but they don't. If mass forms gravity, every planet must orbit according to the velocity provided by mass, which they don't. All planets orbit the Sun equal in maintaining an even pace. Planets disregard mass as they ignore Newton's suggested cosmic laws. The truth is that mass only is the resistance of any independent material to deform. To acquire mass, the individual object relinquishes independent motion. When a falling body is captured by a larger structure, mass forms as the captured body shows reluctance to deform by resisting integrating into the larger structure because of density. Mass retains shape and remaining in form produces the mass. Mass stops the falling by blocking movement, preventing further falling, it does not sustain further falling. Gravity is the moving of the object to the centre of the Earth while falling. I say gravity is movement while mass is obstructing independent movement, which is what gravity is. Mass is not forming the factor responsible for gravity or movement, but prevents further movement. A body falls by gravity. When an object lands on the ground then achieving the status of mass comes through the obstructing that the body causes when preventing any more descending by further falling, while gravity remains present as a factor that brings the tendency or inclination to move or the attempt to continue moving as there remains a direction in which the body attempts to move as soon as the mass restriction is circumvented. Mass hinders movement and therefore mass can't enhance or produce movement or gravity. Mass prevents or blocks gravity. Gravity is the motion that defines the individual identity of any object's structural form by rendering motion while reserving independence in granting free space from other manipulating objects. When having gravity, the object has freedom as to move in a direction, but when moving the structure has not got any evidence of mass influencing or otherwise restraining the movement. Mass and gravity doesn't culminate, but opposes each other. My saying this awards me the cloak of death by Academics that deliberately ignore my correspondence as if I never addressed their mailbox. I keep writing and sending the esteemed academics those letters which is the same letters that they apparently keep ignoring and never reply to.

In the ignored letters I take it to task to try and draw their (the oh so wise and esteemed Newtonian academics) attention to the fact that mass is relenting by suspending motion and by retaining or incarcerating movement and by the arresting of the moving object such capture of free movement is what relinquishes the object's independence in forming a suspending of further independent movement. Gravity is relenting mass by retaining independence in motion. If said by using other words it would read that if the Earth captures the object and reward mass for such a capture, then as a result of mass being awarded the enjoying of the independent movement it had when falling is lost in the rewarding of the mass and such freedom to move independently is relinquished when afterwards free and unrestrained moving seizes to be part of the object's features. By capturing the object, this retaining comes with forcefully commanding movement as prescribed by the Earth and in such prescription mass comes as the result of the capture of the object. Academics know my arguments are solid, but they refuse to read or recognize any possibility that my work is founded. Mass is a fact by human interpretation, creating a usable human standard or a quantifiable norm and therefore my work does not affect general physics. However, in outer space there is no remote trace of mass. The proof I bring of mass and gravity conflicting, trashes other material written on astrophysics, including Steven Hawkins' book that sold tens of millions. There is no remote evidence of mass in astrophysics, except in a despicable manner where science has to bludgeon the truth to present facts. Again, I state that gravity is motion and that motion places all falling objects equal in movement. That is not the way mass influence objects because by implication thereof does mass bring differentiation and falling objects all descend equally. Notwithstanding large or small, objects fall at equal speed.

An Open Letter On Gravity Part 2

The object becomes associated with the Earth and becomes part of the Earth and because the mass gives the object the honour to be part of the Earth, the mass part that the Earth has to supply in the formula $F = \frac{M_1 M}{r^2} G$ falls away as only the mass of the body and the movement of the body holds a relation to the movement of the Earth which is better known by the word gravity. When the equation applies in the format it should then the radius between the mass seizes to be a part of the equation and also the mass the Earth has to provide seizes to influence the formula when $F=mV^2$ or as Einstein equated this $E = mC^2$ which should be $E^3 = mV^2$ where V^2 takes on the value of the speed of the electron which is the speed of light (C) or then as Kepler put it $a^3 = kT^2$ which is all the same thing and hold no apparent, but the final analyses show that Sir Isaac Newton's idea of including the radius or having the mass of the Earth as a factor forming part of the equation never formed part of the entire concept ever.

While Kepler and his work were considered a closed case, at the same time, not one of the following principles was yet successfully proven but I believe I have accomplished that goal. I first started my studies in the field of Cosmology as a spontaneous development of my natural curiosity spawned from childhood interests in the field of cosmology, which I developed even before I went to school. The studies were a reaction (I would imagine) that was part of my personal childhood development in how I was forming a personal concept of a lifelong interest that followed me into my future. At first I conducted all my earlier studying mostly on the basis that inspired me to find out more about what made the Universe tick, with no intention ever on my part to reach a point where I would be writing books on the subject. At first, I was investigating cosmology on a part time basis. This went on, on and off, or the best part of twenty odd years (*as* time and *when* time would permit). Then in later life with my health deteriorating I committed myself to more intense investigation and my effort developed onto involving a study using time that is only permitted by a person when that person is involved in such a quest on a full time basis. That quest has now been going on for the last seven years in full devotion and if one includes all the years invested on my part including the twenty odd years before, part time, then the time I have spent in completing my theory when adding all in comes down to almost twenty eight years. This is to say that I did not come to realise what I am about to introduce on a light-hearted conclusion. I mention this because I wish to ensure the reader that he should have no doubt about my most sincere commitment in producing a cosmic theory on matters concerning the start and the working of the Universe during and before the Planck era.

At first, I began by arguing that there is a "something" that is blocking our investigation in progressing going behind the where the Big Bang prevent further progress. There is some barrier preventing humans passing a threshold whereby our understanding will pass such an obstacle. If there were any way that anyone may break through that barrier which is preventing normal research to go pre-Big Band, it would be accomplished by finding the barrier whereby the vision we use to focus would pass such a limit. If we wished on progress in our pursuit of the very first cosmic moment then we have to find and cross the barrier that blocks our view. We have to look deeper and in another direction should the desire driving us be strong enough to commit us to reach into the very birth of the cosmos. We have to rethink the strategy we use. Max Planck was one of the most brilliant men of all times and even he, notwithstanding all his personal brilliance, accomplished little. There are parts missing in what we have and that which we have at our disposal to use.

If there was no such an obvious barrier then the Wise-Men involved in science would by now have found the way to break through the seal that is locking us out of the critical past which will uncover the origin of the Universe's infancy stage. I went about trying to find what everyone since Adam, (meaning all of the rest of mankind and myself) were missing throughout the ages of speculating and interpreting while philosophising about whatever we find inspirational. The obvious we saw; that was clear. Therefore, I had to find a route that would lead into the not so obvious which all of us were missing, notwithstanding the best efforts of the best qualified to accomplish such a breakthrough. My efforts involved trying to accommodate that which was in the cosmos available to use by the cosmos in all phases of developing. If I had any hope of finding the answer, such an answer had to be simple because I am not very inclined to unravel what is deemed as complicated. The simplicity had to be locked in what was not yet understood about that which was in the cosmos as it formed part of the process used in forming the cosmos. My realising this brought me to focus not on that which we understand.

There is not a lot we actually understand because even gravity is very poorly understood. In fact, gravity is so poorly understood that there is not one person alive that can claim the prestige of understanding gravity and among the dead there is even less that can make such a claim. Several phenomena are presented in nature and acknowledged by science but also discounted by science and therefore not presented as accepted science. By admitting that that what we have available to us to use concerning our research of cosmology in an attempt to better our understanding of cosmology, is useless to use, then one realises that not having what there might be makes what we already have useless. It then is useless to use what there is as part of the big picture we are trying to paint because what we use is not really part of the picture. This leads one to believe that the picture of the cosmos Mainstream science is painting, is being painted without painting a full picture.

In my first attempt to understand the full picture of what science was painting I found so many colours missing there was no picture painted that anyone could appreciate. This is what made me decide to go on researching the 'unknown' in the hope it might clarify the 'known' and as the book unfolds you as the reader may agree that I was correct in pursuing the misunderstood and rejected phenomena. Finding the missing phenomena helped me to place the phenomena mentioned above in a theory where the principles also mentioned above form a part of the overall gravity used in binding the Universe. I believe what is in the Universe is not able to be coincidental because of too many influences contributing to what there is - notwithstanding the fact that this is the manner science uses when they refer to the Bode law. What is in the Universe has a role as it had a role, which is the same role, that phenomena *s* had and in future will have.

This is establishing a very new idea about the working relationship between particles and in explaining it by using Kepler's studies. Redefining the work of Kepler's views brings a new Universe to light involving new concepts that are based on old principles but principles in updating man's view about cosmology are very new in that capacity. Through that new vision I was able to come to realise what the reasons might be why Kepler never saw it fitting to include the measure of Π in his formula. I do not suggest his neglect thereof was intentional, nevertheless the formula he devised without using Π proved that there was no need for the inclusion of Π since his figures brought about a correct answer in the final end result leaving a well concluded fitting answer. The numbers he produced brought about a specific space **a^3** contained in a circle **T^2** at the distance of k from a defining centre thus the calculations did not require the use of Π to find a meaning. In that, Kepler did not see a need to include Π. I would not go as far as declaring with absolute certainty on his behalf that he did it deliberately, however there never arrived such a necessity. It is prudent to agree on whether or not such a need is necessary, because if one is agreeing about such changing not being required a new Universe emerges.

The circle that Kepler discovered came about without ever forcing Π into the frame because it is clear that the circle formation came about as a natural consequence and came spontaneously delivering an equation while he was working. In this book, I prove that the reason for adding Π to the rest of Kepler's formula is unnecessary. This unnecessary addition is because when going one step further in the investigation one will find that k and a and T are symbolising the same value with the only difference being that each one represents a different dimension to our six dimensional or six sided Universe we enjoy. In fact I shall show that Π replaces "**a**" and "**k**" and "**T**" and that Π is the true value that should be replacing each factor as to indicate the correct value to the sides nominating Π. We humans work on a numerical base using ten as a basis where we count to nine and re-establish a new decimal numbering line by adding a nought behind the number in value. This is using the numerical basis of ten, which I suspect we took from ancient knowledge about cosmology and not from using our fingers and toes as the earliest calculating processors. In this letter, there is unfortunately no room to explain my suspicion but another fact I do prove is that the cosmos uses Π in the cosmic numerical basis as a means to measure and quantify.

Therefore in fact the Kepler formula should read instead of **$a^3 = T^2 k$** as it does it must be $\Pi^3 = \Pi^2 \Pi$ where I shall show that Π represents singularity wherefrom the entire Universe sprang from Π and by forming as $\Pi^3 = \Pi^2 \Pi$ it is confirming that space is equal to the motion thereof. Kepler's greatest achievement was showing that the cosmos is space –time **$a^3 = T^2 k$** while time is the motion of space in space. The value of Π is the primeval and most basic of measures applying as an accepted cosmic legal value that the cosmos used exclusively in the very beginning and as it does today. The measure of Π in the Universe, values particle development that brought about all development ever conducted in the Universe. Only after this stage did the rest come including mathematics and went on to freeze spilled singularity into frozen material. Reading this statement may sound suspiciously senseless but as the book unfolds the sensibility will become apparent. The full implication of such a statement will become clear when one dissects different facts coming from studying Kepler. My discovery of this fundamental basis of legal valuing ensured me again that there was no need for someone the likes of Newton to add Π in any form to the work of Kepler because Kepler discovered the ultimate Π in the Universe, the Π giving the Universe form and gravity. The concept of Π that is the only single form of all other forms available that can by duplication of Π assembles the value of gravity. When replacing the symbols with Π the facts of the Universe become self-explanatory because the most basic form that forms the cosmos has a definitive and uncompromising value. But getting this far took me down roads overgrown by ignorance and which I had to uncover myself as if hacking away miles of overgrowth with a machete chopper. All of the disbelief science showed to my work in the past and their refusal to see past Newton made any and all attempts on my part as bad as they could be, strangling and smothering my attempts to announce the uncovering of the newly found insight on my part.

For decades I tried to come to terms with the inability there is in science to explain the cosmos in real terms, when using the science of official reputation. That which there is makes a mockery of science because the undisputable clues left in the cosmos makes what little correct explaining there is available, seem like a comedy of errors, when it is mixed in with all the other near Dark Age errors we still use after so many centuries that provided countless opportunities to revise the old muck. By applying current accepted Astronomy, as such the phenomenon found all over the cosmos is still beyond the explaining ability of Mainstream science. This is true and it is a shame because it also is an undeniable fact in spite of the vast knowledge and progress in other forms of science taken in the manner science uses when it approaches cosmology. Cosmology truly lagged behind while the understanding and advancing of physics, mathematics and chemistry as subjects were flourishing.

By comparison I saw how little there was available in explaining cosmic phenomenon and how much improvement in understanding the other departments such as chemistry, electronics, medicine etc. could offer as results were coming about from research. Even where there is a little explaining available in cosmology it turns out that such explaining is confusing to say the least and at best it highlights the manner in which science is applying double standards. For decades photographs were the only progress forthcoming as an addition to improve the meagre field in cosmology and that improvement was artificially stimulating cosmology. By providing a false impression of advancement, everyone missed what and how much was missing…To the connoisseur desperately looking for more than the obvious stirred in with some out-dated misinformation dating back to the Middle Ages, it all seemed as if it was a picture portraying the ridiculous to make the sublime look good. The pictures only proved the opposite of what progress in cosmology will represent. In truth and as such in cosmology the cover up that was hiding the lack of progress about the science of true cosmology was only forthcoming in the improving of electronic optical telescopic advances and spectroscopic progress. There were only photographs carrying beautiful pictures which pleased the less informed except the photographs did not bring progress to cosmology at any intellectual level by promoting insight. The explaining that the photos demanded about the subject had the opposite effect of installing hope because what it did do was underline what lack in any notable progress there truly is in our understanding of cosmology and laws in the cosmos.

While such Hubble telescopic images might seem to be clear, as daylight it was more than clear there was little academic value to them. To the person in need of more stimulation than being impressed with pictures of God's marvellous Creation and the sightseeing that always accompanies such pictures, such persons always felt very disappointed. The pictures did give satisfaction to those more easily impressed, but the rest of us seeking knowledge accompanied by understanding the images left us despondent. Although they leave the vast majority in total amazement, there are those less impressed about not knowing the 'why' and the 'how' in such amazing pictures. I am aware that the group I fall into may be the greater minority and the majority may only demand the portraying of the images, which is what that "easily satisfied" group demand. The rest of us rouse with anguish at the lack of information about what is known, and what lies behind what those pretty pictures are conveying. Nevertheless, there can be no real progress in scientific understanding about the images portrayed by the Hubble telescope, and others, if no one is able to show the slightest clue of a deeper understanding of what is going on in the Universe. Everyone is almost breathless waiting for commentary by the most informed, which accompanies the magnificent cosmic portraying of God's Creation. When we are portraying the new images, we should also be investigating that what we see the cosmos is at that moment portraying. The lack of actual believable explanation coming from investigating by means of telescopic imaging should impress everyone, but the impressing must not be based on the colours in the images but the sensible information attached to the image investigated. It is that, which we wish to see. What we wish to see must at least be accompanied by scientifically backed information, which provides the proven understanding coming from science. When science is employing new explanations with such photos it should also be discarding senseless baggage carried over from the past.

Most images contradicted Newton and for saying that, every Academic I ever came across in the past ostracized me. That bothers me little! I know I cannot possibly be the only person absolutely discontented with what Mainstream science accepts as science. Here I refer to the out of date theorising Mainstream science still accepts amongst many others as how they suggest stars and planets are forming. One cannot promote cosmology in honesty and advocate scientific fact whilst dishing up such fairy-tale nonsense to students. Moreover, I hold the opinion that amongst Academics, in particular, there must be many if not most that share my personal serious doubts or have an inclination to share some of them. This I say when considering the overall doubtful picture painted about what there is and what one believes there should be. I just cannot believe those forming the most intellectual group of mankind are unaware of the mismatching facts seen over the broader picture because the contradiction and lack of a plan, makes what there is so very doubt provoking. Newton dismissed the formula Kepler presented as all factors forming motion. That is where the apple cart derailed.

Time starts when the top jumps into motion and the dot comes away from the spot. That starts off that which is coming from one spot so small eternity meets infinity within on that spot. One follows another as one comes after the other. One more follows another one that leads as the motion sets in and as the atoms start to move in sequence. The one atom leads the other as it in turn is followed by more atoms coming in an immeasurable stream of atoms grouping as a unit consisting of many forming one. They continue coming until there are a countless number of dots.

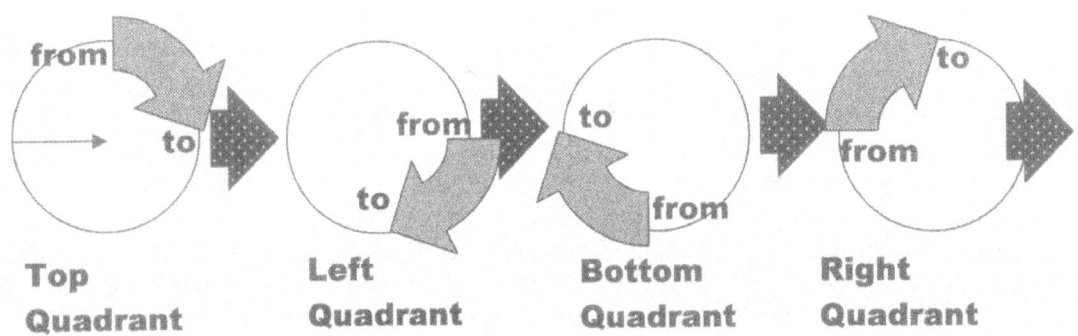

Top Quadrant Left Quadrant Bottom Quadrant Right Quadrant

From this the atom came about and it is the atom that gave the Universe the six dimensional time related structure where every aspect of space had four rotational positions that determines movement of space in relation to movement in positional changes coming in accordance to the centre position. The centre aligns with a rotating Universe that can be the electron moving or it can be the entire Universe repositioning.

The accumulative size of the dots are the same size as one dot because in the true Universe big and small plays no part. The dots are infinitely small and eternally big at the same time because size is a relevancy and without one the other has no size. So in the true perception, there is no difference in size. Without motion parting eternity from infinity, which allows time to be a spot, we can find such a spot within the centre of all spheres. Inside the sphere there is a point holding singularity. At such a point all lines coming from the sides as straight lines that cross in singularity in the very centre and that point can never be not there. That point has to be because the Universe is there and the Universe is what that point is...a holding place for all that is possible to be within the concept of the Universe. The entirety of the Universe is a sphere placing that spot that holds singularity within infinity everywhere. What parts the spot that becomes the dot is motion. When motion applies the spot becomes four dots holding three dots relevant. The seven dots move within five dots on the one side to five dots on the other side and that becomes ten dots.

It starts with the fact that there is no place or part in with which one may associate zero or nothing. There are no room for a number such as nothing. Next to the one dot (infinitely close) one will find the next dot, and if nothing was a factor then that is precisely what one will find between the two dots.

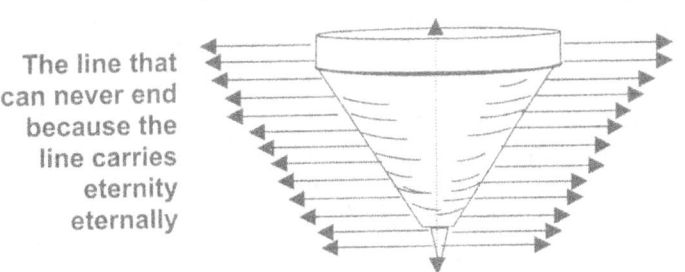

The line that can never end because the line carries eternity eternally

Nothing of space, a non existing entity, taking up no space, and much more important, no time, therefore the dots are infinitely close to one another, being the same space, eternally big as much as infinitely small. If we as humans cannot find a manner in comprehending this notion, there can be no manner ever understanding the cosmos as much as the start to the cosmos. Every dot is a Universe in its own and the accumulation is a Universe. The Earth in itself is a Universe as the moon is a Universe, because rules applying on Earth do not apply on the moon and visa versa. When in the ocean, another set of rules applies; therefore being in the sea places a body in another Universe.

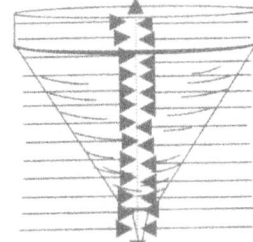

The line that can never start because the line comes from infinity with no starting point eternally

The number of Universal entities is still countless, as much as it was at the beginning. Every dot insignificantly small as it may be, is a part of another Universe as much as it is part of the accumulative Universe and every dot in the infinity holds singularity, which we translate as " nothing" being " darkness". There cannot be "nothing" just as much as there cannot be "darkness". There cannot be something big or small, but is holding relevancy of perception, and then the relativity of perception becomes the question. There cannot be hot as much as there cannot be cold. The Sun FREEZES hydrogen to a liquid at six and a half thousand degrees Celsius and the Universe boils over in the form of the Hubble constant at the temperature (we presume from our vantage point) at minus 273 degrees C. If we Humans cannot or will not abandon our human perception and our manly perspective, we may as well return to astrology for all its worth.

Every point in the infinity we may observe is not merely part of the Universe in not being nothing, but is the point where the Universe started representing singularity. It is the very first point where everything began so many eternities ago, because after all, how can we ever determine where the first point was, as they were very much equal and alike at the beginning. Every aspect of the Universe started with the fundamental fact that no point in the Universe can represent "nothing" as a number, because every aspect in the Universe represents singularity in what ever form it may hold in that specific spot forming space-time. Every point represents the start of a line and all lines start at infinity. If man does not reach a conclusion where that conclusion is matching the Universe and stop to match, the Universe with man needs to survive and maintain the cosmic centre where man wishes to be (and man's incapability of realising man is not the reason why there is a Universe in the first place), we may all go back to caves and become starving hunter-gatherers again, because we will never find a way to progress to the ultimate understanding of the Universe. After personal research on an extensive basis (when leisure time permitted) the past twenty-seven years, and in depth study the past six years, I completed the version of my book, which I named **THE THESIS**. I have first tried the commercial route in order to introduce my work with the book I named **THE THEORY**. As there is little commercial value in such an academic book, I have not been able to find a publisher willing to publish the work or so I thought at the time. I then tried to publish the book I renamed **THE HYPOTHESIS** with a greater emphasis on proof than the previous book, but the only route I can follow down that alley, is through the Internet. This route is seemingly impossible because the people in search of information on the internet, is not capable of understanding the work and those that can understand the work does not search for information on the internet. After the disappointment that this route proved, I decided on an academic version with all the proof I hoped would satisfy academics.

This forces to find a promoter willing to allow me to follow the route through the academic institutions, should I ever have any chance to gain recognition. I have now completed the book version that I named **THE THESIS**. As I am of the opinion that the work in research may prove valuable to science, as material for further study potential, I feel it worth the effort to promote. By it self, it does not change the view science holds on cosmology, and on the other hand, nothing science at present views about cosmology is correct and therefore it lays groundwork for a change in the direction of further study. I submit a summery, indicating the line of argument this book takes on cosmology. I do realize the method of calculation is new to science, but it proves many aspects that science does not understand.

This book **"MATTER TIME IN SPACE: The Thesis"** proves the following:

1. The Universe is in a constant balance between positive space-time displacement and negative space-time displacement. The heat distinguishing matter's particle individuality has a definite and crucial position in this motion of gravity. All bodies' particles down to atoms apply both positive and negative space-time displacement in relation to each other's position in occupying space in the duration of time.

2. There is no force, other than "life". I have managed not only to bring prove of The Big Bang and indicate that the Universe at that point went from singularity, but also how matter and space came about from the original value of time. In conjunction to two newly defined, but very well known cosmic principles, I manage to take the Creation of space-matter-time back to previously unproven territory. Through that evidence it becomes clear what time is and how time changes in accordance to space-time developing.

3. Space has no value, other than that which time lends it. Therefore, time determines the value to space. This also is a deduction, which I prove from the

formula $a^3 = k\ T^2$. Under normal circumstances, the major component having the biggest proton value also carries the circular displacement value or positive space-time displacement value. The bonding value of gravity runs through the heat separating particles, and that mass by number of particles alone does not carry gravity. Gravity comes about from the density that particles have in their proton mass. Hydrogen and helium contributes to almost no space-time intensifying, and iron $_{56}$, in association with Cobalt $_{59}$ and Copper $_{63}$ that holds the Universe in space-time.

4. The space-time, which, is to the "outside" of the electron, is as much part of the atom as the space-time between the electron and the neutron, which, again is as much of the atom as the space-time between the neutron and the proton.

5. Time located in space is moving to time located in matter. Matter is growing while space is diminishing in terms of density. That which the growth matter obtains, are depending on the amount of protons forming the concentration of time in that particular location of space forming the concentration of space-time.

6. All stars developing in this era has to have an iron inner core which forms the last link a star has to space-time, before becoming a proton star, consisting of the single dimension also known as singularity.

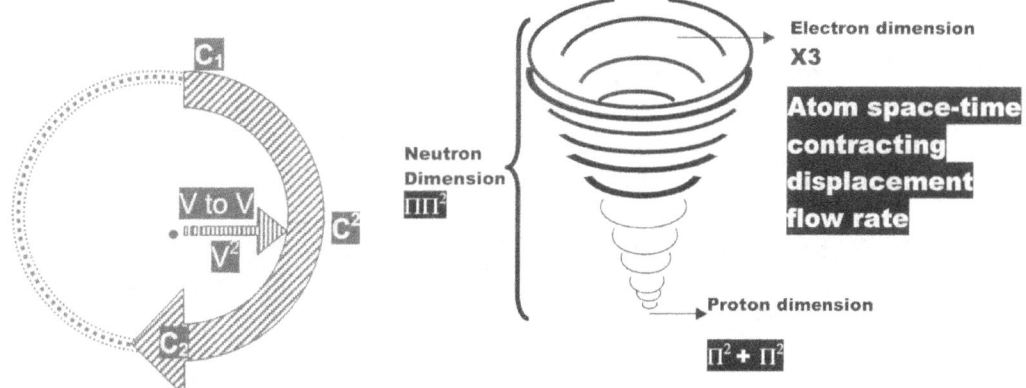

7. Time is located in the proton and not only in the Universe as a whole. Time is everything space has to offer in between such points. Through this I show how the atom influences the cosmos by means of the dimension changes 3 to Π, Π to Π^2, Π^2 to $\Pi^2 + \Pi^2$. When multiplying all these dimensions the answer is 1836, the mass difference between the electron and the proton.

8. The speed of light is not a constant, but depends on the ratio between space and time, whatever the value to that might be in any location of space allowed by time. The speed of light also has a linear component $3\Pi^2 = 29.6$ (the photon) and a circular component ($3^3 = 27$) the wave. The combined value shows the limit of light's photon displacement potential within the most intense gravity.

9. The Doppler Effect is a manifestation to the true cosmic law the Titius-Bode principle that I renamed as the Titius-Bode law. The Titius-Bode derives its space-time value from the 7^0 inclination all spheres hold to space-time, forming the linear contingent to space-time and Π^2, which is the circular displacement value to space-time.

10. Electricity is the revaluing of space-time from a linear stance to the speed of light in a circular stance, forming an alternating of space-time between the linear and circular space-time in and between atoms of the conductor. Therefore, lightning is not electricity, but lightning is moreover wind in its most advanced form or gravity in the most intense form it can be.

11. The Hubble constant is not a universally applied single confirmative value to time's expansion, but each galactica performs to its particular time value to space. The growth rate is in accordance to the Titius-Bode number arrangement of space-time (7/10) (10/7) and in its three dimensional value is 3; 6; 12; 24; 48; 96 etc.

12. The Universe came about not from one "Big Bang" event, but seven individual cycles of light and darkness, and (**whether you like it or not**) just as the Bible indicates. The seven different periods (that are mathematically proven) from the **inner-Core-value** confirming the time related to each era where the value of Π changes every time.

13. The Earth and the solar system also came about from seven individual separated periods of light and darkness, positioning every planet in accordance to the Titius-Bode rule **NOT APPLYING** WHERE AS IN NORMAL CONDITIONS, IT SHOULD APPLY.

14. Gravity is **strongest** when and where **space is least**. That, I admit is old news but I came to realise why that is a reality. Where **space disappears** and space **finds singularity** in the **centre** of the **sphere, that** is **where gravity** is at its **ultimate.** That is the location of the strongest gravity. It is in the place that the heat is the most, which is in that centre area of any sphere. That is where Einstein saw gravity draw space flat.

15. Singularity is present in motion as motion charges space-time.

This is only a few, of the findings I publish in this book; however, **IT IS PART OF A BOOK NAMED MATTERS TIME IN SPACE THE THESIS**. Armed with this information and still very impressed by the honesty being one of the admirable qualities Academics hold, I set out to find an audience. How naïve was I not.

At first I wanted to find any academic willing to listen, and introduce such a person to see what I saw and then allow that informed academic to take the matter from there. I was ready to hand over my ideas and disappear into the vastness of the crowds because I was very aware that being amongst the common folk that placed me amongst all those persons that was with me in the place meant for me because I am a qualified mechanic and no physics professor. I was also very aware that I did not have the insight and the means to press the issue to the end and was only on a quest to find the right person to whom I could hand the material over. This hand over I was prepared to do and would be the first academic I could find that was willing to lend me an ear as I explained the issue.

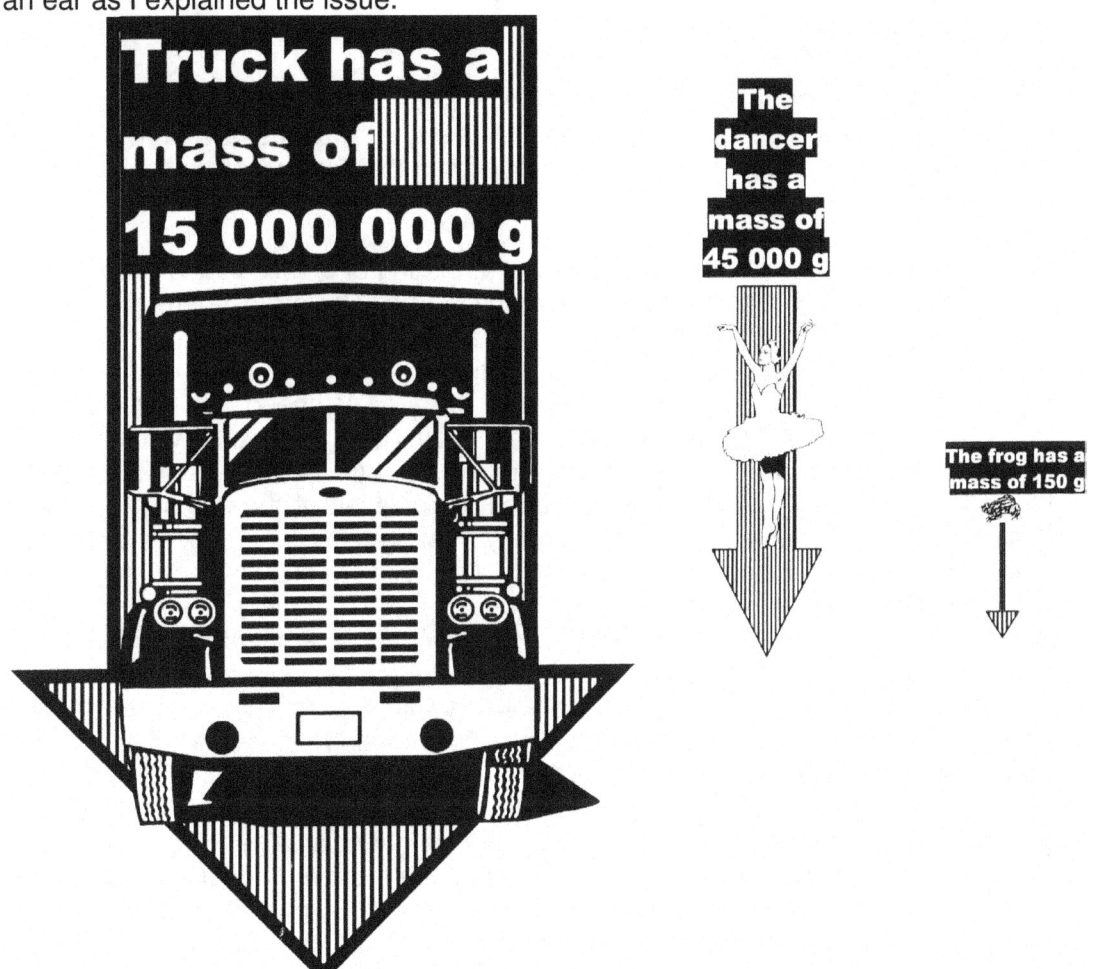

I wasn't that naïve as to believe academics were not aware of the predicament about Newton not matching reality, but I was under the impression they used that fable until something better came their way. I thought if they had a look at what I presented they would immediately catch on and take everything from there…after all I was dealing with the most powerful brains that the human race had to offer. I went looking high and low; far and wide and the more I went on the pursuit of finding one academic willing to read my work, the less I came across any academic that would face the music that was prepared to admit to the reality of what Newton left his audience. Not once was there one academic that was prepared to listen to my plea.

At first I was filled with respect for such noble gentlemen as the ones filling the shoes of the Academic physics profession. But as my quest brought less and less fruit because I ran into more and more rejection from the academics that I contacted, the more my suspicion rose about the true nature of the affairs. It later became apparent that the resentment was aimed at me in what I present to them and their dismay was not about the quality and the correctness that my work delivers. Yet I placed all the failings I attributed to my success in convincing the academics at my door. Every time I went back and seek a new approach to the concept as to find better meaning to my work.

It drained my funding totally and I got nothing back for all the effort and the money I spent in trying to charge a reaction. The only response was just a vacant silence; they did not even E-mail me in response acknowledging that they received such a book that I sent their way.

I say what I prove and I prove that gravity is a rotating of a solid that is moving through a liquid space. The Earth is a solid that much is true. The atmosphere is regarded by physics to be a liquid and that much is also accepted as true. Gravity is the Coanda effect and the Coanda effect is where a car tire spins through water and the spinning wheel gathers the water onto the surface of the tire. At speed, the tire picks up the water and secures the water around the tire. The motion of the tire, which is the solid, contracts the water, which is the liquid onto the solid tire surface notwithstanding the thrust of rejection of the liquid coming about as a result of the spinning momentum. The contracting of the liquid onto the solid by rotating motion produces the gravity that attracts the liquid to the solid. That is gravity. The solid tire can cover the surface of the tire by a layer of water where the water is as hard and as sturdy as what the solid tire can be. But the water might be hard and sturdy, yet it remains a liquid with all the characteristics attached to liquid. That is why driving in the wet is so dangerous. The solid tire can surround the tire surface with as much as one inch of water. That is gravity. When the solid tire spins the rotation coming as result of the spinning, contracts the liquid water onto the solid tire. The water being a slid in the form of ice can't perform in the way the liquid does when the liquid is surrounding the tire. An inch of ice will never be strong enough to carry the weight of a car and to allow the car to drive over it but in the case of the Coanda effect the motion of the tire allows the water (still in liquid form) to be much stronger. While when being in the position where the water is surrounding the surface of the wheel through the spin of the wheel, that contracting gravity then makes the water as strong as the tire which enables an inch of water to support the entire car running on the water. The spin of the tire produces a gravity contracting by rotating motion, which turns the density of the liquid water to the same compactness as the tire surface being a solid, will have. The spinning solid of the tire turns the density of the fluid water into a solid equal to that of the solid tire. The tire asserting a rotating motion does the producing of gravity by motion. The expanding of the rotating action produces a contracting of the liquid space it moves through. By expanding, the solid tyre that is rotating, takes away some of the space that the liquid holds and that gives the liquid in that space solid characteristics. The tire rotating is expanding the space it holds. That is called fleeting momentum. The matter tries to move away from the centre in an effort to gain more space. As the tire spins the tire, acting as a solid, it tries to capture more space and the tire can thrust this so hard that the tire does go oval. The tire tries to hold more space than it has because it is capturing more space in an effort of expanding.

While the tire tries to capture more space, the tire also reduces space that the liquid water holds. By capturing the space that the liquid water holds, the tire is capturing water, and in the effort the tire is trying to gain the space that the water has and in that the solid tire is making the liquid water solid. Therefore, we can drive on an inch of solid water while the water is liquid. By the water contracting, we find the density of the water changing where it meets the surface of the solid. The solid tire expands and in the expanding it makes the liquid water solid. Then where the liquid water finds itself being reduced and being made denser, the liquid water gets so dense it becomes a solid water area. The tire wall turns the water into a solid by the rotating action of the wheel. That is when motion applies to the wheel. When considering the wheel at speed and the Earth at speed we tend to think of the Earth being still and motionless. The Earth is spinning at a far greater speed than the tire would ever be capable of. The Earth is spinning so much it is concentrating air to become a liquid. This I try to tell the esteemed Newtonian's but their time is far too important be lost on a soul like me. What can a motor mechanic that does not understand Newton tell them?

When objects fall they have no mass and this is in spite of all the claims the Academics in physics try to produce. Galileo said all objects would fall at an equal pace and hit the Earth at the same time when falling the same distance through the same air under the same conditions. Newton said mass is responsible for that which produces the gravity by which objects fall. That means the object being more massive must fall faster than the object being less massive. If mass brings on gravity, mass must distinguish the amount of gravity by applying a bigger, more or a lesser falling pace. If that does not happen there is no evidence of mass applying because then all objects hold equal mass while descending to the Earth. We see frequently that the object can fall at a specific rate depending not on the size it has or the shape it has but the distance it travels through the air. We see so many times that a car drops from an aeroplane with a human falling next to the car. There is a car advertisement where the human that is falling has a parachute in a bag falling next to the person that is falling next to the car. The car is around twenty times more massive than the human and the human is about twenty times more massive that the bag containing the parachute. If the falling process depended on mass that instigated the gravity action as the intellectuals wish to declare, then the lot cannot fall at the same rate. The car must fall twenty times faster than the person and the person must fall twenty times faster than the bag. We can see that the lot are falling at the same rate and the descending bears no implication to any mass that shows differences. That is what Galileo said when Galileo said all things fall equal at the same rate and land at the same instant and a massive object will land at the precise instant that a very light object will land, on the condition that they are dropped equally and that they fall through the same space at the same time. That statement excludes mass from any part that gravity has.

From that, one can see that the falling has no implications brought on by mass. Mass has nothing to do with gravity but to restrain any further gravity effort putting individuality to the item falling. The gravity is produced by the moving of the object and while mass restricts the moving, the moving still applies as gravity because the moving remains as a tendency to move when mass applies. Even where mass stops the gravity moving, the gravity remains as a tendency to move towards the centre of the Earth. It is not that the mass is pushing but the mass is stopping the moving of the item to the centre of the Earth. Mass only applies when any further motion of objects are restricted. If the restriction of the larger object that inflicts the mass suddenly also starts moving, the mass turns to gravity that very instant. While the larger object retains a position of preventing the independent object from any further individual moving, mass comes in as a factor that stops further motion. Mass counter acts gravity by stopping gravity. The object only obtains mass when gravity becomes no longer applicable. When the objects all fell and they all hit the surface of the Earth at the same time. Only then, after contact is made with the surface of the Earth, is there distinction about size and mass. That makes mass a factor of the Earth holding a restraining on the object and then mass is not part of gravity because gravity is part of the falling or the moving of the objects. When objects fall they have gravity and the gravity is equal applying to all because the gravity is the moving of the objects without restriction. When the objects hit the ground, the objects lose independent motion and with the accepting of mass the objects retain the motion that the Earth provides. The mass renders the objects the motion of the Earth and having the motion of the Earth they move at the same pace as the Earth. The mass then makes them having a relation to the Earth where the mass puts them in a restricted part of the Earth. They move at the rate that the Earth move because they then are part of the Earth by the provision of mass. Mass shows how much the object that then landed on the Earth and no longer moves independent of the Earth, then became Earth.

The gravity is the Earth forming the solid that rotates and the air is the liquid through which the Earth rotates. The liquid is contracted onto the surface of the Earth by the solid of the Earth trying to expand into the space the liquid holds and this happens due to fleeting momentum or rotating motion. Our distinguished academics filling the upper class of thinkers and forming the super league of the intellectual supremacy live in a sphere they create where everything in their Universe is tailored to be subject to their needs and their requirements and what will suit their comfort. The one academic gives homage to the greatness of the other academic while the one giving the homage receives it right back from the one getting the homage. They lick one another and smell each other's importance and admire one another's prominence while enjoying the comfort in the knowledge that only they may groom one another since all others may wish to have, but has not have such a privilege to touch their greatness. They are the supremely blessed that are living in a class that they share with no mortals. Only they are filling the class where one may relate to the others also filling that class and only those being in that supreme class of society know what intellectuals think or may think is true. Only they are filling their norm as the top estimable centre of the privileged Estrada. They have the capacity to know that only they can form an opinion in the knowledge that only God will fully understand the correctness of their opinion. They are the only intellectuals and that gives only them the ability to think whatever supreme beings are able to think. From the cloud the ride, they have reached the limit of the super intellectual capacity, which they share with others of the same esteem and stands apart from other mortals that they have to share an Earth with. In that sphere, they are also being the only ones that have the prerogative to share such space with the other of their class that has franchised intelligence to their class alone. They have the same secluded world of the advantaged because only they having space in that sphere are the ones being liberally blessed with mind and thought because of a much superior intellect. Only they are the cream of the thinking class to raise an opinion on any matter. It is a small wonder they put their view in line with God Almighty and that is the only entity they find worth sharing a thought with that is equal in comparison. No other being on Earth has the right to an opinion. No other being may think because thought is reserved to their class only.

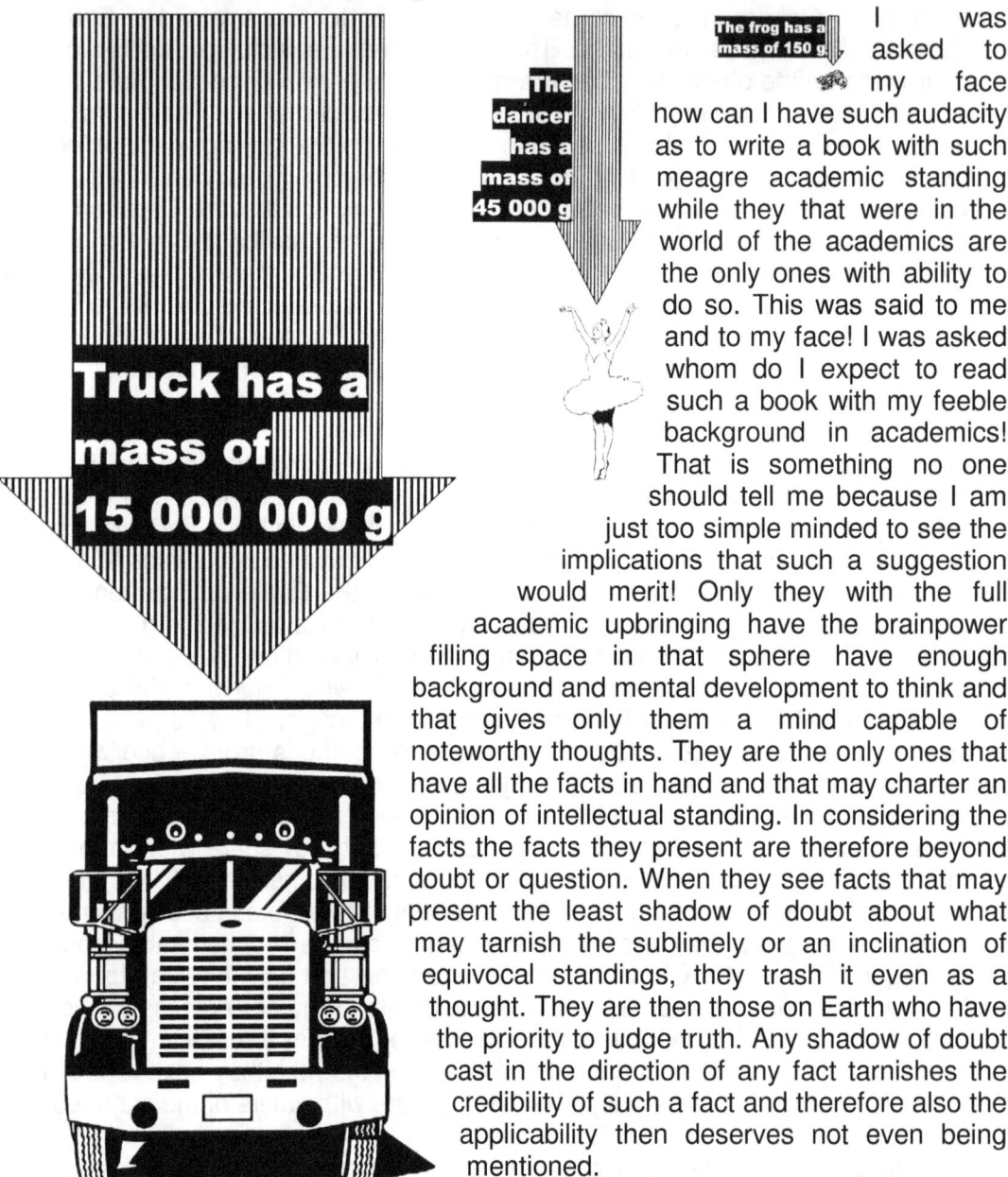

The common drop height from where the three fell

I was asked to my face how can I have such audacity as to write a book with such meagre academic standing while they that were in the world of the academics are the only ones with ability to do so. This was said to me and to my face! I was asked whom do I expect to read such a book with my feeble background in academics! That is something no one should tell me because I am just too simple minded to see the implications that such a suggestion would merit! Only they with the full academic upbringing have the brainpower filling space in that sphere have enough background and mental development to think and that gives only them a mind capable of noteworthy thoughts. They are the only ones that have all the facts in hand and that may charter an opinion of intellectual standing. In considering the facts the facts they present are therefore beyond doubt or question. When they see facts that may present the least shadow of doubt about what may tarnish the sublimely or an inclination of equivocal standings, they trash it even as a thought. They are then those on Earth who have the priority to judge truth. Any shadow of doubt cast in the direction of any fact tarnishes the credibility of such a fact and therefore also the applicability then deserves not even being mentioned.

When I say mass is not supporting gravity but it is restricting gravity, it then is I that is without merit because I gave birth to the idea in the first instance and they were not responsible for establishing such an idea. Mass is the resistance that a moving object shows when stopped by a larger object that blocks the smaller object from moving further as an individual object, where by showing mass the smaller object resists the effort the larger object asserts on the smaller object to compromise the form the smaller object holds as a unit and not to accept the form the Earth imposes on the object.

This entire sheet of paper, no, this entire room formed as one piece of paper, could not produce a scale that would even closely resemble the falling differences there has to be, between the three falling objects if mass was a consideration in the falling process. The truck would have to land on the Earth at much more than the speed of sound if the girl was hitting the ground at normal falling velocity of between $(7(3\Pi^2))$ =207.km / h and $(7(\Pi^3))$ = 217.km / h and while all this commotion is going on, the frog will have time to grow wings, take flying lessons and circle the Earth before the frog will have to land, that is if mass was the cause of the falling of the three objects because mass will render much differences since mass distinguishes and differentiates according to size and compactness.

In contrast, every one knows the frog and the girl will hit the Earth the very instant that the truck hits the Earth. So what has mass got to do with gravity if it is the gravity that is making them fall and bringing on the movement, while it is clear that it is mass that stops the falling and it is the mass that is making them stick to the Earth as something that is motionless or in motion with the Earth. If mass was responsible for gravity then all gravity will not be equal but distinctly bias towards size and density.

By having mass the object no longer holds independence but retains some form of individuality by not compromising the unit it forms as an independent structure. When the object lands on a larger object and the larger object is part of the Earth, the larger object halts further motion. The larger object removes the individual characteristics that the moving object has.

But since I say that mass has nothing to do with gravity and considering I am uneducated, thus not a Newtonian, it is not worth their bother to investigate or to consider the idea. Then furthermore, since it contradicts Newton, it is despicable. Those rejecting my point of view do so because they only work with proven facts, or so they claim.

I wish to bring to mind some of the facts that academics work with when academics as scientists only work with facts. Remember they are the ones boasting that if facts are not proven then it is fables and those very important academics don't waste time with fables because they only work with facts. Students, it is your liberty to ask them to explain what they say is such correctly proven facts. They maintain it is a fact that we have to have mass in order to produce gravity. Mass is responsible for gravity. If you don't have mass you're not going to have gravity. Mass is equal to gravity and gravity is only where mass is. If mass is anywhere it should show its presence otherwise mass is absent. If a body falls it is the mass that allows the body to fall because the body receives gravity by ratio of mass and mass is what produces gravity in relation to the mass available. It is mass that drags you down because the mass is in charge of the gravity and the gravity finds the value from the mass available.

Is it mass that drags you down and does mass have some input in you're falling? Does mass pick up some responsibility for the speed and the manner in how a falling body falls? Charged with these questions it is the responsibility of the tutor to find answers when students start to pepper those tutors with questions. It is the responsibility of every tutor to have the answers ready when students start to ask questions because at the end of the month the tutor stands ready to collect his or her salary for teaching and the teaching includes all questions that was put to the teacher. Some questions are as obvious as balloons flying and parachutes falling.

No one of the most intellectual Newtonians ever express their opinion about the parachute. When a body is in free fall the body travels at around 207 km / h straight down. The body in mass falls in a direct line with the Earth at a specific speed. When a parachute opens the decline of the body slows down to a speed that puts the person falling well in control of the fall and under normal conditions and with an experienced parachute diver the likeliness of injury is very low.

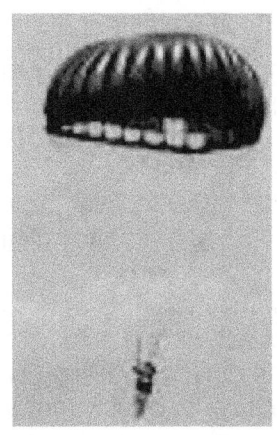

In this, questions arise about mass and gravity. Since the body falls, the question comes to mind about gravity and mass. If it is mass that has the body falling, then does the body become more massive when the parachute is open because the parachute plus all the air it gathers must add to the body mass. Does the mass become less because it is obvious that the body is falling slower and if it is mass having the body fall then a slower falling body must have less mass? Is the gravity becoming more because the gravity has to have to show some effect in the altering of the falling tempo? If the gravity is less, then the question is, how did it happen and if the gravity is more, then how did that occur? How does the parachute alter the conditions in mass because with mass responsible for falling then the mass has to reduce to slow the descending down…that is if mass is responsible for such falling? Then in what way does the parachute reduce the mass and how does the parachute reduce the mass? If the mass does increase with the additional adding of the parachute plus all the air particles that the parachute gathers, and the mass is responsible for the falling, how does the adding of mass reduce the falling tempo of the skydiver? It is obvious that with more massive objects falling, such as vehicles and army battle tanks, there is more than one parachute required. Then the parachute must be there acting as anti-mass because with more mass one requires more parachuting but if that is the case does the anti-mass render anti-gravity or not? Remember, it is mass that is responsible for gravity and it the gravity that gets the body falling (according to Newtonians).

An Open Letter On Gravity Part 2

So what happens to balloons? Have they got anti-mass or anti-gravity? They are moving up when the air is heated. Mass pushes you down by the gravity it forces onto you. If mass drags you down then what is lifting you up in the balloon? If mass gives the gravity to drag you onto the Earth then why would the hot air lift you up? Is the hot air causing anti-gravity or anti-mass because gravity and mass drags you down or so Newtonians say? The balloon is lifting the passenger and all that is in the bag plus the bag plus the balloon into the air. So what is then pushing the lot up, if it is mass that drags you down. Has the air not got mass because then the air can't have gravity and then the air must escape into the blackness of outer space because by going up it shows a resilience of either mass or gravity. We have seen that it is mass that pulls everything onto the ground. Why would the air defy mass and allow the balloon to go anti-whatever? We find mass being the equivalent of that which brings the object to the ground. The object has mass to produce gravity. Why then would hot air allow the balloon plus everything in the balloon to lift into the air? The balloon lifts in relation to the hot air that blows into the sack. The more hot air and the hotter the air is, the more lift and the swifter the lift will be that the balloon provides.

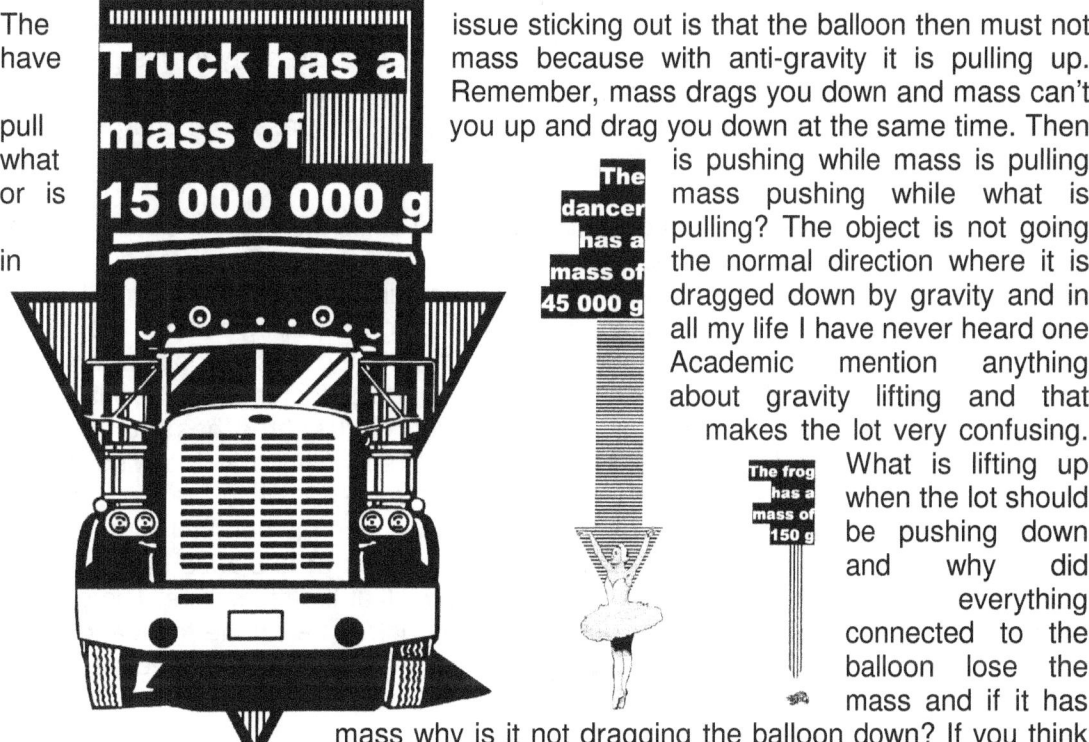

The have pull what or is in issue sticking out is that the balloon then must not mass because with anti-gravity it is pulling up. Remember, mass drags you down and mass can't you up and drag you down at the same time. Then is pushing while mass is pulling mass pushing while what is pulling? The object is not going the normal direction where it is dragged down by gravity and in all my life I have never heard one Academic mention anything about gravity lifting and that makes the lot very confusing. What is lifting up when the lot should be pushing down and why did everything connected to the balloon lose the mass and if it has mass why is it not dragging the balloon down? If you think this is a little confusing try what is to follow.

The wise teach you that it is mass that produces gravity and gravity makes you fall because while gravity makes you fall, mass drags you down. You have heard this before and I suppose you believed it. It is because those mind controllers are lying through their teeth with a menace in which they are the experts, as they know just how to pull cotton wool over you eyes. Take a truck of 15 tons into an airplane. Put next to the truck a petite little dancer weighing 45 kilograms. Put next to her a frog weighing 150 grams. There is a fifteen-ton truck next to a girl of forty-five kg holding a frog of

150 grams company. Then get this lot into the air by airplane and let them jump. They have to jump the same instant under the same conditions. Take note that you are told by the wise amongst us that it is mass that produces the gravity that pulls you down. We have just had a lovely debate on how it works and how mass drags you down and wondered if it then is anti-mass or anti-gravity that lifts you up with the hot air balloon. Well take note of this as your airplane reaches 11 thousand meters, which is eleven kilometres straight up into the air.

We'll leave the debate about how you get there in the plane for some of the other books because space is a little cramped but I can assure you the Newtonians got everything up side down there too. Now we drop the truck and the girl and the frog at the very same time from the airplane. They are coming down as a team of skydivers and the dancer is dancing around the frog that is jumping in and out of the truck cab. The frog then pretends he drives the truck and in the next scene he is dancing with the girl while the truck is falling as fast as a truck can fall. Who do you think is lying? Is it the academics or is it me that is not telling the truth in this case? Am I telling the untruth or are they telling the untruth. Remember only one group can tell the truth and the other must be lying. Have you thought why one party is lying while the other party has to tell the truth? Why can there be only one believable party in this picture while the other is untruthful? They say the lot is falling by mass driving the lot. The academic Brainy Bunch are telling students all over the world that mass is in charge of gravity and it is mass that's pulling you down.

Then the mass is pulling the truck of 15 tons down since the mass produce the gravity and the gravity produces the fall which is three hundred and thirty three times more in a down direction than the mass of 45 kg is pulling the dancer down. The mass providing the gravity that pulls the truck down is doing the pulling down of the truck one million times stronger than it is pulling down the frog. If the mass is doing the pulling by establishing the gravity, the truck must fall 333 times faster than the girl and one million times faster than the frog. It is either that or the three has the same mass because they are falling at the same rate. If the Brainy Bunch all too wise is correct, the frog can fly to America and have a pizza in New York while the truck has a few micro seconds to get down if the girl is going to fall during the normal falling duration of a minute or so. Everyone has seen skydivers jump out of airplanes next to cars and trucks and bags. Every one has seen how they all fall at the same rate. The girl can do tap dancing around a jumping frog on top of the truck or below the truck and they can be inside the back of the truck galloping on fresh air inside the truck because the lot is falling at the exact same rate.

Let's find mass where our Super-Educated Brainy Bunch say that they find mass. No one ever seems to directly link orbiting planets to the falling of objects to the Earth. It is as if everyone parts the two ideas of $F \alpha \frac{M_1 M_2}{r^2}$ and $F = G \frac{M_1 M_2}{r^2}$, while Newtonians cover both concepts under one formula $F = G \frac{M_1 M_2}{r^2}$. I thought it wise not to keep the two ideas apart and investigate both on equal terms. They are ardently supporting the idea that man can now determine the force there is that is between the Sun and any planet that has gravity, which is mass driven.

There are nine planets in all revolving around one star in the centre. Being typical we gave every planet a name and we called the star our star and also the Sun. We realized after some sole searching that we couldn't uphold the prehistoric concept that the Sun and the lot of planets are circling around the Earth just because we have the Pope on the Earth that liked the idea that he holds the centre of the Universe and from there rules the Universe by control on behalf of God. There came a concept that there has to be more than the Earth in place than to have the Universe spinning about the Earth.

Then someone came up with the idea that we lot spin around the Sun and that some planets are far away from the Sun while others are closer to the Sun. From those planets far away, the Sun would seem small or either very small and others being closer the Sun will seem very large. The Sun will appear differently from various stances because the distance in relation varies considerable according to the position and in some cases the location of the planet relative to the Sun.

Then another person came up with the novel idea of locating and mapping the planets and positioning the lot according to their locations.

What a job that turned out to be because it took the one person all his life and then his lieutenant had to surpass his lifespan to complete the job. A little later the lot got really clever. Some other clever guy got the idea of determining a force and found a way to calculate what the driving power would be needed to have a planet go around the Sun in a circle.

This was the epitome of human brilliance because from then on man had a window through which man could enter the mind of God and calculate how much force God released to have the Earth spin around the Sun. The calculations made science experience God not by religion but as equals…or then as semi-equals…or if you will, then on sharing some level. Science found a way to imitate God. Science was now with the ability of mathematics with which man could enter where only God could work in the past and that made science extremely powerful.

Previously I spoke about the top circle of the academic intellectual that rises above the level of the mortals they find below them being those that fill the streets and

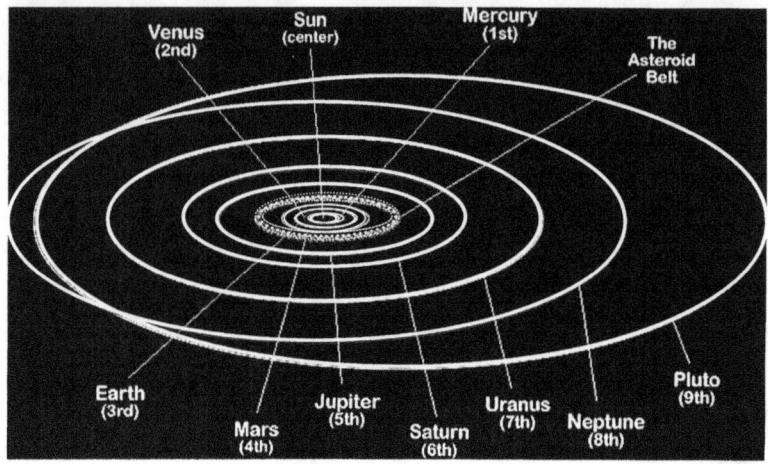

doesn't understand Newton. From the cloud of supremacy they enjoy, they do tolerate the mindless mortals filling the plain below their supreme level of thought but it is without doubt that in their superhuman thoughts they have little grasp for those in the world below them. Those living in the world they do not share are divided from the rest on the grounds of them being common mortals. They have the ability to think and grasp Newton while those common mortals are incapable of having thoughts. Those mortals without minds have no ability to think for those common mindless cannot understand Newton and not understanding Newton leaves them sub-human.

Those that don't understand Newton can never understand God and they are there with the good fortune of having the super intellectuals to think on their behalf because those that have the ability to understand Newton are the only ones that may move into the Universe where God plans and creates. They represent God to those with minds as small as not even able to understand Newton because those thoughtless creatures not with a comprehensible ability to be able in understanding Newton are just machines with a capability to breathe and to breed. They that understand Newton can follow the thoughts of God because only God and they can understand Newton. Newton saw that God gave mass and in giving mass, God gave gravity and with gravity God created the Universe. All that was required was to measure the mass of the planet and from the mass they could measure the force there is required from God to drive the planets around the Sun. If mass was used in the Creation of the Solar system one would expect the layout to be more or less in the order depicted in the sketch above. That small piece of evidence everyone seemed to miss because no individual ever gave that thought much notoriety or prominence of any description worth mentioning. They discovered the formula that God used to design the Universe. They found that God gave a dash of mass here and a dash of mass there sprinkled at random throughout the Universe. They saw that God gave mass the ability to establish a force called gravity and God gave gravity the opportunity to destroy a nothing found where they placed the gravitational constant. This nothing held the Universe together, but God gave gravity the pulling power to destroy that nothing called the gravitational constant and convert that nothing into the normal nothing or (I suppose) your ordinary garden variety nothing found all over. God's trick to create, they found, was establishing mass that could nominate a force called gravity and this force could convert the nothing we have in outer space to nothing we don't find in outer space.

An Open Letter On Gravity Part 2   Page 410

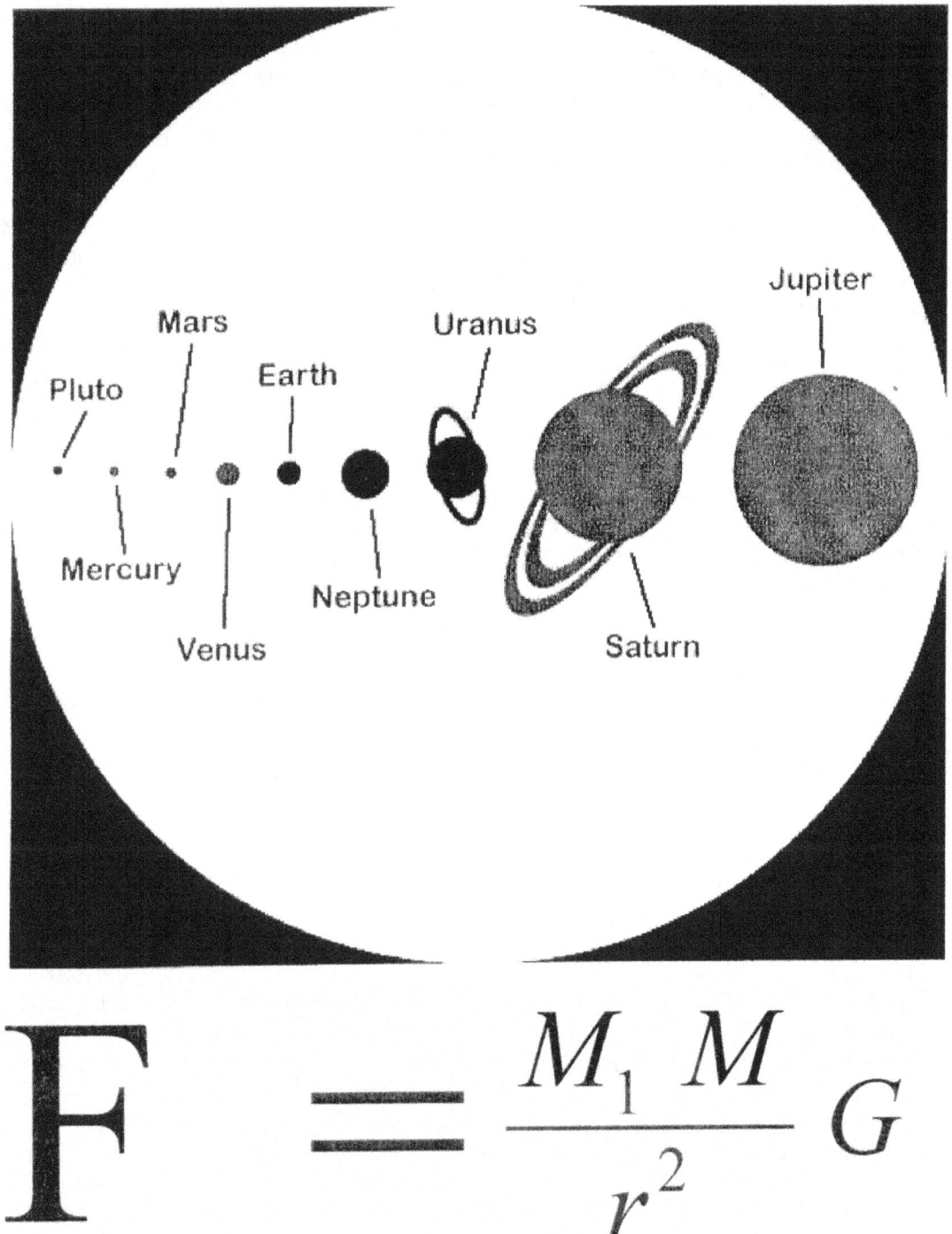

$$F = \frac{M_1 M}{r^2} G$$

They had the recipe around which the entirety of what is, revolved. They copied God's formula He used as He created the Universe. They found the foundation of Creation and they got so bewildered with power they then had, that from then on, they no longer thought of it as God's creation because they now were fully armoured with the formula by which they then could also create the cosmos. Now it became not God's creation but just any one's creation and therefore was no longer the creation of God. They had the formula and now they could see how it was done. They no longer needed God to create because with the formula in hand they had. They had the key with which to invent a Universe to their taste and with that to solve all the riddles and mysteries and unknowns there just might be in God's Universe.

What was required was finding the mass because the mass was critical. There were some other measurements they had to get around and out of the way because it was not supporting their formula but that was easy because they had the formula and now they had to implement the formula. What can be that difficult if you have the formula that unlocks the Universe? Now with the formula they could find the engine that drives this simple devise. With little effort someone somewhere could even calculate when this lot will end in one lump of material. The person just needed enough time on his hands with little else to do but since they now had the formula there was no such a person available with spare time or with time to spare as to calculate when this lot will conclude because everyone was too busy imitating God by playing with all sorts of various and wonderful calculations. It took a few fanciful arguments and some reinventing of principles but eventually everyone came around and saw how the Universe worked.

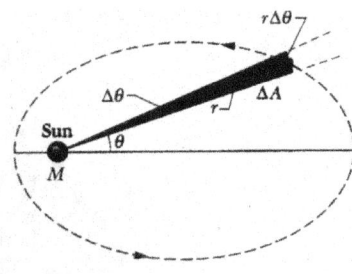 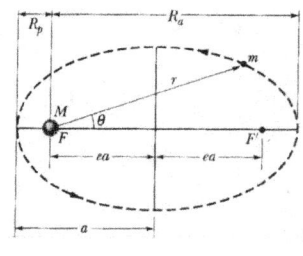

Every one with mathematical skills could apply the formula and with the formula find the answer to whatever question they had because they had the formula that was the key to being God. They still have the formula that they use as the key by which they become the masters of the Universe and no one actually needs God anymore because everyone can now be God by using the formula that made everyone God. They even invent space whirls and be blessed so that they can rule out any affect time may have on them…and all that because they can use the formula in some or other form! They could now devise methods to jump Universes and fool time by disallowing space to have any space. With the aid of the formula they can swim through two Black Holes and still be unscathed! Where light cannot go and not be destroyed their formula will take them and bring them back with no harm done to them in any way. This achievement is all contributed to mass being given. If they use mass then no harm can come to anyone because they have the key that unlocks the Universe: it is mass that is the key. No further investigation is required because with the gaining of God's formula, they gained the powers of God and with such powers comes the security that whatever they decide will be the correct decision made. God played the same games as the Romans did, with using a somewhat larger arena. He put in the arena planetary gladiators according to mass and had these gladiators then in the cosmos to fight it out by reducing the distance between the planets according to size or mass. This showed their atheistic point correct that God has a human side, which made it true that they could become God in the human side. Right or wrong was never in the discussions about decisions because with the formula that mass holds, man is now God.

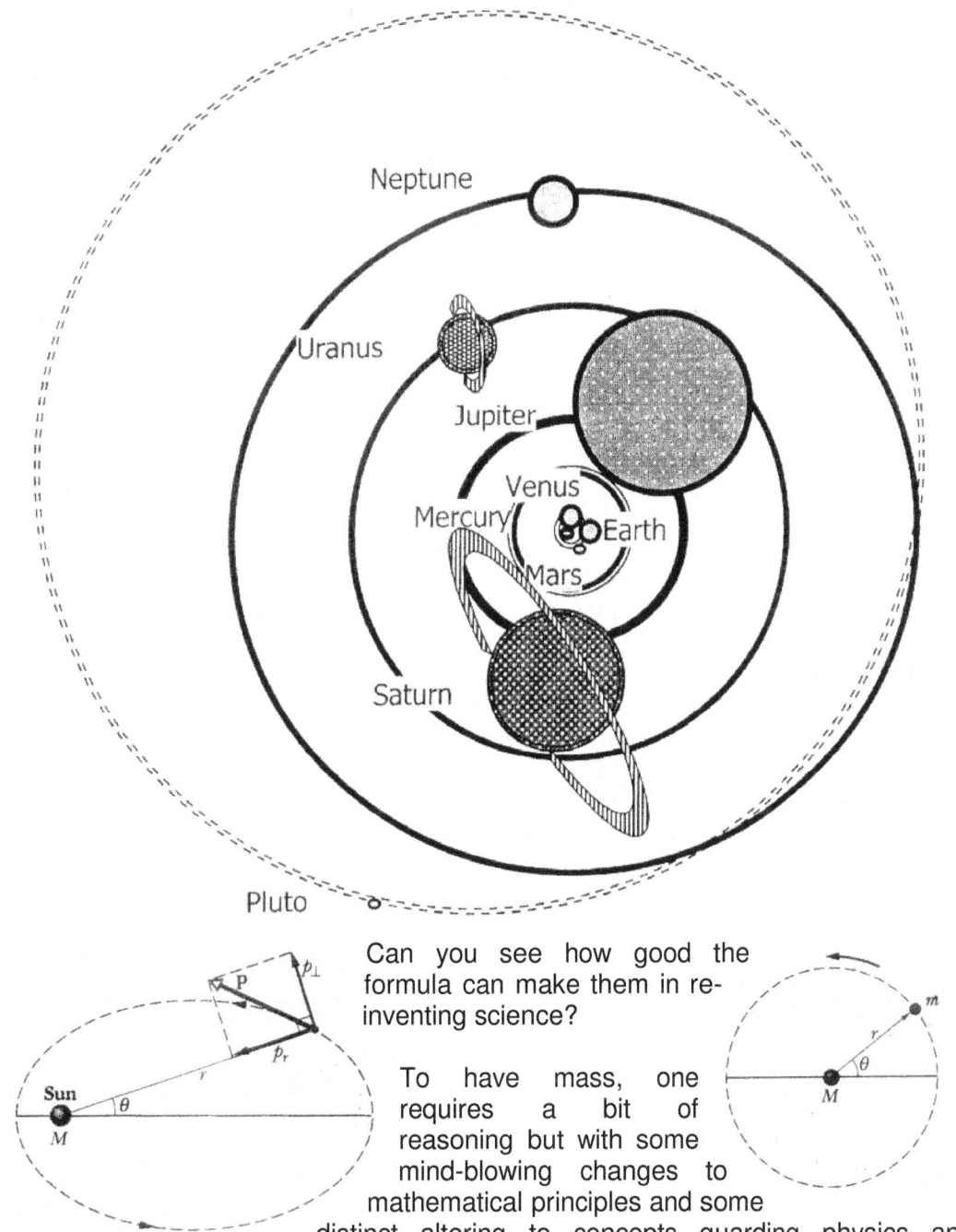

Can you see how good the formula can make them in re-inventing science?

To have mass, one requires a bit of reasoning but with some mind-blowing changes to mathematical principles and some distinct altering to concepts guarding physics and changing all laws applying to mathematics, then the rest was rather mundane. It was a bit of cheating here and a bit of corrupting there and ignoring laws on purpose by misplacing common sense. By changing and applying mathematical science and corrupting cosmology, they managed to force the formula into having a force and having a force that runs gravity per mass, you have all the requirements in place that God uses to create the cosmos and then by applying mass as a usable force one can begin to replace God. All you needed was having mass as a measurable value and fabricate estimations of what the components might be that will give the mass that was needed and the rest was good enough. Then the estimated mass that was precisely calculated was used as the driving force in the Universe. Let's be honest about it.

This was rocket science and nuclear know-how of the day back in the seventeen hundreds or so. This is what made men great and allowed some to look down their noses at others and that principle still applies to the letter to the day.

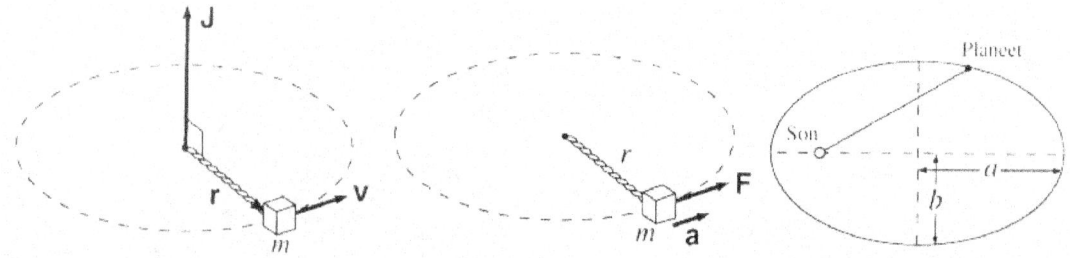

I find it a little surprising that they did not find it a little surprising that after measuring all the mass and calculating all the forces floating around, that the planet layout was not from large to small or from small to large since the mass was so crucial in the whole process of calculating the forces flying throughout the night sky and between all the planets that was in the process of turning around the Sun. At least with mass being so important one should wonder why God forgot to arrange the planets according to mass, but then I guess someone remembered that with the formula no one needed God any more. Without God the planets arranged themselves and who knows what planets use when planets arrange and allocate various positions to their fancy. So then even that was sorted and understood.

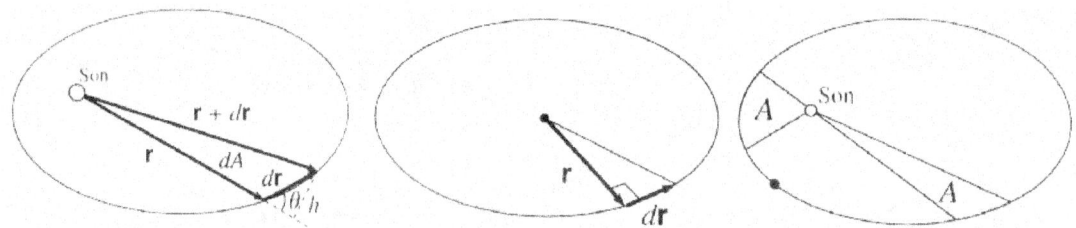

The information filled the night sky and he that was wise could see and calculate and be God. The requirement was mass and if you had the measured mass then the rest of the calculations were simple enough to be student labour. The mass of the Sun pulled on the mass of the planet and with a little help from the gravity constant, the force will eliminate the distance between the two objects.

That force that was in place to eliminate, was also the force the engine used to drive the mass as it eliminated the other mass and in the process the two ends carrying mass was devouring the distance parting the objects by the square.

Physics had the formula. They stole the formula from the clutches of God and now every person in physics can surpass God by inventing and re-inventing whatever any one was less satisfied with in the Universe. Science now knew all the nine planets by heart and by mass and by size as well as location. They had the distances. They had the mass. What more could they ask. It was a question of letting the games begin and finding out who could be God of the week for this week coming. The one that could calculate the best had the best chance of being God for the week. Not needed was a sound and clear mind or a drop of logic because now they had the formula to calculate the gravity going around and that feeling made them drunk with power.

According to the formula that they now have and that they use which enables the academics in physics to be so powerful and smart and bright and…think of any name that is complimentary and it will describe this lot in some way because they have the formula. The formula is putting them in such a commanding place, they are beating God every round and doing it so much they can even declare they have beaten God out of existence. And all this they accomplish by having the formula. With the formula and bestowing mass throughout the Universe, they now could be atheists

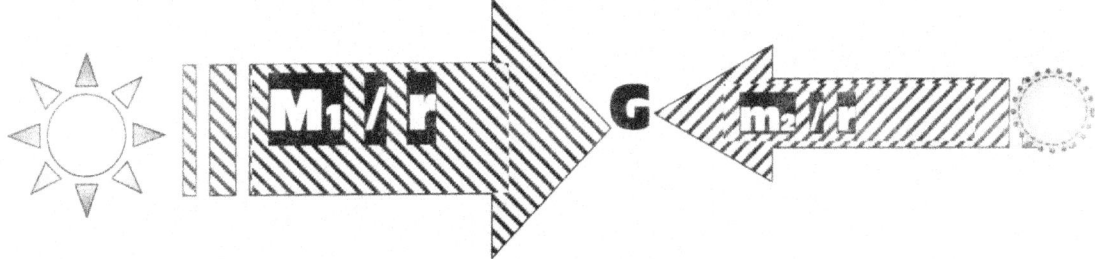

But undisputedly true is the fact that a human, a wheel borough and a train will descend at an equal pace!

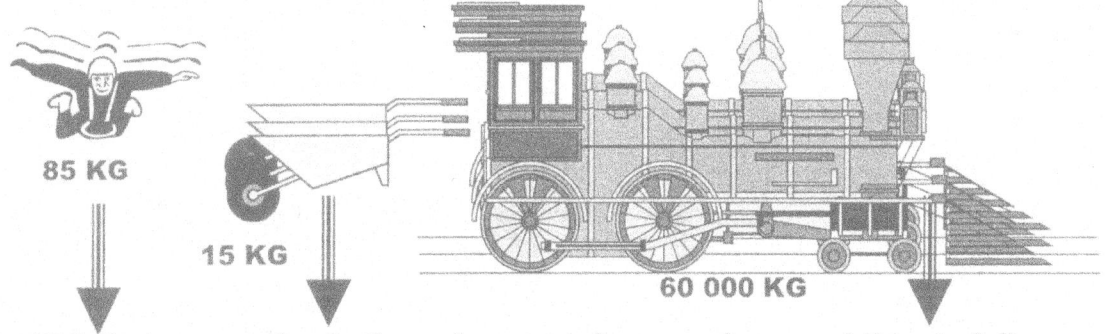

All fall at an equal rate through an equal pace using equal time to fall

If the train and the person as well as the wheelbarrow were pulled by "mass" how much quicker will the train hit the earth than the wheelbarrow would and how much slower would the person be in falling down to earth. To hell with vacuum and feathers and the moon's atmosphere because all that is just to cloudy the issue and draw the attention away from the fact that without vacuum the lot still falls equal under equal circumstances. …And that is what Galileo said happens and Galileo would know because he was the true master.

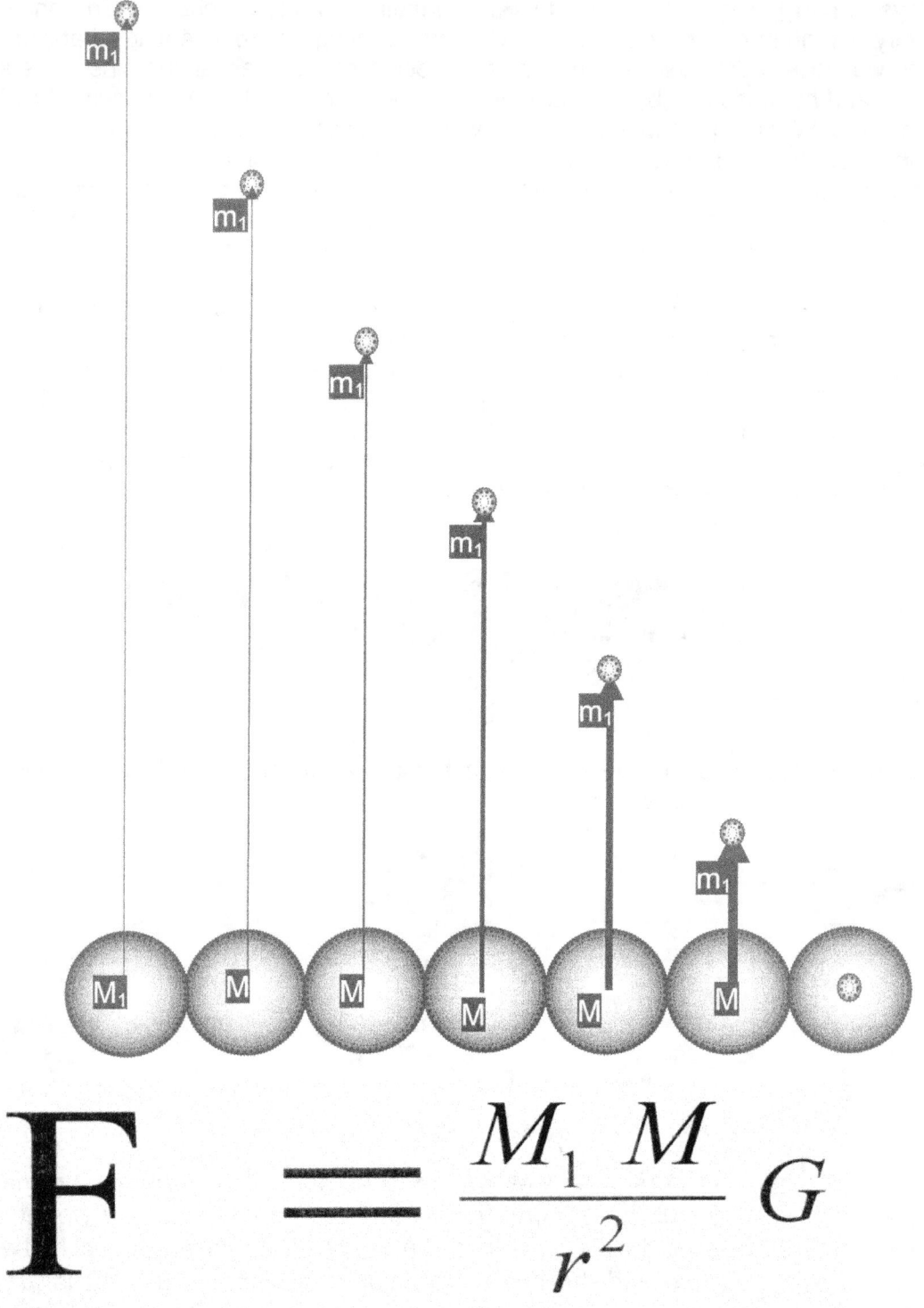

$$F = \frac{M_1 M}{r^2} G$$

That is how mass destroys the gravitational constant, which is filled with nothing in the square (from both sides) by producing gravity.

Today the formula works so well that it is estimated in some official circles that more than 90 % nine (9) times ten (10) out of every hundred % ten (10) times ten (10) are atheists. (I'm giving the mathematical wonder-children some mathematic medicine so that even they can understand what I am saying) are totally defiant of all religion and most persons of those forming the numbers of persons in the ranks of physics are atheists…and all that because they came by the formula that enabled them to become gods in their own right.

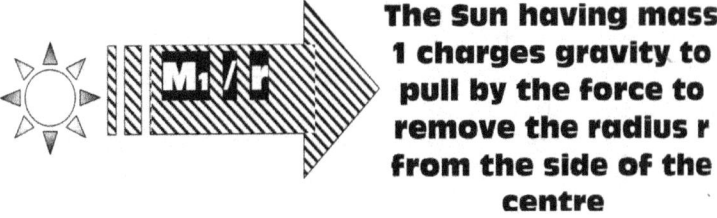

The Sun having mass 1 charges gravity to pull by the force to remove the radius r from the side of the centre

Let's see how this wonderful formula works that can elevate physicists to the level where they are equal as gods. With them having the formula and the power to bestow mass they now see their potential as equal to God. The requirement is mass and that they can award without God helping them.

The formula insists that the mass of the one puts gravity at work to eat up all the nothing holding the gravitational constant in place that there is between the Sun and any given planet. This mass inflicting gravity is at work so that the radius diminishes to yes you are right…more nothing and that it does from the other side too. It turns the nothing there is in outer space to nothing that is not in outer space because the nothing that fills outer space becomes nothing that does not fill outer space leaving outer space filled with nothing. See how clever mathematics can make those in

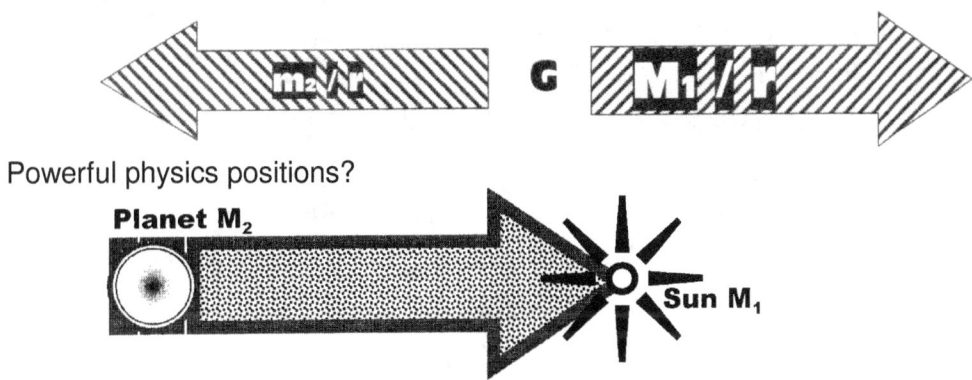

Powerful physics positions?

The planet having mass charges duplicated as gravity pulls by the force to destroy the radius from its sides

It seems that they can use nothing for whatever pleases them most. We find nothing between the Sun and the planets where it is disguised as the radius covered in and pretending to be nothing and then also we find nothing when the radius is not there and has been eaten up by gravity where we then find nothing in place where the radius has nothing in place since gravity devoured it. The gravitational constant that is according to Newtonians "a nothing" that is allowing the participating mass factors unrestricted access on the condition that the force destroys the radius to nothing. That same nothing that is also incidentally what gravitational constant, comprises of when the constant has a radius and when the radius is resolved. This mass is that which applies gravity, which removes the nothing and replaces the nothing with nothing. This removing of nothing and replacing that with nothing establishes gravity that acts as a force that is bent on destroying the radius comprising of gravitational constant which means the lot is in self destruction all the way. What a morbid lot does this picture turn out to be?

The mass has it at task to produce a force of gravity that is hell bent on destroying the gravitational constant that is hiding in the radius because it has no other place it can be hiding but in the area that r holds and that r is what everyone wants out of the way.

Apart from mass bent on destroying the radius to get to mass, there are a number of other forces also at work. There is a circular motion that has the planet spinning at a distance and they say because it is spinning at a distance the planet is idling and doing no work but to sit and spin.

They have even gone as far as comparing this lot spinning, to a top. It has nothing better to do than spin in the manner that a child's top spins. The turning extends as the distance from the outside border to the middle at the centre of the top is keeping the top up straight. However, they say this constitutes to no work being achieved by the spin and this results in the top leisurely doing nothing by spinning idly. In that the one motion not only cancels the other motion but also totally annihilates the other motion to render the motion completely absent and invalid. It renders the other not only being incapable but totally absent. It removes the one from the Universe completely by removing the value of the other one to zero. $\frac{dJ}{dt} = 0$

Should you ask me which one is rendering the other one out of existence, I have to be honest and say I don't know because this in mathematics I have never come across. It is the privilege of our most honourable academics in physics to be able to go and divide a measured value by another measured value where they render both values legal and with some value while also at the same time being competent to have a measure and then they remove the measured value by having the outcome of the ratio to be non-existing. This is new to me and I have seen this occurring only once and it is in this case. See what I mean when I stated before that they are able to use "nothing" for whatever pleases them. They can use nothing even to be their God because then they call their religion atheism. Is that not because they have nothing to use as mental powers…I mean that is then the reason why they manage to use nothing so liberally.

I have never seen that one value of a specific measure divides into another in which the two forms a ratio and in the process the one removes the other from the division. You ask me which is removed and I have to tell you in all honesty from where I see the top spins that both must be there because the top is not removed from our side of the Universe and both factors are applying while the top spins.

One can see the presence of the top throughout its spinning… so where did the nothing take the top to work? However, this is all little mathematical detail and why is a law a law if one can't break the law and come out of such a law that is confining and restricting the abilities of the mathematician and especially if the persons are the very ones that have the appointed position not only to uphold the mathematical law but to interpret it and to see how it applies. But since they now have the formula that puts them in the bracket that previously was reserved for God alone and they now, as acting gods, can change some detail about mathematical principles especially when those principles stand in the way of the formula and hinders some aspects of the formula not proving that efficient.

After all, what is required is the mass and the lot will come together. The one mass jerks the other mass and while the other mass is jerking right back, the one jerking mass is jerking the other jerking mass around in a circle. The formula proves unsurpassed by any intellectual standard and once any person accepts the formula by declaring that, then such a person understands the formula and then that person is immediately promoted to the ranks of also being god.

There are no questions to be asked. There are no comments to be passed. This is a religiosity and as it goes with al religion, one only must accept that it is true and their soles are saved. Well it happens in this case as well. Don't ask questions and the lies will become instant truths. One needs to know the mass and the rest is of no importance. To test mass and find how mass applies by implementing it into the formula; I have made tables to hold values.

The explanation will pose certain factors that seem important.

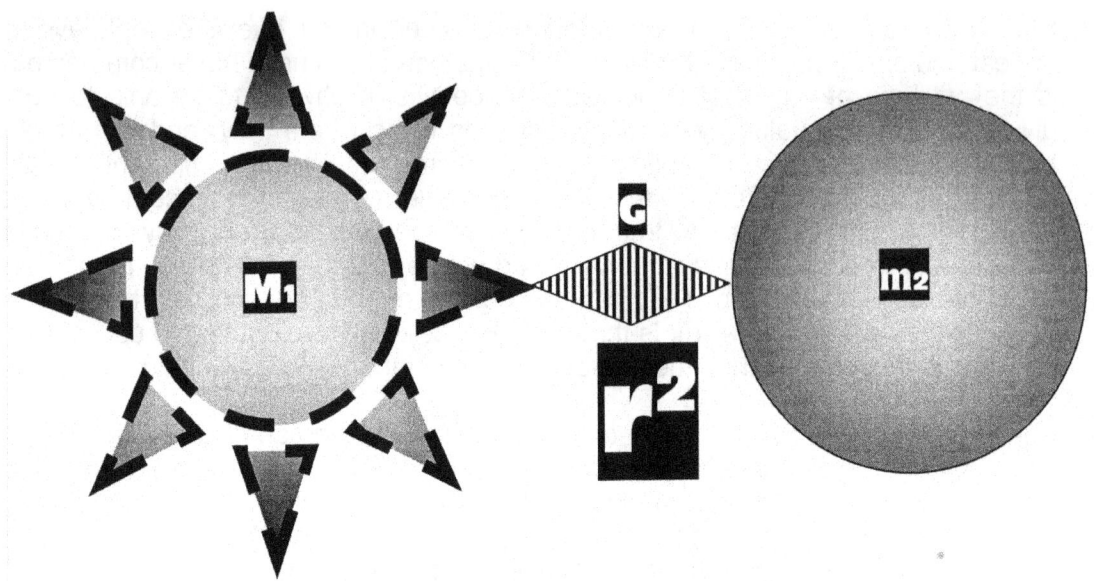

**Planets name
Planet number from the Sun
Mean density 5.43 g /cm3
Mass = is taken as the Earth is 1
Volume = is taken as the Earth is 1
Mean distance from the Sun is taken as the Earth is 1**

All measurements are taken in relation to what applies to the Earth as if the Sun and all the planets in concern are still revolving around the Earth. How far did the art of cosmic physics not evolve and how much did they not break free from the previous interpretation that the Earth was the bona fide number one around which all spun.

We follow the presumption that mass does apply as Newton stated it would and that mass will be responsible for the force of gravity that will form the driving force propelling the planets around the Sun each to it's individual mass. We also presume that every planet is orbiting the Sun at the rate and the pace set by the mass that produces the force of gravity that propels the planet in accordance to the mass awarded to the planet by the distinguished Newtonians.

These are the smaller but denser planets being to the inside circle of the Sun and has no real gravity potential since these planets don't show much mass to work with. The mass can only generate weaker gravities that bring on the slower rotating speeds and that could explain the smaller circles there is found with the small mass and weak gravity used by the small inner structures. We have to consider those that come from this point on and circles in orbit paths much bigger because they are the true mass carrying giants and they should travel much faster than the smaller ones on the inside.

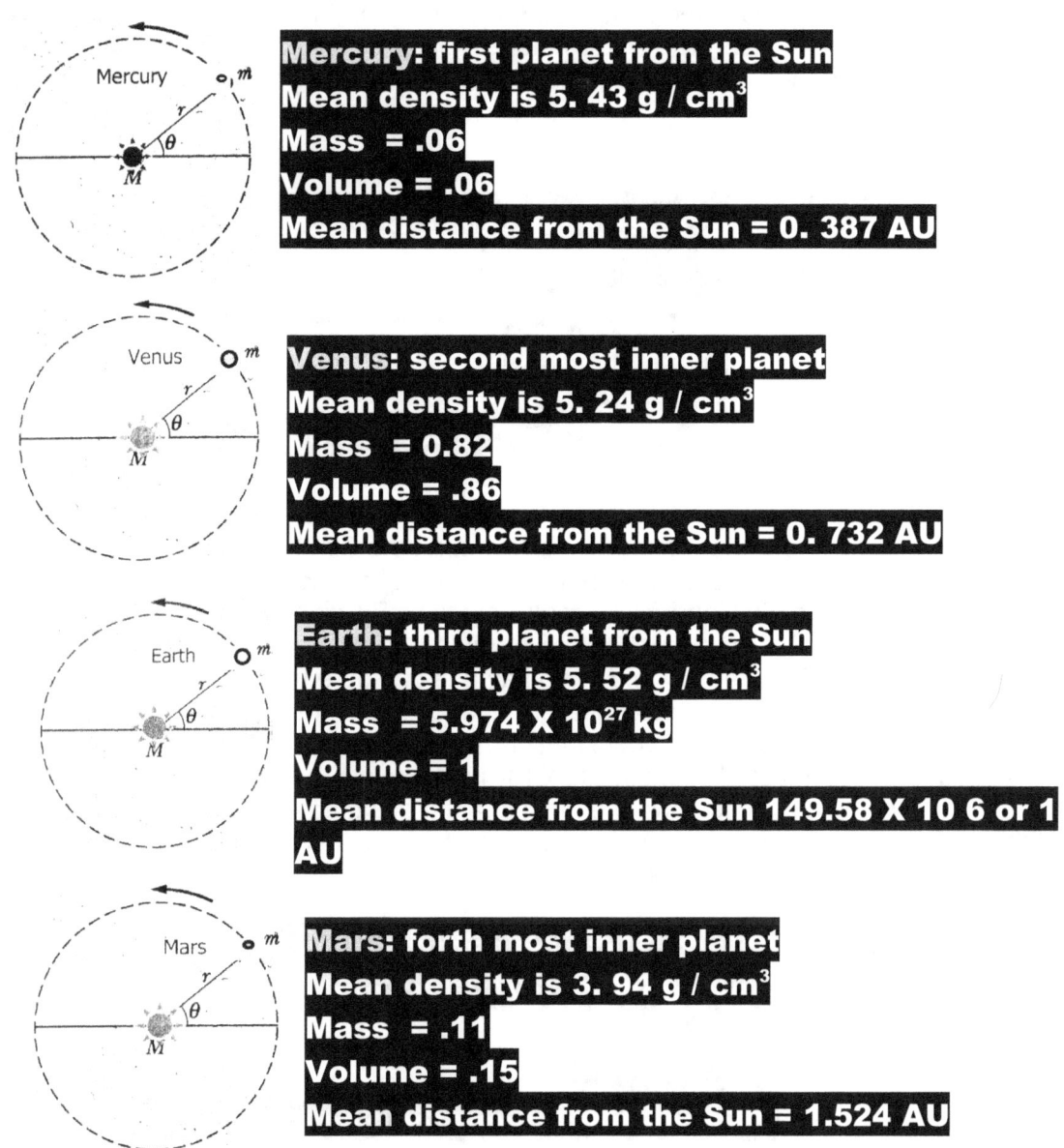

Mercury: first planet from the Sun
Mean density is 5. 43 g / cm^3
Mass = .06
Volume = .06
Mean distance from the Sun = 0. 387 AU

Venus: second most inner planet
Mean density is 5. 24 g / cm^3
Mass = 0.82
Volume = .86
Mean distance from the Sun = 0. 732 AU

Earth: third planet from the Sun
Mean density is 5. 52 g / cm^3
Mass = 5.974 X 10^{27} kg
Volume = 1
Mean distance from the Sun 149.58 X 10 6 or 1 AU

Mars: forth most inner planet
Mean density is 3. 94 g / cm^3
Mass = .11
Volume = .15
Mean distance from the Sun = 1.524 AU

The planets running from most inner to the third planet shows distinct increase in size most of all when adding the mass of the Moon to that of the Earth. There is a distinct patter of mass increase, which must cause an increase in orbit velocity in accordance to the mass growing. The growing mass factor awarded to each planet in accordance to size will establish an increased orbit velocity in line with the increase the planet has in mass and such an increase in mass will be beneficial to increase the planet as the gravity increases the propelling ability of the planet.

But immediately after savouring such a thought, Mars comes along and burst the gravity bubble. Mars is twice as far as and even further than that in some cases from the Earth as well as being smaller than the Earth and moreover, Mars has one small rock in custody that is not even comparable in size to the Moon of the Earth. It seems the sooner there is a rule no quicker nature breaks the principle that might support such a rule. Let us see how mass pushes gravity propelling when the real big boys begin to play.

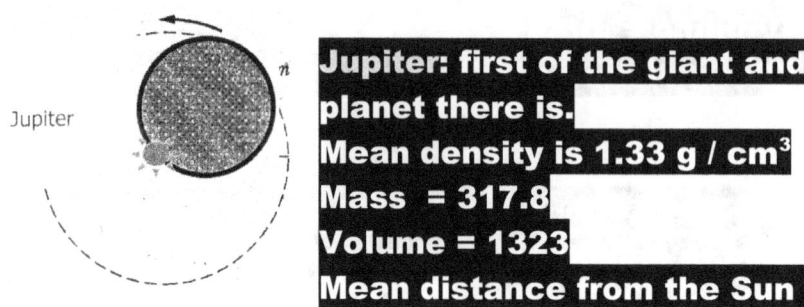

Jupiter: first of the giant and the biggest planet there is.
Mean density is 1.33 g / cm³
Mass = 317.8
Volume = 1323
Mean distance from the Sun = 5.203 AU

If mass had anything to do with gravity, this giant should be flying because this giant has a lot of mass to pull the Sun and to race around the Sun than any other planet can dream of. This giant can contain most, if not the lot, of all the other planets and debris that roam about the Sun and fill the solar system. By planets going in motion this planet is a racing car and should be the formula one of planets speeding. Take the size and the mass that Mercury and Pluto has and compare that to Jupiter and you will see why Jupiter should be breaking every speed record there is to break. This one should be the Schumacher of planets that can fly. It should have the shortest year, that is because of the mass that must propel it in a circle around the Sun. Think what speeds it can do as it races down the main straight going past the Sun at a speed that the Sun can hardly see where Jupiter is coming from as it is racing past going on at a neck breaking speed.

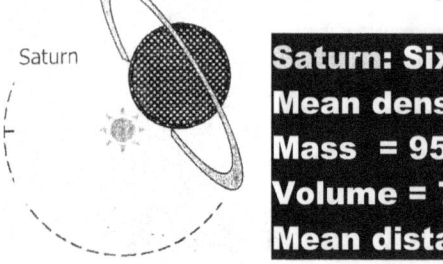

Saturn: Sixth planet from the Sun
Mean density is 0.70 g / cm³
Mass = 95.16
Volume = 752
Mean distance from the Sun = 9.539 AU

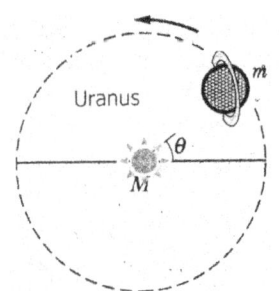

Uranus: Seventh planet from the Sun
Mean density is 1.39 g / cm³
Mass = 14.5
Volume = 64
Mean distance from the Sun = 19.182 AU

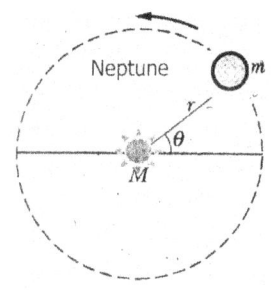

Neptune: Second last planet and the last gas giant
Mean density is 1.79 g / cm³
Mass = 17.2
Volume = 54
Mean distance from the Sun = 30.058 AU

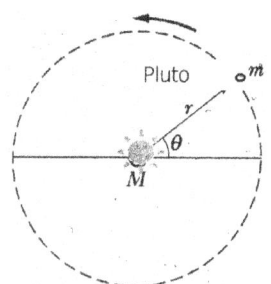

Pluto: The very last planet and almost a double planet
Mean density is 1.1 g / cm³
Mass = 0.00252
Volume = 0.1
Mean distance from the Sun = 39.44 AU

This is what five hundred years of cosmic research unveiled. Persons put in years of dedication and painstakingly researched all the above information. They went to the ends of the Earth and further to obtain this information. The dedication and resolve of the persons researching this vast network of information puts those in research in a class far superior to others and having the outer space vastness as a laboratory only exemplifies their admirable perseverance. Their achievement is supreme amongst all that is supreme. What does all this information then bring to mankind? What has this information helped us divulge the formula that gave us the ability to enter the heavens and push God aside with all the wisdom man now has accumulated? It helped us in no way getting nowhere in a big hurry except if we wish to start playing God by lying through our teeth as all the physicists do about their formula.

The following table lists statistical information for the Sun and planets:								
	Distance (AU)	Radius (Earth's)	Mass (Earth's)	Rotation (Earth's)	# Moons	Orbital Inclination	Orbital Eccentricity	Obliquity	Density (g/mc³)
Sun	0	109	332,800	25-36*	9	-	-	-	1.410
Mercury	0.39	0.38	0.05	58.8	0	7	0.2056	0.1°	5.43
Venus	0.72	0.95	0.89	244	0	3.394	0.0068	177.4°	5.25
Earth	1.0	1.00	1.00	1.00	1	0.000	0.0167	23.45°	5.52
Mars	1.5	9.53	9.11	1.029	2	1.850	0.0934	25.19°	3.95
Jupiter	5.2	11	318	0.4111	16	1.308	0.483	3.12°	1.33
Saturn	9.5	9	95	0.428	18	2.488	0.0560	26.73°	0.69
Uranus	19.2	4	17	0.748	15	0.774	0.0461	97.86°	1.29
Neptune	30.1	4	17	0.802	8	1.774	0.0097	29.56°	1.64
Pluto	39.5	0.18	0.002	0.267	1	17.15	0.2482	119.6°	2.03

An Open Letter On Gravity Part 2

If we start to interpret the formula correctly by instituting the mass and the mass in relation to the gravitational constant that has the task to divulge the radius as soon as possible, it means the following: It means absolutely nothing but to give the idiotic atheist with no brains a means to boast about what they cannot even engage with a clear vision about what serves as fact or as fiction. This senseless formula brought the mindless into physics by placing the idiots into positions where they can boast how clever their godlessness is and why in their stupidity they can prove there is no God because the formula indicates that there is a chance of randomness in selection of building the cosmos. How idiotic are the wise in physics not to see how incisively incorrect the formula is that proclaims them as godly wise.

With the use of their mass creating the force of gravity, the following motion and planetary speeds of motion should be in place being in accordance with every planet having altogether different mass. The relative mass each one has should propel it in relation and according to the mass it has in contrast to the mass the Earth has, because not one planet has either the mass or the radius from the Sun which is the same as that which the Earth has.

KEPLER'S LAW OF PERIODS FOR THE SOLAR SYSTEM			
PLANET	**SEMIMAJOR AXIS** $a\ (10^{10}\ m)$	**PERIOD** $T\ (y)$	T^2/a^3 $(10^{-34}\ y^2/m^3)$
Mercury	5.79	0.241	2.99
Venus	10.8	0.615	3.00
Earth	15.0	1.00	2.96
Mars	22.8	1.88	2.98
Jupiter	77.8	11.9	3.01
Saturn	143	29.5	2.98
Uranus	287	84.0	2.98
Neptune	450	165	2.99
Pluto	590	248	2,99

Mercury ○ Mass = .05 and that means it has to revolve half the speed of the Earth.

Venus ○ Mass = .89 it means it has to revolve point eight nine times the Earth speed.

Earth ○ Mass = 1 bang on target and it is the only one they got correct

Mars ○ Mass = .11 this one moves so slow it must be standing still most of the time

An Open Letter On Gravity Part 2

Jupiter

Mass = 318 which means we can hardly see this one running. Shooting a rocket at it is impossible, because take the speed of the Earth from which the rocket has to launch and multiply that with 318 times the motion of the Earth and to do that, one will surpass the sound barrier countless times, that is if the measure of mass applies.

Saturn

Mass = 95. It has to be circling almost one hundred times faster than the Earth with that enormous mass charging gravity all in its favour.

Uranus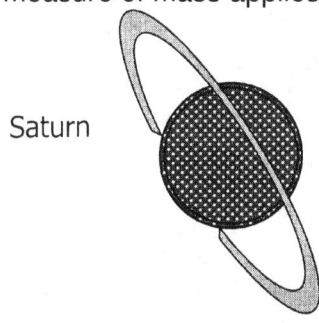

Mass =14. Uranus is moving 14 times faster than the Earth

Neptune

Mass = 17 Must run at least four times slower than Uranus because it is twice as far away

Pluto ○ Mass = .002. Pluto is not moving. Pluto is frozen in time and space and only once in a blue moon can Pluto move an inch. But can you remember what we spoke about in the previous article about things falling and not having mass?

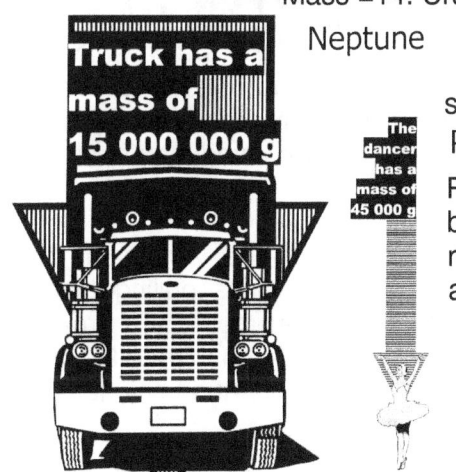

We can't just bring in mass because sometimes mass can't just be "brought in". Looking back at the truck, the girl and the frog, it was clear that mass as a factor was invisible as the three components fell to Earth. The girl was dropping as fast as the truck while the frog was coming with, as it would not be left behind on its lonesome self. The lot was falling equal because there just is no possibility for involving mass, since mass does bring distinction in motion efforts and there is no distinction in motion efforts when the three objects are falling together. This made Newton's initial concept of

$$F = \frac{r^2}{M_1 M_2}$$ total rubbish.

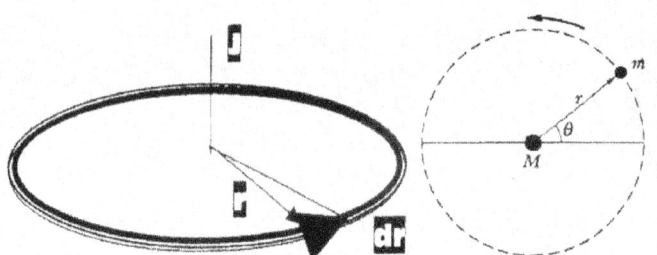

KEPLER'S LAW OF PERIODS FOR THE SOLAR SYSTEM

PLANET	SEMIMAJOR AXIS a (10^{10} m)	PERIOD T (y)	T^2/a^3 (10^{-34} y^2/m^3)
Mercury	5.79	0.241	2.99
Venus	10.8	0.615	3.00
Earth	15.0	1.00	2.96
Mars	22.8	1.88	2.98
Jupiter	77.8	11.9	3.01
Saturn	143	29.5	2.98
Uranus	287	84.0	2.98
Neptune	450	165	2.99
Pluto	590	248	2.99

With the table to the top of this writing being figures calculated from Kepler findings we can see $T^2 / a^3 = k^{-1}$ Making $k^{-1} = \pm 298$ and $a^3 \neq T^2$. That proves that Newton knowingly and deliberately committed fraud to validate his invalid and illegal presumption

Mass has no part in this equation at all.

Kepler said the cosmos said that the space is time $a^3 = T^2k$ but time comes in two parts being T^2k and not as Newtonians state being one part T^2. According to mathematical principle one cannot throw k out the window. This Newton did when he came after Kepler and changed what Kepler said to what Kepler never said which is that space is equal to time in rotation $a^3 = T^2$ and that T^2 only represents the circular movement. If $a^3 = T^2$ then that will place time equal to space in Kepler's figures being $T^2 / a^3 = 1$ which one can see from the table it is not. Clearly Newton did that to defraud the truth because we can calculate from Kepler's findings that he found the calculations present $T^2/a^3 = \pm 298$. Since Kepler gave the formula as being $a^3 = T^2k$ then $T^2 / a^3 = k^{-1}$ and since Kepler devised the formula Kepler holds authority over how the formula is read and is accurately presented.

When Kepler placed three factors in a specific relevance as $a^3 = T^2 k$ and which Kepler concluded after researching the cosmos for two life times, no one in his sane mind have the tenacity to change $a^3 = T^2 k$ to $a^3 = T^2$ and just merrily drop k because he has some criminal ploy driving him to do so, even when such a criminal carries the name and the status of Sir Isaac Newton.

This proves the cosmos is moving $k^{-1} = T^2 / a^3$ and also at the very same time it annihilates mass and everything Newton ever said about mass pulling mass and making mass responsible for causing the contraction between objects. It proves that the Sun is contracting outer space and all the planets are COUNTERACTING equally notwithstanding mass differences. This proves Newton was the fraud I say he was.

If Pluto ○ being almost part of another system because it is so distant, moves $(T^2 / a^3) = 299$ as fast as that of the most inner planet Mercury ○ how would mass then come into play?

If Venus ○ that almost has the same mass as that which the Earth ○ has but is a third of the distance closer to the Sun and yet is moving $(T^2 / a^3) = 299$ as fast as what the Earth is moving, then what role has mass to play?

If Mars 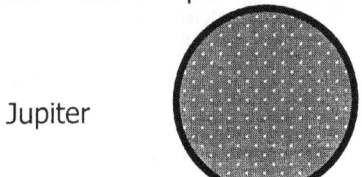 has so little mass compared to the Earth but is so close to the biggest

giant of the lot being then where is the explaining as to how this attributed mass comes into play seeing that both move about equal?

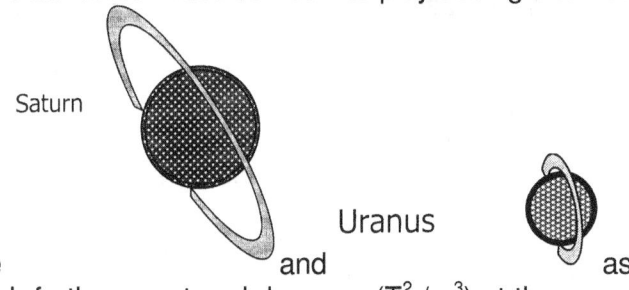

Where and as well as is so much further apart and do move (T^2 / a^3) at the same pace as all the others what is the effect of mass?

This made Newton's later concept of $F = G \dfrac{M_1 M_2}{r^2}$ total rubbish.

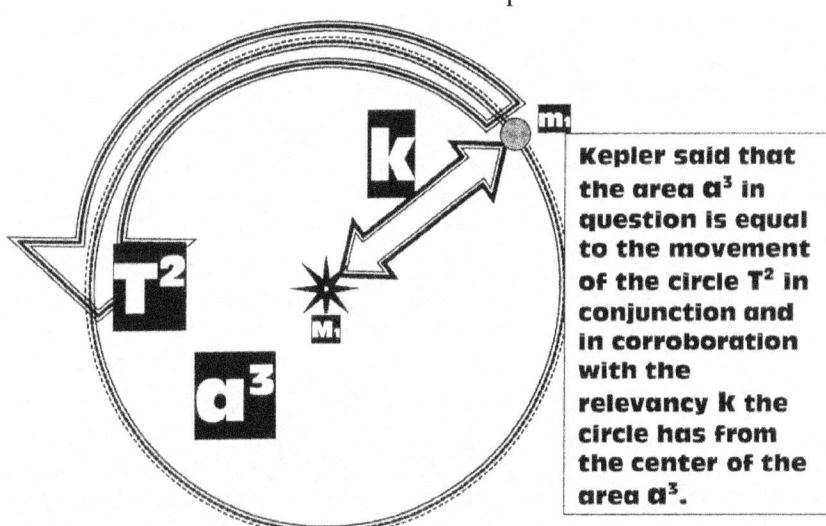

Kepler said that the area a^3 in question is equal to the movement of the circle T^2 in conjunction and in corroboration with the relevancy k the circle has from the center of the area a^3.

Again I am going to repeat what the honourable Academics say by never saying what they are saying. They don't disagree with Galileo because that is not possible to do when facing evidence in support of Galileo. Since the time of Galileo all time chronograph devises are used to using the principle that Galileo's theory lays down. They will never admit that if Newton is correct then by that implication Galileo can't be correct because Newton doesn't back Galileo up, it contradicts Galileo. Even saying both Galileo and Newton is equally correct becomes one argument they are bound to lose, as Newton contradicts Galileo and does not support Galileo. Galileo, unlike Newton is well proven and there is not a trace of doubt. From the chronograph Galileo devised he declared that all pendulums respond to time in the same way notwithstanding whatever size differences there might be. Newtonians say it is mass that delivers the gravity that swings the pendulum that makes the object falls. Galileo said all things fall equal...then how the hell can mass produce that which makes everything fall equal when mass puts specific distinction in the mass factor?

I just have to remind you again that Newtonians insist that a feather and a hammer falls equal under equal conditions through equal distances using equal time in distance (space) and period (time) to travel. That is what Galileo said and he proved that with all the numerous clocks that were on the market since then up to now. With that proof there is little aptitude for confronting this matter. Then by the same measure Newtonians insist that gravity is in force by the measure of mass because it is mass that finally is responsible for gravity. If mass was the responsible factor, there has to be variation in travelling time because mass will inevitably have the massive one travel faster than the slighter object will travel. Do you get this one together because I can't!

The space holding the truck being next to the space not holding the truck is falling as fast as the space holding the girl and the space next to the girl which is not holding the space or the frog where this lot is falling just as fast as the space holding the frog. The space is falling. The space is falling whether it is filled or whether it is empty and that means the mass has as little to do with the falling, as the colour of onions has to do with the depth of the sea or the temperature of the shining Sun. If it is Galileo that is correct and if all things fall equal then mass has no part in gravity. If mass is the inspiration behind gravity, the truck must fall a million times faster than the frog and in fact the frog should almost land in another country because that is how slow it must fall. The fact of the matter is that I don't wish to be near the place when any of this lot hits the ground because the truck will cause a quarry and the dancer will be a splash of red fluid while the frog might not be that worse for wear if the truck or the dancer don't land on the frog. The hole the truck will make will be big enough to hold a million or more frogs and that I admit. Then aspects change. As soon as there is contact between the falling object and the Earth, mass comes in bringing dispensing of the motion. But that is my point about Newtonians being incorrect and Galileo being correct and the fact that no marriage can be between Galileo's principles and the Newtonian vision of mass interfering with the fall of objects. This matter would be less serious if brilliant Newton did not extend his vision into outer space and helped God find a manner to modify the Universe. He put his nose where his mind was too small to walk. Newton cheated blindly to give value to what he fabricated and in deceiving the world he thought himself more brilliant than all others. That is a common trademark all criminals have.
Criminals always think they are smarter than the rest of the population. Newton was just another common criminal that changed Kepler's formula from $a^3=T^2k$ and rewarded k the value of zero $\left(\frac{dJ}{dt}=0\right)$. Then after this, Newton cheated even more and in three hundred and fifty years of study not one Newtonian, notwithstanding their obvious mathematical genius, was able to see or detect this fraud. After awarding **k** the uncompromising value of 0 by initiating an absolute unlawful mathematical principle, he furthered his criminal ploy by assessing that from the formula $a^3=T^2k$, k can become zero and in the meanwhile also a^3/T^2 can be **299**. But also typical of all

criminals he (Newton) pushed the blame quickly onto Kepler by naming the fraud Kepler's law while poor old Kepler had nothing to do with the laws in the first place. Newton laid all the blame at the door of Kepler and that, too, is another very common criminal ploy, to blame others and not accept responsibility for your personal devious actions, therefore pushing the blame onto the innocent party that was never involved in the first place. Newton not only changed the value of **k** by corrupting Kepler's work but then he threw the lot away by dishonouring the existence of **k** altogether in $T^2/a^3 = 299$. **The fact that $k^{-1} = T^2/a^3 = 299$** proves that outer space is moving relative to the Earth by a negative direction or then by contracting.

We have the same case scenario in this case. I can't find one shred of physical proof that there is mass in the whole case scenario because al the planets are not moving by mass but by equality. Look at the table just below and see the lot is moving at similar rates. If any of the Newtonian corruption was excusable, then the corrupting of Kepler's work is beyond repute. It took two men two lifetimes of toil and in a blink one corrupt Englishman destroyed the lot.

How can they say space $a^3=T^2$ is equal to time while at the same time declare that the motion of space in time is $T^2/a^3=299$. Has he (Newton) no respect for other person's labour? Was the man so criminally insane he had no respect for the life tasks of two outstanding researchers? Moreover, was the criminal intent spread across the entirety of all Newtonians that followed Newton in his footsteps and never blinked when reading the corruption their master committed? This is something that is going to upset the record books. In spite of all other magic Newtonians are capable of, they now can accomplish to move nothing by lengths. If it is nothing that fills outer space then nothing is moving quite rapidly because where we have Kepler stating $a^3=T^2k$ we have to presume that a^3/T^2 will bring about a negative factor to **k**. If what Kepler said that $a^3=T^2k$ is true, then $k=a^3/T^2$ and that any schoolchild doing mathematics in a classroom will testify to in this mathematical explanation. This then means space is growing by time developing. If space is declining by the cycle of time, we find the indicator of growth will be negative where that shows why there is a decline in space around all cosmic structures. It is space declining by time reforming space and space is what time transforms into as time moves on $k = T^2/a^3$.

The only time mass can be awarded is when **k = 0** but that is when the movement of the object is the movement of the Earth and the object lost all independent existing. There, in that case Newton is correct. Mass does become a factor when **k** is not zero as Newton professed, but $k^0=1$ as Kepler claimed. By disallowing k any influence it puts $a^3 \div T^2k = 1$. In that case where $a^3 \div T^2k = 1$ and also $k^0=1$, then mass becomes valid. But in those terms mass is hardly a cosmic reality with $k^{-1} = T^2/a^3 = 299$. There is no mass and the planets are shouting it but no Newtonian is prepared to lose his or her claim on performing as God and allow the truth to come and destroy their favourite lie. Those God forsaking nihilists can deny God existing on the condition the god they profess is covered with besmirching fraud. Only in the case of $a^3 \div T^2k = 1 = k^0 = 1$ can any factor such as mass find validity, and then it only applies where life intervenes in normal cosmic motion. Mass is something humans invent to make their life more calculated but mass in the cosmos is a non-existing entity.

The Brainy Bunch in physics academics awarded mass so generously to all falling objects while there is no evidence of mass interfering with the fall or is favouring those with more mass while penalising those with little mass, then the evidence of mass is totally absent. All the planets are falling towards the Sun as they drift away by the same measure and that is why Kepler said his formula is the moving of space a^3 that is equal = to both the rotation T^2 in conjunction with the linear movement **k**. Afterwards Newton went and wished away **k** by declaring $a^3 = T^2$ and that is fraud. If (T^2 / a^3) is = equal to anything from 300 to 296 then **k** is anything from 300 to 296 just because Kepler said that $a^3 = T^2 k$ and that gives **k** a value of anything from 300 to 296 and not as Newton so fraudulently said $(T^2 = a^3)$ putting at a value of **k = 1** or something as detesting as that. It is then little surprising that Newtonians stubbornly refuses to acknowledge the Titius Bode law. Then the four cosmic principles found in the cosmos and is a reality unlike mass, Newtonians discard since it does not fit into their Newtonian fabricated facts. The most obvious one is the Titius Bode law, which Newtonians claim as "incidental" where mass is awarded.

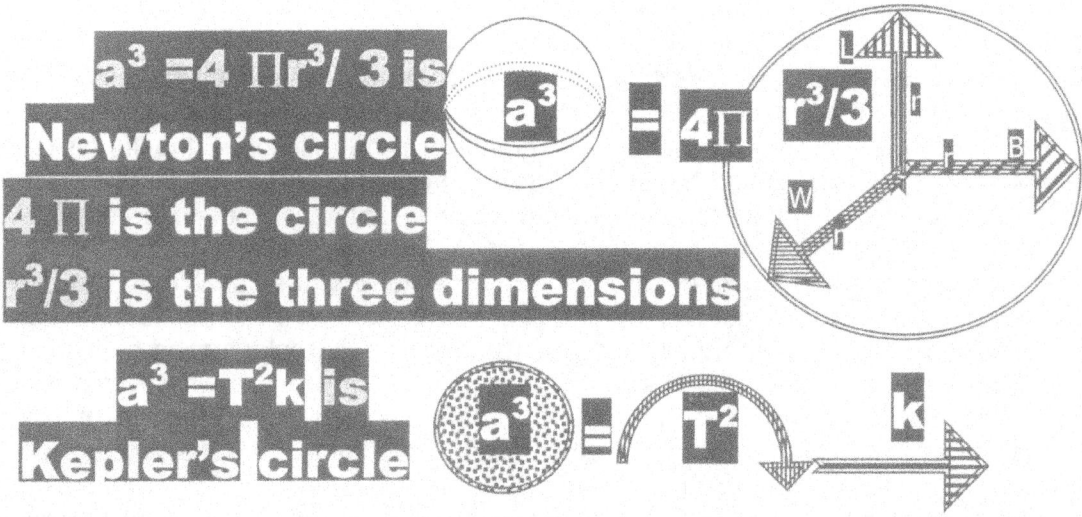

$a^3 = 4\Pi r^3/3$ is Newton's circle
4Π is the circle
$r^3/3$ is the three dimensions

$a^3 = T^2 k$ is Kepler's circle

a^3 is the space that time moves and that moves through time

T^2 is the time covered by rotation

k is the time covered by positional replacement of material

Newtonians should by now have realized that Newton and Kepler were not sharing the same views on the same matter or issues. Newton also should have realized centuries ago that Newton and Kepler did not have the same principle in mathematics in mind when Newton saw the work of Kepler as a mistake on the part of an ignorant Kepler. In the manner that the two applied the method used to measure the volume of a circle was different and the formula that Kepler used was unknown to Newton. But notwithstanding, Newton saw his ability suitable to change what he did not understand to his liking and that by the way is also part of his fraud...to tell Kepler what Kepler found by changing what Kepler found to the likings of Newton.

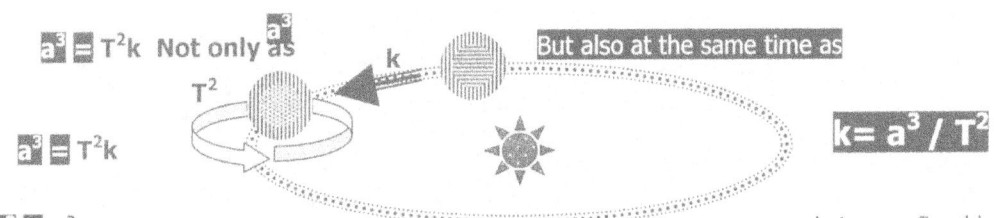

$a^3 = T^2k$ does represent space being confined by the circular motion reserving the in rotary motion of space to a specific location and acknowledging a specific center point to which the motion refers

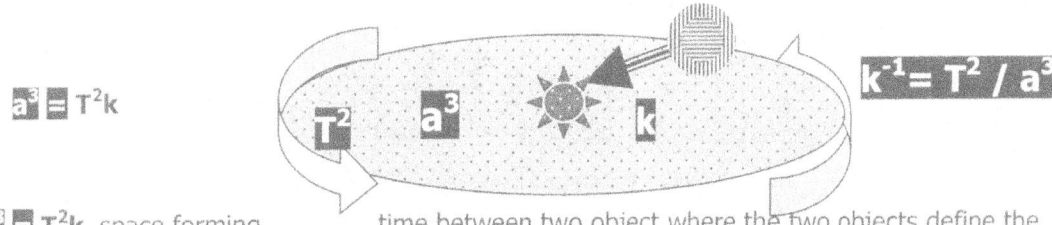

$a^3 = T^2k$ space forming time between two object where the two objects define the definition of the space that serves as time between and time allowing rotation

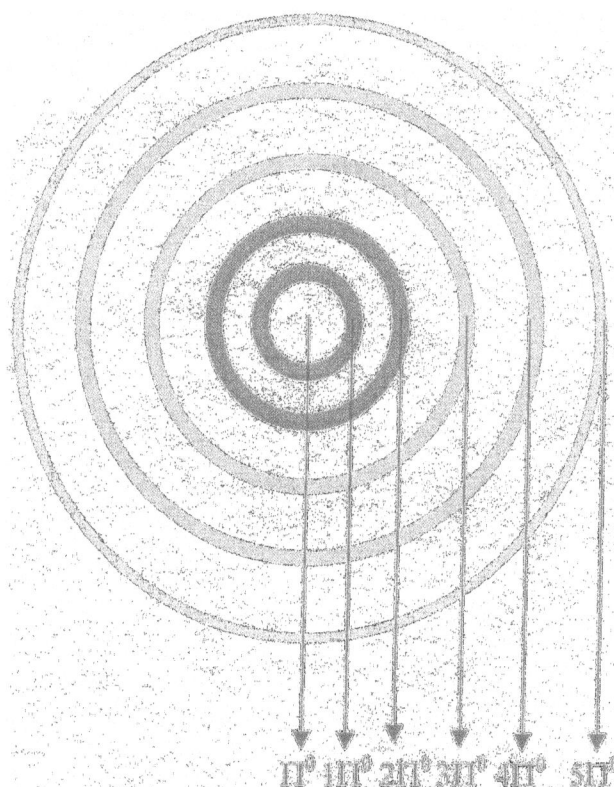

And that is why Newton and all his High priests could never understand the Titius Bode law. This might be excusable, but what is beyond any excuse anyone can offer is the blatant fraud they cling to, to not be able to understand the Titius Bode law.

The Doppler Effect is the Coanda effect that is the Titius Bode law in relation with the Roche limit.
When an object stands still, time moves space along in relation to the point standing still. This we see as a ripple effect and the ripple effect always moves away from the centre point by using a circle. This applies when an object holds k in relation to k^0 by duplicating the position of k^0 as k repeats the allocation of k^0. This Newtonian science knows as the Doppler Effect and for almost two centuries science could not move past this because there is no mention of mass and without mass mentioned, Newtonian brains come to a halt.

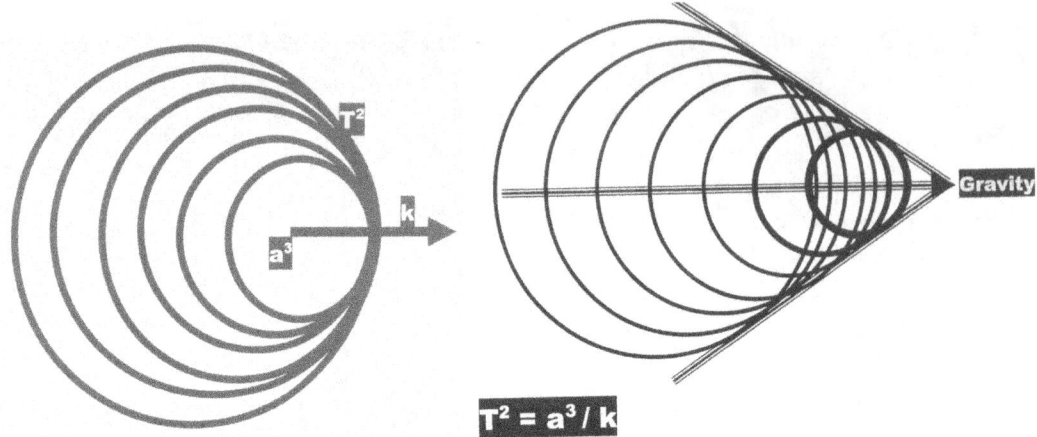

$T^2 = a^3 / k$

Position allocated to sound traveling according to the Doppler effect.

Π^0 $1\Pi^0$ $2\Pi^0$ $3\Pi^0$ $4\Pi^0$ $5\Pi^0$

Position allocated where the aircraft is positioned in relation to k.

When the object moves independently, as only an object can move in relation to the manipulation of motion using space-time in time, we find relevancies change to an unnatural state of affairs. By using motion, life surpasses cosmic time as only life can produce motion and motion is time, therefore life is reproducing the movement using time in space- time. Time moves by placing time in the spot k^0 or then Π^0 in relation to time producing space in space-time. This producing of space is too complicated for this part of the letter and I leave that to another point in time. When the aircraft is moving through **k** in space, such movement coming from life exceeds the time in space that the earth can produce. Therefore there is another star within the Earth and that star found independence by half the Roche limit $\Pi^2 / 2$.

According to Kepler's facts it is clear that the Sun is holding space a^3 and in the space there are planets held together as seen in the designated or allocated relevancy or distance placed represented as k and every planet is placed at a distance to orbit at a rate of T^2.

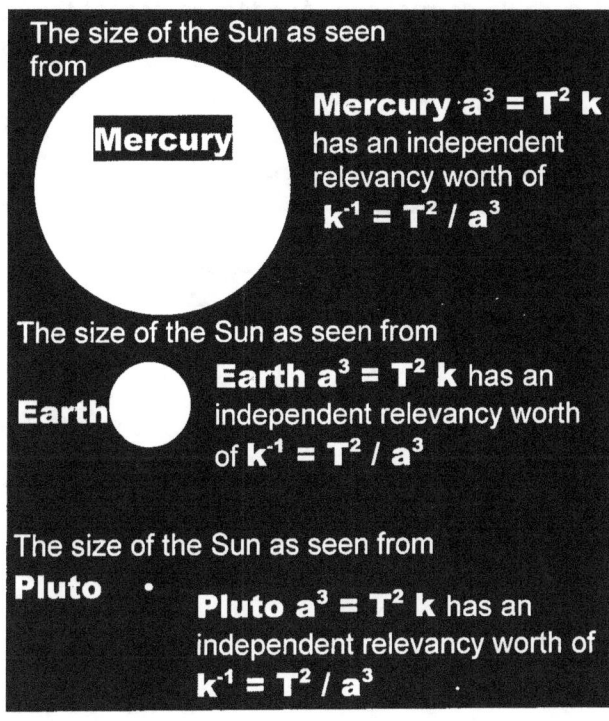

In spite of this $a^3=T^2k$ mathematical truth that even scholars know, Newton changed Kepler's $a^3=T^2k$ to an unbelievably corrupt $a^3=T^2$ by just blatantly wishing away the relevancy factor k stating that the circle T^2 divided by the distance k was equal to a total of 0. Newton gamely further corrupted this rule in the mathematical principle whereby he accomplished more fraud.

$\frac{dJ}{dt}=0$. This statement can never be and is blatantly corrupting mathematic principles.

This picture in its entirety condemns Newton's statement of the rotating distance nullifying the motion totally. $\frac{dJ}{dt}=0$. This is quite ridiculous in mathematical terms but Newtonians never once admitted to this being fraud.

Newton and science made one enormous blunder from taking this stance. It is as if they took the idea that when a wheel spins, the radius of a wheel has no influence on the wheel. In doing that, they removed the very fact that keeps the wheel at a radius and size relevancy and then announced that there just is no relevancy applying between the two. When on divides into another there is an irremovable ratio in place. Removing that ratio is breaking the most fundamental mathematical principle.

$\frac{dJ}{0}=dt$ or $\frac{0}{dt}=dJ$ This disputes mathematics. DJ/dt can have any number except the only number not possible to have is zero. One cannot place two factors in a relation of dividing and then proclaim the ratio as zero by not committing the relation to be forming a relevancy. By placing the two factors in a ratio of division, such relevancy then puts two factors in place that at the smallest must be one, showing absolute equality. One cannot say $\frac{1}{1}=0$ and find such a statement still acceptable, notwithstanding your academic position. It has taken me more than seven years of constant effort and continuous but fruitless communicating with Newtonian academics in physics, with no results to show. Thus, I finally am convinced that I now can report that I could not convince one Newtonian about this invalid ness, which implicates that every Newtonian is participating and furthering all Newtonian fraud. Their spreading this indicates their deceit by total corroborating with Newton's recklessness in mathematics that misrepresents the correctness of Kepler by corrupting what Kepler stated in the first place. This carries the hallmark and trademark of nothing less than corruption and blatant fraud. They'll not correct this because they are as corrupt as the Master they serve in upholding Newton's corruptions. Academics accept this total rape of mathematical law and defend the statement until death sets in but sneers at the Titius Bode law

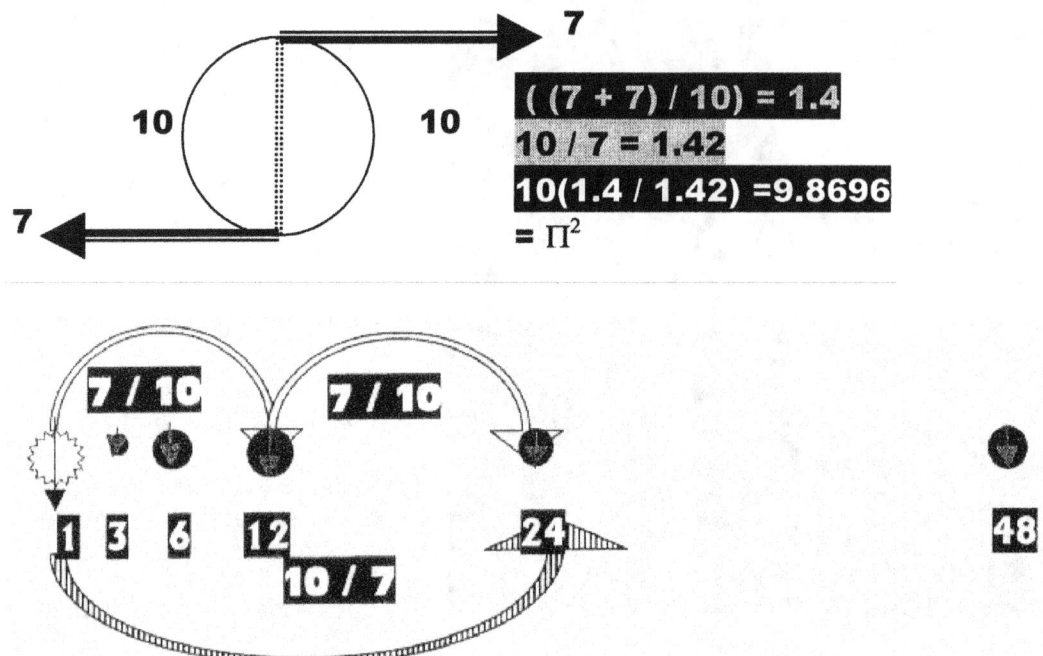

The culmination of the Titius Bode law is the turning of time into history and that is what space is. Space is the history of time in the present going to the past as the future arrives forming the present. We are living by the flow of time **(k)** forming the movement of time **(T^2)** that becomes the history of time **(a^3)**

What I do find is a lot of twisting of rules to that has no evidence in the evidence of underhandedness for the unspeakable Academics further cheating as Newton's deceit to a level in order to found the absolute policy of furtiveness slyness. The words that requires to accurately

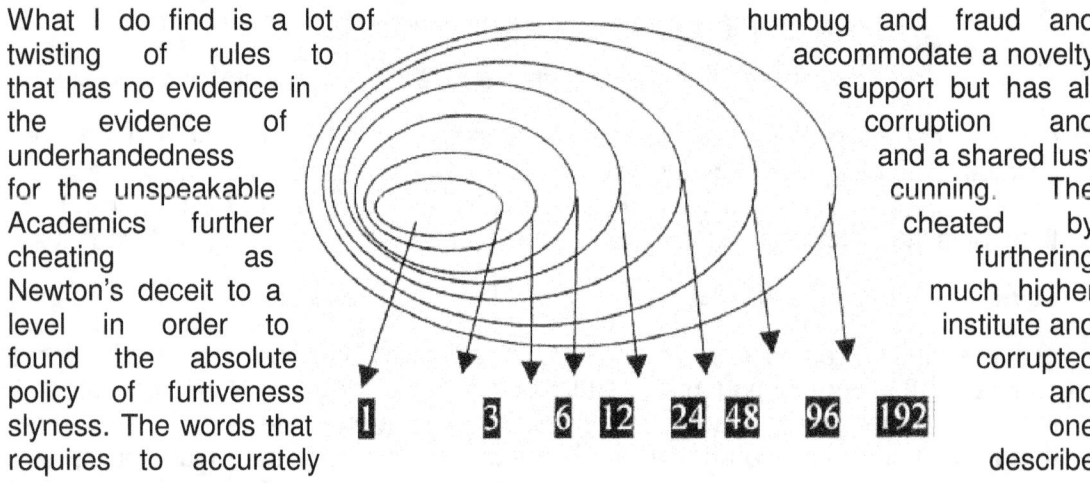

humbug and fraud and accommodate a novelty support but has all corruption and and a shared lust cunning. The cheated by furthering much higher institute and corrupted and one describe such fraud still have to be brought to paper since no language can describe the evilness behind such deception. The fraud and defrauding of facts are beyond comprehension to the non-criminal mind.

However, in the cosmos there is evidence of a law those in deceit refuses to acknowledge. Why do they refuse to recognize the formula? Because these phenomena don't show evidence in support of Newtonian fraud, it is rejected by Newtonians outright. If it is not corrupt and falsified it is not worth investigating!

Since the Bode law does not use or incorporate mass and thus does not support Newton in his fraud, therefore anything that renders any doubt in their formula such as the Bode law does is from the devil and should never be discussed amongst civil persons with highly graded education. In the cosmos we find planets being allocated specifically in a seemingly pre-selected allocation.

It is called the Titius Bode law and is named after two persons that both did the discovery at the same time although not at the same location or where the two was aware of the existing of the other person.

Bode's law: A numeral sequence announced by J.E. Bode in 1772 which matches the distances from the Sun of the six planets then known. It is also known as the Titius Bode law, as it was first pointed out by the German mathematician Johann Daniel Titius (1729-96) in 1766. It is formed from the sequence 0, 3, 6, 12, 24, 48, 96 and 192 by doubling of the distance after every number to each number. The planets were seen to fit this sequence quite well – as did Uranus, discovered in 1781.

The Roche limit and the Roche lobe.
We think of the Universe as being $7/10(\Pi^6/6) = 112$ that give us the nice six sided Universe we came to enjoy so much. The real Universe where gravity forms time we have a linear and flat Universe where only lines crossing space less singularity have any validity. The line flowing crosses singularity to form $\Pi \times \Pi = \Pi^2$. To do this crossing of the divide the four points where the extending goes is in division of four where time alternates the divide to form the wave. That makes the limit that singularity sets to be a part of one unit $\Pi^2/4 = 2.467$. That forms the Roche limit and everything inside the Roche limit will form liquid which is the Roche lobe. These are facts where mass is a lot of fantasy.

The combination of the four cosmic pillars form the movement of time that forms space-time through which time moves as time builds space as a spoor time left behind while walking along the route time is destined to go.

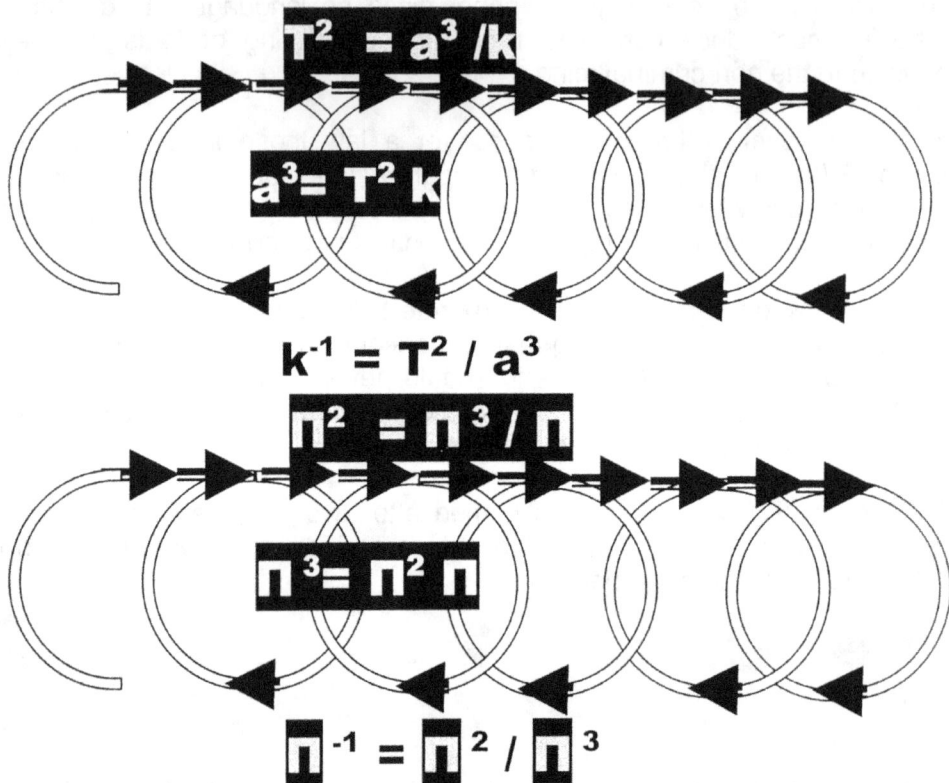

As time forms space-time it leaves a trail of space. By retreating in space as space-time forms, time moves back one position (k) as it extends two positions (T^2) to become the Universe (a^3). This repeat in motion by countering direction is very obvious in the way time (or if you prefer to use the other name as gravity) moves through space to form space-time. In that we find the principle behind the Coanda effect.

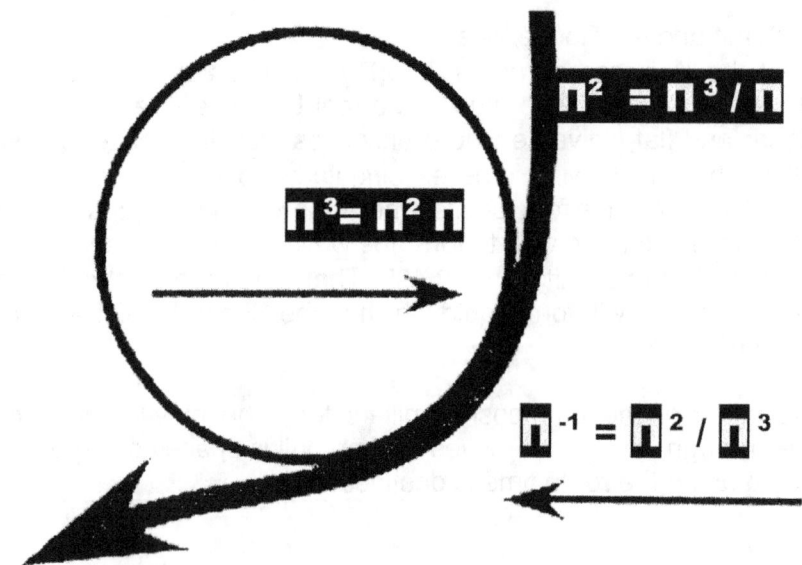

The Coanda effect is totally time related and implicates the motion of time moving through space as time as no other example in the Universe can equal. It shows that moving of space ($\underline{T}^2$) through space ($\underline{a}^3$) is the doubling of time ($\underline{k}$) or ($\underline{k}$) in the double ($\underline{T}^2$).

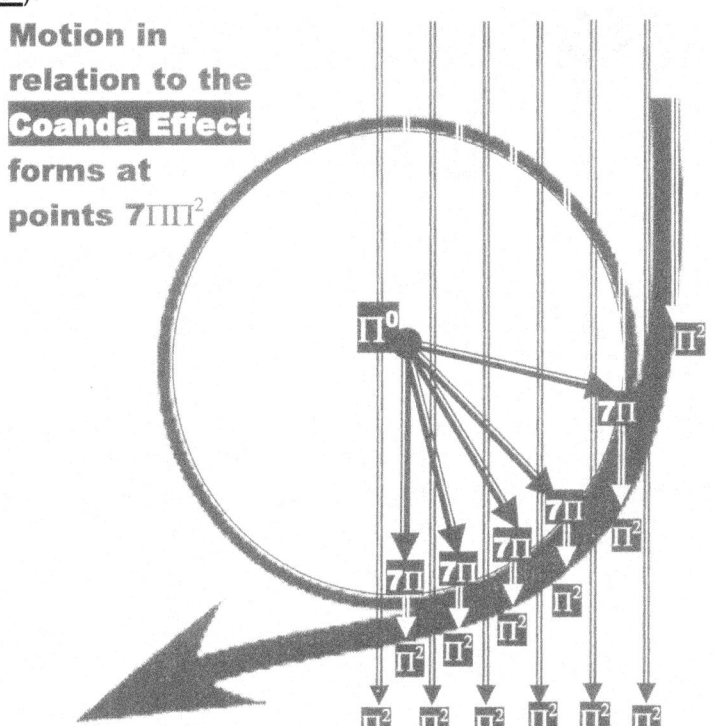

Motion in relation to the Coanda Effect forms at points 7ΠΠ²

Lines Π that connect Π (or also known as mass if connecting to the Earth surface) to singularity Π⁰ at intervals of 7⁰ (the circle in the sphere). As gravity or the motion of the Earth in rotation forming Π² crosses the marker Π at an interval of 7⁰ a point in time is established in accordance with the gravity of the Earth

This relation we find in motion takes direct value from the Titius Bode law and that is the speed at which gravity allows time to descend to Earth. Mass has no validity in all this except in the manner that time as a liquid transforms the object holding mass from s moving object to become part of the Earth.

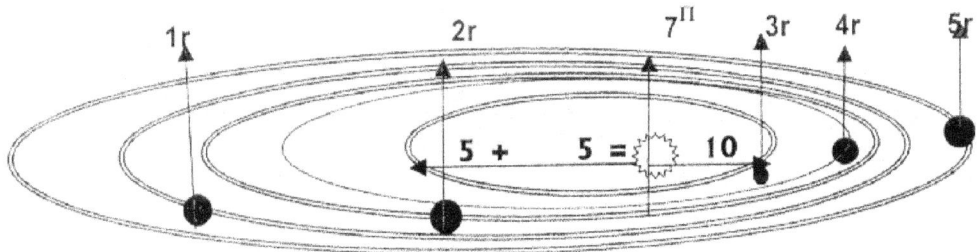

Newtonians make a big issue about Neptune and Pluto not conforming to the law. This explanation is so obvious that when reading it in ***The Seven Days of Creation*** even children will laugh about the simplicity there is in the explanation. Bode's law stimulated the search for a planet orbiting between Mars and Jupiter that led to the discovery of the first asteroids. It is often said that the law has no theoretical basis, but it does show how orbital resonance can lead to commensurability.

An Open Letter On Gravity Part 2   Page 437

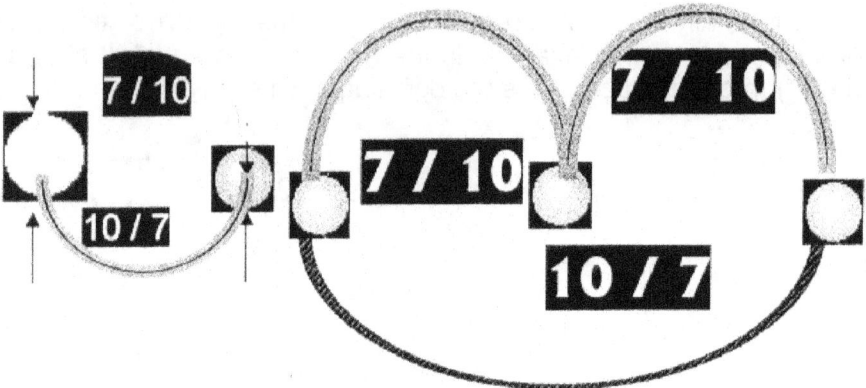

Normally it is expressed as Bode's law but I decided to use the full name in all my books, as the Titius Bode law is one of the four pillars on which gravity rests. This is the pillar that assembles the solid with the liquid but we will get to that later on.

There are three other cosmic pillars on which gravity is founded and inspecting any or all of the others there is not a single bit of evidence of mass playing any part in the overall concept. They are the Roche limit, Lagrangian points, and the Coanda effect.

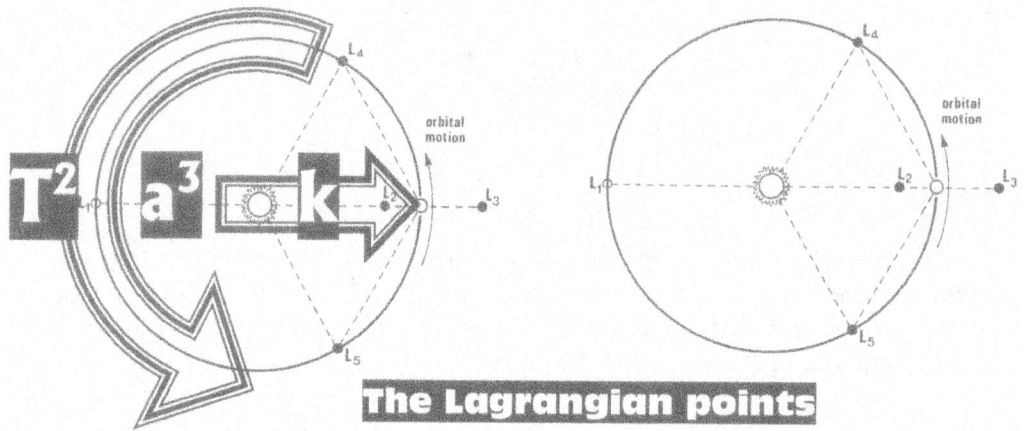

The Lagrangian system is in its full compliment the Coanda effect that derives from the Titius Bode law, which presents the best example of the Coanda Effect establishing gravity by lining up liquids in relation to solids rotating.

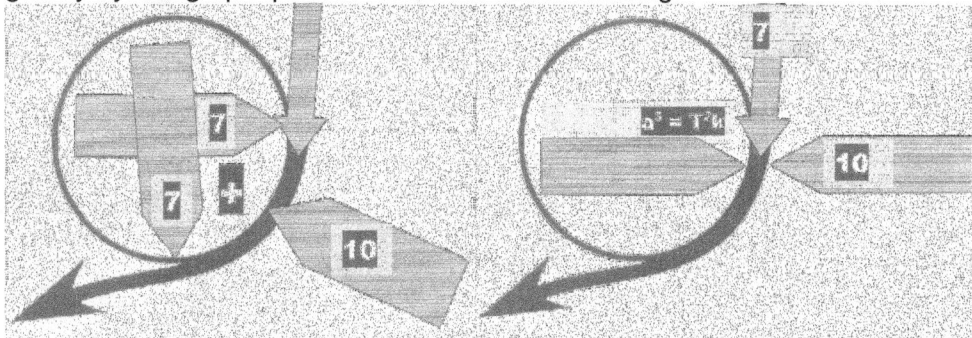

Lagrangian points: Five points in space at which a very small body can remain in a stable orbit with two very massive bodies. The points were first recognized by Joseph Louis La Grange and are rare cases in which the relative motions of three bodies can be computed exactly.

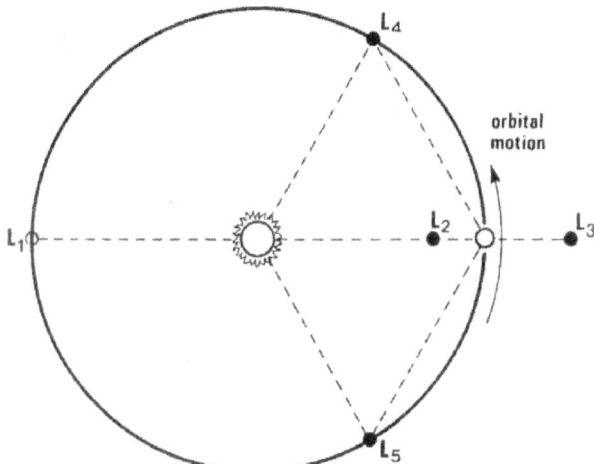

In the case of a massive planet orbiting the Sun in a circular path, the first stable point (L_1) lies on that orbit diametrically opposite the large planet itself. The points L_2 and L_3 are both on the Sun-planet line, one closer to the Sun than the planet and the other farther away. The remaining two points, L_4 and L_5, are oppositions on the planet's orbit where each forms an equilateral triangle with the planet and the Sun. Clouds of gas and dust are believed to collect at the Lagrangian points in the Moon's orbit around the Earth, and the Trojan asteroids are found at the Lagrangian points in the orbit of Jupiter.

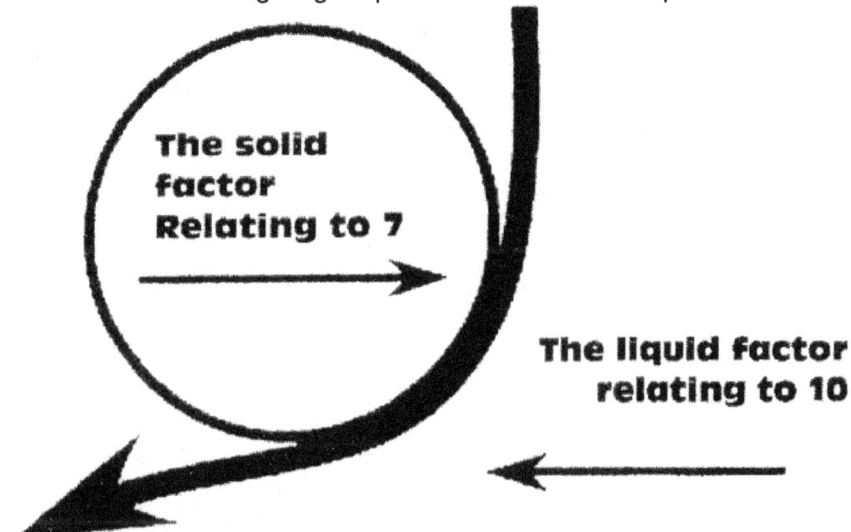

Then the Coanda effect is the glue to the three phenomena where each contributes in a specific manner to form a unit in movement that combines as gravity. The Coanda effect has the other three principles so tightly woven forming part of the Titius Bode law and not giving that much distinction as recognisable and individual factors in the unit but is more forthcoming as the culminating of the factors where the one is being part of the other but because of the tightly woven form the independent distinction is not that obvious. The Lagrangian system is also another part that implements the Coanda effect as a duplication of the process.

Since the Titius Bode law in particular doesn't support any corroboration with the formula of deceit, Newtonians are attacking the concept as they try to deny the existing of this law. They break it down to a coincidental occurrence because it threatens their deception of mass being the result of cosmic forming.

They are unable to explain it and since it does not support Newtonians corruption in the fabricating of mass, therefore they find no reason in supporting the law. Newtonians deny the Titius Bode law being a law or even being valid, although it is there and it is so accurate that Newtonians applied the law in the past to detect at the time yet undiscovered planets. Because it is there but not supporting Newton's fabrications, Newtonians turn on the laws as if it is the cosmos that has gone criminal and it is the cosmos that is breaking Newton's laws. Breaking Newton's laws then make the cosmos criminal or so do Newtonians wish to portray the situation. They report that it is the cosmos that stands in corruption because the cosmos produces insignificant factors such as the Titius Bode law. The way they go about in trying to diminish and tarnish even the possibility of law applying outside the Newtonian spectrum of possibilities, such laws are referred to in the climate of suggesting that the cosmos commit the fraud by producing an unlawful practise. This sounds terrifyingly arrogant but scrutinise the defining explanation and one would see that they accuse the cosmos of being un-substantiating and deceitful in creating a false and non-existing image. However coincidental it might seem in their view, still they do not doubt or question the accuracy of the coincidental principle, because they use the principle as a law to great effect when in the past they used the Titius Bode law when they detected the missing planets! It is very surprising that they still believe it is coincidental, because they used it to detect the missing planets. But this is not the only time they accuse the cosmos of fraud and underhand dealings and when we discuss the critical density dogma, we will see just how ridiculous Newtonians get when their formula comes under threat.

These are the facts that the atheistically motivated group of supreme intellectuals regard as so well proven they rank it higher than religion. They doing the ranking are those that find it funny when someone asks about Creation and the Bible. They are those supremely gifted that find us believing in the Bible to be rather feeble and they think that only persons with a childlike mentality living in a fantasy world and those not very skilled in intellectual standing will associate with such trivial unimportant issues as we find in the Bible. They are the ones holding a posture they think represents the absolute epitome of human intellect while they are being so superior in intellectual standing that they carry the mentality of the Godly sane.

I gave a glimpse of how I show that in their world of pipe dreams with their unsubstantiated formula that puts such acclimation on mass, they are the actual ones that never can read and who never will read into what is true. The formula does not inspire greatness but proves how little is the measure in mentality that those atheists see while holding their position as the super educated and how wise they truly are. As all other persons that are not blinded by the Newtonian greatness can see they founded their lives on fraud. The corruption is so basic to realise that after reading this letter with the view I introduce not even a child will find them much convincing any longer. They promote the absolute fable that the glorified formula represents but in truth it is a myth based on a corrupt fallacy of changing facts by the hypocrites to suite the needs of the stupid. By changing mathematical principles they create the fallacy no other deception can beat.

Changing science and rules governing science that applies does not establishes but it destroys. Can any one imagine what the intellect must be of persons thinking their position to be so absolute they can deny the existence of a living God and yet they are too feeble to see there can be no mass present in outer space because mass has no proof of being present in outer space. What is not there is not there because there is no evidence of it being there. Placing something in a position because finding it in the location is required in formulising a theory is what constitutes to fraud and not creativity. I am sick of atheists' high and mighty cloud of supremacy under which they sport the banner of being beyond reproach because they hold the supreme proof and yet the supreme proof is not worth the ink on the paper.

The formula they so grandly hold, places science above any believers because in the science they prove the absolute confirmed. In that they see that the intellect that they are so privileged to have and which makes them so supremely gifted that they fall in the upper-class where they are the only ones that may converse on the level equal to what God can enjoy while holding conversations on intelligentsia in an intellectual capacity on par with only God Supreme and therefore they will not allow such common beliefs as the Bible presents to interfere with their skills. To them and to those I challenge to prove my theory wrong after reading the entire works. Try to disprove me where I show not only how but also why the only way the Universe could start is by the interpretation we find in the way the Bible indicates and only the introduction the Bible gives has any validity.

My books have no religion of any sorts and only find support in real mathematical explanations that bear real mathematical laws. On occasions in my other books I do use one verse or so to show how the start could not have been any other way but Creation had to have started the way the Bible declares. It had to start with light and then I show where the light comes from all that is otherwise darkness. I show mathematically where the darkness is and I show mathematically where the light that the Bible refers to at present is and that has to be the light that formed the Universe because that light is the Universe. But being so blind that they fail to see that planets can't have mass, leaves no reference to their insight. If there was mass then there was distinction brought about by mass, as the big planets will move faster than the smaller planets move. It is then hardly surprising that the atheistic fools can see only their formula and not what their formula misrepresent. Their formula can't take them past their own stupidity. I challenge the atheist to disprove my mathematics that I present when I show how the very first moment came about exactly as referred to in the Bible and why the moment it is carried with us during all this time. The book is written and the book is waiting for funding to be published.

It is not lightly that I call them stupid or fools but if those with corruptness in their hearts and deceit in their minds that come and criticize my religion from a tower of what they think is strength, where they try and belittle my God, then I have no quarrel to show who they are. It is with incompetence raging out of control they did then trod on my religion. The dismay they show is as a consequence resulting from their stupidity caused by their blindness, leaving them with what is coming from incoherencies their stupidity can't realise and their mindless blindness can't foresee and it is with pride that I enjoy the moment where I uncover their corruption and blow the cover of what covered the size of their little minds to reveal the fools underneath a blanket of fraud. Just this one question: if they are so stupid that they can't trace mass to where mass belongs and remove mass from where it is not present, how on Earth will those being that blind understand what is truly Biblical and holy? If you read on you will see how shallow their deceit really is. Venture with me as we trace the path of their criminal deceit and we journey to see how the formula $F = \frac{M_1 M}{r^2} G$ truly applies. I explain gravity in all the books. The issues about gravity are very complex although understanding it is not very difficult. This achievement is possible because I saw a way to break away from invalid concepts which Mainstream physics hold. It is all the misinterpretations Mainstream physics holds that is resulting from corruption that they portray as truths that makes cosmology not understandable. Cosmology is so simple that even I can make sense of it but the way their inadequate presentation presents cosmology is not coming as result of what they can see but is a result of what they do not understand when explaining what they cannot see. I recognised the impossible double standards Mainstream physics apply to promote their much shady explanations.

The inconsistencies brought them double vision and to compensate for their incredible theories they simplify issues to a level where everything is becoming meaningless and in comprehendible. At this junction I can mention the senselessness of Einstein's declaration that it is the entire Universe that draws flat. Einstein presumed he new what the Universe was and never went in search of what the Universe is that he saw was drawing flat. What they say can't be supported and authenticated in any investigation even in simple terms. It is as if they never read with interest, that which they explain and they never scrutinise that which they advocate. They give values that are senseless and make that which they say meaningless. Mainstream science finds stars on the edge of a limitless Universe. Mainstream science advocates that the limitless Universe expands its limitlessness. It sounds degradingly stupid and irrational but that is what Mainstream science is doing. They expand that, which is already everything that can be into being more while it is impossible to put more into anything that is already everything. Mainstream science can fill an already overflowing universe with nothing where the nothing lets the Universe overflow while they place that nothing in between stars and galactica to form space by filling space with nothing. Then they create mass that creates gravity that creates a force that removes the nothing and replaces the nothing with nothing being not there. The list of stupidity and irrational senselessness goes on and on and on…and they are the Brainy Bunch, the so wisely educated!

An Open Letter On Gravity Part 2

In this letter I am going to further investigate how much truth there is in mass pulling by the force of gravity. To most if not to all of the persons reading this where all those being with Newtonian corrupted minds have the view that just the thought about me embarking on the investigating of Newton making mistakes, that the issue then is totally senseless and a waste of effort to investigate. To every person that went to school investigating Newton from a neutral stance is senseless because the concept his work carries became accepted as household practise and life science.

Do you think of astrophysics as the department that is run by the wise and the level minded, the sober thinking and the absolute trustworthy? If you are a student there is no other choice you have but to think of the faculty in those terms because your tutors brainwash you into accepting all the senseless stupidity they present. If you think those in charge of astrophysics are the pillars of trust, then get wise and read the following. What you are about to read is simply mystifyingly simple and yet to this day I have not challenged one academic anywhere that had the honesty to admit to the fact of Newton being wrong. After you have considered the following you might agree with me that even small children can reach a higher level of clear-minded logic and find more sensibility than what those scientists promoting astrophysics have because science lives in a make believe fool's paradise. If you are a student then ask your Educated Masters to please explain the following abnormalities and inconsistencies they promote. It is those whom I present in this letter to become wise instead of brainwashed. I say brainwash again because they force-feed you fabrications, as you will come to see. They can't explain the facts as the facts but hide the fact that the facts are in fact untruths. Tell them to prove that planets have mass. Tell them to prove that it is mass that generates gravity that pulls the planets. Ask them to explain gravity in detail.

In the book named "an Open Letter Announcing Gravity's Recipe" I bring the solution to the mystery behind gravity. I tried in vain to introduce the principles I find valid to the academics in charge of astrophysics. Facts that Science presents as being the uttermost explicit and unwavering truth, fails to bring any logic answers to so many questions that it should address. It fails to have substance in addressing the most basic and simple questions about gravity and physics. Yet, to every question science can't answer, my approach do bring many solutions. The presentation and the delivery of my answers that I reach are understandable and simple where it serves both logical science and the truth. However, since my answers do not match Newton and his misconception about gravity and supports the idea that mass generates gravity, those in charge of science don't even bother to read my work.

With their affixation on the corruption they portray and as a result of their uncompromising position they hold, I can do little to the giants where they are in the mighty positions they have and just because of that they can go about to sideline and ignore my work and this is notwithstanding the correctness that my work delivers compared to the utter failing that Newton's work shows. When they are confronted with my evidence that my work delivers and where they then have to match my work with the hypocrisy and misleading nature of Newtonian cosmology, their defence in substantiating their claims are to ignore me. Since I do not applaud mainstream science and the clear fraud they embrace and fraud it is that they embrace, I am silenced. Why is it that my work is going unrecognised or even in the least goes never debated and never commented on…it is because it will then trash every article anyone has ever written about astrophysics and general science forming cosmology.

As academics with high standing such as they supposedly should have, then as such high standing academics, they are obliged to show a lot of integrity in the way they conduct their work and do it in a fashion that we all clearly can see …but in contrast to their surmountable integrity that should be surrounding them like a well fitting cloak, in stead they will rather protect the blatant and obvious fraud of their Master, while at the same time saving their skins than would they seek the truth.

It is the perpetual fraud that results from the choice they make when they protect Newtonian untruths that becomes paramount. It is the result that comes from their decisions they make that gives them the integrity they claim and when they begin to read my work their in discrepancies force them to disclaim the truth I report. It is when they disclaim the truth, instead of choosing to back my work that they become participants in crime by promoting their fraud. If they would back my work they would have to commit to the truth, although doing that it will come at a price, I know. Then in doing so they will trash all the previous fraudulent work in cosmology and everything that was delivered this far.

They will condemn the fraud to the waste paper basket and render all previous work they produced on the subject of cosmology as invalid and void. Doing such condemning of past flaws will require a very brave person that shows the utmost integrity. But it will put all the Newtonian's bias and fraud into the place where it belongs. However, I realise that it is a very expensive virtue to have since considering that such acting will lose them money. Those academics in controlling positions then will rather miss to subscribe to the truth in order to benefit financially from continuing to corrupt student's minds further than put the students on the correct road. So, to keep benefiting from the wrongs in the past they have to justify their inconsistencies and that brings them where they rather have to attack my work and disprove the accuracy of my work than forcing the real corruption into the daylight. That they cannot do.

Therefore they then rather ignore my work because they cannot attack my work since my work is rather indisputably true. I am not merely saying this but charge anyone to prove I am wrong when completing the reading of both parts of **An Open Letter On Gravity**. In that sense they also place their work beyond my approach, as they can simply ignore me as if I represent the plague while they carry on with little consequence to bother them. I challenge them or any other person to prove Newton correct and not just declare Newton being beyond reproach after all has seen the evidence I bring. After reading this all students must challenge them to prove everything what they cannot defend or prod them to get honest. After the entire work is read, see if you still can defend the Newton formula as it reads when it says that

$$F = \frac{M_1 M}{r^2} G$$

$$F = \frac{M_1 M}{r^2} G$$

This is the basis that Mainstream science uses as the foundation of all physics anywhere. If this is wrong then everything they have got to work with goes out the window. They put mass and the distance that parts objects in a relevancy, in other words the one is a ratio to the other. The one factor brings a measure to the other factor's value. The one cannot be without the other. The increase in one becomes the reducing of the other and the other way round also applies.

When the distance is large, the influence of mass will be small and when the distance is small, the influence of mass will be overwhelming. Then they state we are in a Big Bang expanding of the entirety. Why then, when considering that if it is mass that produces an inclining force of contraction as Newton says and this is all what still there is that is going on then…why didn't the expanding stop before it started when the Universe was little developed and still very concentrated in comparison with what applies today?

Today after we are able to use the luxury of hindsight through studying Hubble's discovery even better, it becomes even more apparent that the lot of everything is not imploding, as Newton would have us believe. Instead reality shows that the fact is the exploding Universe is instead of imploding, the entirety is expanding just as Hubble proved. The radius at the time of the first instant back then was no factor, which makes the gravity at the time a totality of unrivalled force. The radius being that insignificant leaves the mass unchallenged in asserting power in relation to the non-existing radius it had.

I say the following and while I say the following they ignore me while they remain as tight lipped as they ultimately can in defence about my challenge. Mass is the resistance that any moving object shows when stopped by a larger object that blocks the smaller object from moving further as an individual object whereby showing mass, the smaller object resists the effort the larger object asserts on the smaller object to compromise the form the smaller object holds as a unit and not to accept the form the Earth imposes on the object. By having mass, the object no longer holds independence but retains some form of individuality by not compromising the unit it forms as an independent structure. When the object lands on a larger object and the larger object is part of the Earth, the larger object halts further motion. The larger object removes the individual characteristics that the moving object has.

Let us have a good look at mass. When there is a ship we see the painted waterline of the ship indicating the load in mass the ship can take. The ship is lighter than the water because the ship floats on top of the water. By loading the ship, the ship can take in a lot more mass than what fills the volume of water that the hollow ship displaces. That is because as long as the hull displaces more water than what the mass of the water is in terms of the area the ship claims, the ship will float. The ship is less dense than the water when taking the area it holds in relation to the density of the water it displaces.

When the ship has an equal density to that of the water, the ship will float in the water while being buoyant. The ship has to accept a certain percentage of room the water normally holds to allow the ship to float inside the water and prevent the sinking of the ship to the bottom. This is an act of precise balancing. The ship then being somewhere inside the water has no mass but being equal to the mass that the water holds that the ship displaces. Then the ship is submerged and floating in the water and not on the water. The ship has to displace as much water as what the area it holds will be in mass when being only with water. When the ship goes submerged but not sink to the bottom more of the mass will have to be equal to the water in order to achieve the total buoyancy. Then the mass of the water is the same as the mass of the ship and therefore the ship can hold ground inside the water without sinking or floating. In other words, the ship is just more water. If the ship again wishes to float the ship will have to exchange some density of the water with the density of air in order to displace some water it has in its hull and exchange that water for air.

Then, having more air than water will enable the ship to have less density per volume of space and less density will render the ship in having less specific density than the water and therefore the ship will float once more. When the ship sinks to the bottom of the water and rests on the bottom, the ship has more mass than the water it displaces. The ship has a bigger specific density than what the water has and the mass of the water pushes the ship to the bottom of the water. At such a point when hitting the bottom the ship then has mass. It is the mass of the ship at the bottom of the water being more than what the mass of the water is that floats over the sunken ship that puts the ship on the bottom of the water. In water or in liquid the ship or the solid either has buoyancy or it has mass. It either floats in buoyancy or it sinks whereby it receives mass when the sinking ends. When it floats it has motion by buoyancy that the water provides.

Displacing liquid space as air

Displacing solid space as water.

The buoyancy of the ship is a mixture of the density of the steel that displaces the water and is taking up the density of the water that then is added to the density of the air that the steel that is above the water level displaces. The steel displacing the water combined with the steel displacing the air form a unit that holds density that is in ratio with the water density. The mass of the displacement density must be lighter or equal to the mass density of the water that the ship displaces to gain buoyancy. That is the water displacement as well as the air as a combined unit that the total density should be less than the water.

While floating the ship per overall volumetric space used has less mass than the water has that has equal volumetric space in use. While being completely submerged, the ship has the same mass than what the water has per volumetric measured unit. The ship has more mass per volumetric measured unit than what the water has when the ship sinks. The mass is a relevant factor because of the fluid aspect. The mass is considering the density per volumetric measured unit of space. When floating the ship has little mass and all floating objects has mass that is less than what the water has. The density of the air reduces the density of the total volumetric measured unit of the total combined density of the ship plus the air the ship displaces. When placed in relation to the density the water holds per volumetric measured unit of water the ship overall should have not less mass but must have less density. When being submerged it does not matter if the ship is a big ship or a small ship because the ship then in a submerged state has the same qualities as the water and therefore a big ship will be as buoyant as a small ship will be. This is very important to note. When in a liquid there is no mass factor. It is all about density. Only when the liquid suppresses the object by pushing the object onto the solid, there then is no distinctive difference between the motion that the object has and the motion that the solid has does the body turn from being a liquid to being part of the solid. Only then does the mass factor enter the equation. A big ship will float submerged next to a fish and from the mass aspect the two would be equal. It is depending on motion.

That is gravity but that is the motion of liquids in relation to solids and the solid try to secure the liquid as an extension of motion. Only when the substance attach to the solid does the attaching piece form mass. The mass has no ability to charge the motion but in fact is there to stop any further notion as being a liquid. To remain a liquid, the object has to maintain the formlessness of liquids and reform its form at the first point of touch. By not reforming but maintaining its shape, the object separates from the liquid and retains the state of a solid. A solid will not dispose of form but a liquid will relinquish form at the slightest provocation to do so. I am about to show that mass has no implication on orbiting planets and therefore mass has no implication on the cosmos. With no mass playing any part all the rules applying to the cosmos will fit liquid and density instead of mass forming part of solids.

This is why planets circle the Sun. The explaining of planets and therefore the action of the comet is the same. The planets are in the liquid motion and that we can see because using Kepler's formula shows that the flow of space is towards the centre where the Sun is. The explaining is not difficult but is tedious and takes time because it involves interpreting the four cosmic principles being the Lagrangian system, the Bode law, and the Roche limit where the three interact to form a balance between what forms a liquid in relation to a solid. That relation then becomes the Coanda effect and the planets are buoyant in the liquid space that spins towards the solid Sun. But by own individuality, the planets too are solids that spin in the liquid and the solids react as independent solids relating to the Sun. From that we find that the building of space or the expanding which forms part of the Big Bang expanding uses the form of 10 in relation to 7 which was what Titius and Bode saw. This forms gravity and gravity is the process whereby liquids move in relation to solids. There is no evidence of mass anywhere in the cosmos and that criminal deliberate misinterpreting of the truth has to stop. Mass is a unit Newton that came up with but only has merit when applied to movement of material that has individual qualities and that can move above and beyond the movement of the Earth. To that however, there are also limits and one of these limits in stretching the atmosphere is the sound barrier. If you grant an object mass, the mass you give only applies when there is contact between the object in mass and the Earth. To have mass the object must become part of the Earth singularity and not as Newton presumed reach a point of $\frac{dJ}{dt} \neq 0$. Instead in reality by touching the surface of the Earth, the object becomes on solid part of the Earth singularity $\frac{dJ}{dt} = k^0 = 1^0 \times 1^1 = 1$ and become one with the Earth. By having such contact mass is granted. Such contact can be direct where the object touches the Earth or indirect where the object rest on or hangs from a body that touches the Earth and through that distribute the mass it has to the Earth. But as Newton, said mass can only apply when the moving of the rotating body equals the linear moving of the rotating body. However, when this happens it is stopping the gravity and not promoting the gravity. One may say that mass forms anti gravity because it prevents gravity by repealing the individual object's independent motion or gravity and replaces such individual motion with the motion of the Earth.

When the body becomes part of the Earth, it finds mass by the compressing of the Earth's movement forming a resistance in space and that compressing becomes the atmosphere. The rotation T^2 of the Earth a^3 shears through the liquid atmosphere by extending **k**. The atmosphere then resists the Earth motion and with the resisting of such motion by the liquid that is outer space compresses into the atmosphere, the liquid pushes against the Earth just like the water pushes against the spinning tire. By not being on the ground a body holds buoyancy in liquid space and in that Galileo's principles find truth. While any body falls the falling body is part of the liquid thus it is in buoyancy. Because it has buoyancy, it has no mass and therefore it fall, the same whether the object is big or small. Big or small is just more liquid when travelling through the atmosphere. When the object lands on Earth, it switches relevancies and become part of the solid Earth. But since it strives to retain form and refuses to compress into a denser compact state where it then would submerge into the Earth, it retains a part of its individuality by maintaining form but most independence it loses by accepting mass. The more this liquid pushes the object in mass that places the unit into a smaller or less spacious environment, will the liquid compress the object by becoming more compact or denser while it claims less space. The object is going denser and that is why the object gets many times smaller but also many times more massive when the object will have mass on a much larger structure.

When the structure, that was falling and landed, the landing alternated the applying relevancy and the object becomes part of the Earth by holding mass It then swaps sides with the solid the object having no longer a relation with the

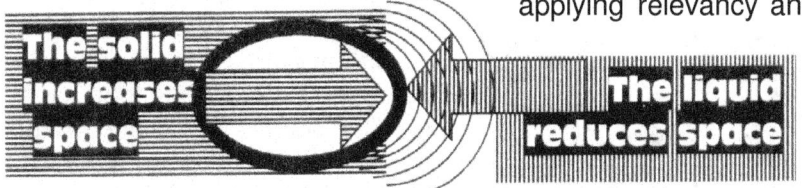

liquid that it before formed a factor within the liquid and therefore holds buoyancy but then when it made contact with the Earth, it then stands related to the solid extending by claiming space that the liquid holds. As it connects onto the Earth ($k^0=1$) it no longer flows with the liquid but it relates to the solid. It is when the solid makes contact with the liquid that the liquid suppresses the liquid factor's claim on space by granting the solid a claim to more space and thus compressing the space the liquid holds by reducing the space the liquid claims in favour of the solid's space expanding. By moving faster there is more liquid resisting space claimed by the solid and therefore the solid can claim less space. Because the flow exerts the mass that the contact in resisting provides, the compressing of space claimed puts a depressing of space granted and such depressing of space granted brings about a more resilient claim on space.

Newtonians regard size and mass being directly connected but reality tells a different story. Earlier I showed the information in terms of mass but it also shows a tell tale picture in terms of concepts. Where from do we get to the "giant" planets with the "massive" gravity because they are so "enormously big" we find that also to be total a fabrication of facts. When comparing the gravity produced in relation to the space used to produce the gravity, we find Jupiter producing the feeblest gravity and only the Sun eclipse Jupiter as a useless generator of gravity. For every cubic meter of space used, Pluto and Mercury are producing some gravity that should make them become Black Holes, while it is not very clear what keeps Jupiter intact with the feeble gravity that it produces in comparison to the smaller Pluto and Mercury.

Let's place the Earth at 100% gravity efficient meaning for every cubic meter of space we use for filling space within the Earth; we have one unit of gravity produced. Since Venus is slightly smaller, Venus holds an efficiency rate of 111 % compared to the Earth having 100%. Looking at Uranus and Neptune we find them generating in comparison only 6.6 and 6.3% per cubic meter of space occupied when placed in comparison with the Earth. If these ratios of gravity charged per cubic space applied, were placed in context then the following would apply:

$$F \propto \frac{M_1 M_2}{r^2}$$

That means if one puts the gravity Pluto generates into the mass that Jupiter have, there has to be some sort of Black Hole coming about should Pluto eventually gain the mass Jupiter has. It is obvious that with the smaller planets being dense and the larger planets being "gas" (which they are not) our Newtonian genii have something screwed up as usual. Size and mass does not seem to bring the resulting gravity production into play as Newtonians so admirably advocate.

If mass got more then gravity must get more, because gravity is the result of the mass that generates gravity. Then every cubic meter of space that is filled with material must get evenly stronger in the ratio of generating gravity, since it is the same material generating more gravity. According to our ever so wise Newtonian masters, it is the amount of mass becoming more that increases the gravity applying. According to their wisdom it is the gravity that is the result of mass and if more mass is filling the structure by the square or the cube, more gravity should be coming about per square cube. They say It Is not In ratlo to a centre but it is individual particles packing the object to capacity.

Again the reality defies Newtonian misconception. Pro rata stars become gravity active and apply less gravity per size when the radius becomes bigger, Newtonians were supposedly correct, then the star that held the most space had the most mass and the more mass has to produce the most

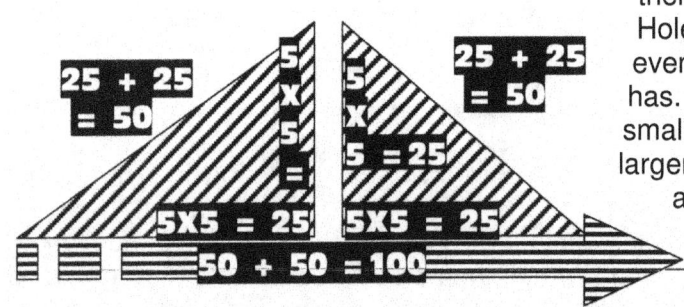

gravity bringing about that the escape velocity would be proportionate to the radius of the planet or star in terms of the mass and the radius holding the mass $v = \sqrt{\dfrac{2GM}{R}}$. Fit this idea into the idea of which sizes apply when a Neutron star and a Black Hole is in place and we find Newtonians double-talking as if they are not even using the same language.

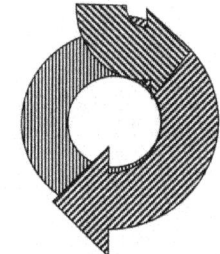

The liquid forming motion in terms of the solid

However, if gravity came about in relation with motion of time serving a relevancy by placing infinity (that which allows movement of material within) in ratio with eternity (that which allows movement to be), we get a better perspective of gravity. By placing material (the solid aspect) in a smaller unit, it will spin faster and through that come into contact with more liquid, which is making the ratio there is between 10 and 7 more.

The faster that material spin in relevance to what is not spinning, the more it will become in ratio with what is not spinning. The material will become more by being in contact with more liquid. We will find that this argument finds validity in cases where motorized boats go faster in the water. When going faster the boat lifts from the water by placing less material inside the water because by moving faster the material becomes more or bigger in relation to the water it is in contact with per time unit spent in motion.

The claiming of space is forming mass because the particle holding mass is siding with the solid as then being part of the entire solid, the entire solid is claiming more space. With the entire solid claiming more space because of more resilient motion of the solid claiming space from the liquid, the resenting of the liquid is per time unit much stronger. This is the same as a boat rushing through water. The faster the boat moves the faster water will be rushing past the hull. By moving faster, the boat claims more space per minute used and the more space the boat claims by moving faster in space during time, the more the water will resist such accelerating claims by the boat on the space the water holds and therefore the less will the water allow space per time unit to comply to the request of the boat on space that the water has. By moving faster, the boat relent space per volume per time unit and will concede to less space given per time unit because the boat as the solid in the water, which is forming the liquid, will use more space per time unit in the overall use of space by the boat per time period. We see the boat leap out of the water due to faster motion but the faster motion renders the ship a higher relation to more water and therefore more water pushes the boat higher out of the water giving the boat less space in the water. The boat becomes a lot more forceful or massive but also has al lot less solid in the space, which the liquid holds. That is the Coanda effect.

The explaining involves a lot more facts before the argument becomes completely clear. However, in this piece there is no room for an elaborate explaining such as I do in the other books dealing with particular matters dedicated to every book.

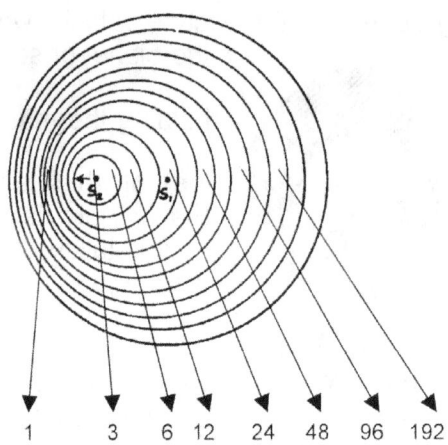

Planet	Mercury	Venus	Earth	Mars	Ceres	Jupiter	Saturn	Uranus	Neptune	Pluto
Bode's Law distance	4	7	10	16	28	52	100	196	-	388
Actual distance	3.9	7.2	10	15.2	28	52	95.4	191.8	300.7	394.6

The Bode law is there and any amount of their denial will not tarnish or remove the Bode law as a fact. The Bode law is the indication of how space grows in relation to the expanding of space in the Big Bang. It shows how the expanding progress by the interacting of the solid holding seven and the liquid being five on both sides of the Universe. There is no evidence of mass anywhere because all planets rotate at an equal pace. There is the Bode law and the Bode law cannot be coincidental. Coincidental might involve two of the nine planets. Coincidental might involve a few of the moons forming satellites to the planets. In truth the prediction derived from using the law is so accurate that one can predict the position of undiscovered planets by using the law and that can hardly be coincidental. This was used in the past.

What is coincidental is the mentality of the smartest there is when they do not understand anything and then transform that which they do not understand to what they understand and that is nothing. They understand nothing because they wish to live in a Universe where they can create mass and if their mass does not create, that which they then are able to create, is nothing. The way Newtonians become impaired in mental capacity is directly linked to atheism. They think they remove God from denying the existence of God.

In that manner they also think they remove the Titius Bode law by denying the Titius Bode law any presence in science. But the way the atheistic minded scientist reason, puts the Titius Bode law in the same bracket as the way they approach religion, because as with God, the simpleminded atheist thinks like in the case with him not believing in God, the atheist remove God by denouncing the presence of God. That is so simpleminded Newtonian thinking as one can ever find. Whether the atheist denies God existing does not kill off the fact of God being or not being there.

Denying the Titius Bode law does not remove the Titius Bode law but does remove the Newtonians' chances of ever understanding the cosmos. All that is not there is the understanding of the atheist concerning God. It does not remove God from the ranks of the atheist but it places the atheist in the ranks of what animals can understand about God. No animal can understand God, therefore to the animal the existing of God means the same as if God is nothing. The atheist comes by virtue of stupidity and lack of insight into the ranks of all animals, because like all animals the atheist too cannot understand God and therefore they share the understanding of the animals. The same argument goes concerning the being there or not being there of the Bode law. They have to reduce the Bode law to be coincidental, but that does not remove the Bode law from forming a crucial factor.

However, it does form a nothing they understand about the Bode law and therefore remove their understanding from the ranks of the Universe. Then the only nothing found in the Universe is the nothing they understand about the Universe and that nothing they wish to transform to everything they do not understand. But it is the nothing they understand because by understanding nothing they remove their understanding from the cosmos. The Bode law is there and is so accurate that it was implied to predict the locations of the last number of planets that were discovered. The nothing they wish to put between planets is the very same nothing they understand about the Universe. Because they understand nothing about the Universe but they have to understand the Universe, since it is their job to understand the Universe. They relocate that which they understand to what they believe is what they understand about the Universe. They understand nothing therefore the Universe has to comprise of nothing.

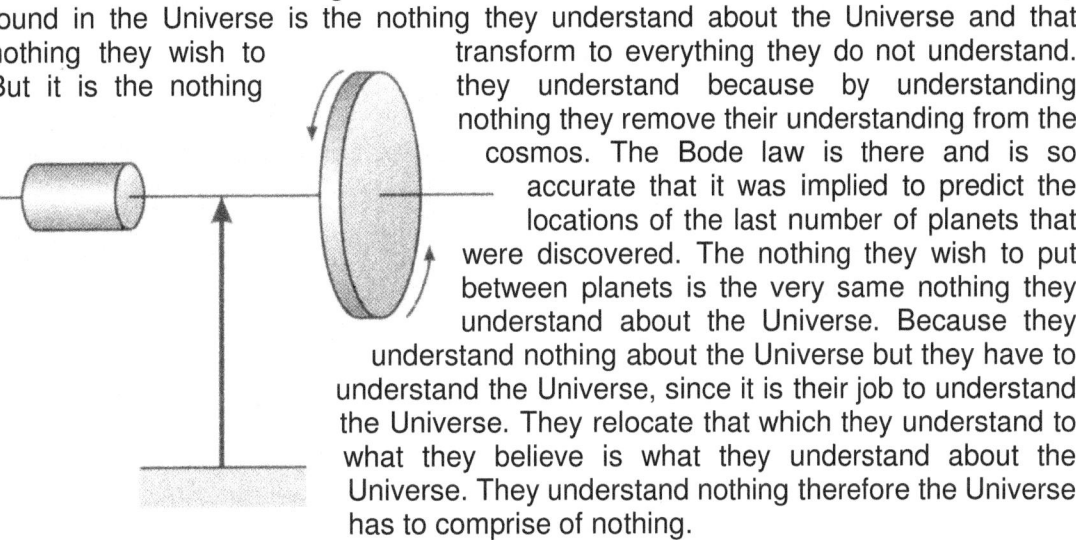

Newton used the top to prove his incoherent arguments about work done and not done. According to Newton it is said that the gyroscope and the top spinning does not work. There is a difference between the gyroscope not spinning and the gyroscope spinning. There is a difference between the top spinning and then stop spinning. The difference is motion. The top not spinning has lost the independence it gained from the Earth when spinning. When spinning, the gyroscope does generate a balance by such spinning that is not present when it is not spinning. That balance comes from some form of input that otherwise is not present when it does not spin and therefore the motion does not generate such a balance when the gyroscope is not spinning. By generating an upright stance when spinning is a fact to become understood and not to be wished away by the Newtonians' simple-mindedness. The Newtonians' disclaiming thereof by refusing to recognise this fact does not remove the input of the motion in establishing the balance but again removes the Newtonian's understanding of the concept. By claiming that such a spinning gyroscope does no work does not do the comprehension of the Newtonian's effort to understand, any good. The Newtonians commit corruption when they teach that anything not spinning is equal to that same thing what it spans. The fact of difference proves that the one in motion does deliver work and the one not in motion does not deliver work. There is a Universe of difference coming about between the two options where the top can exert motion and strive to gain independence from the mass the Earth inflicts or when the top succumbs to the mass and lose all independence when it loses independent motion and therefore independent gravity.

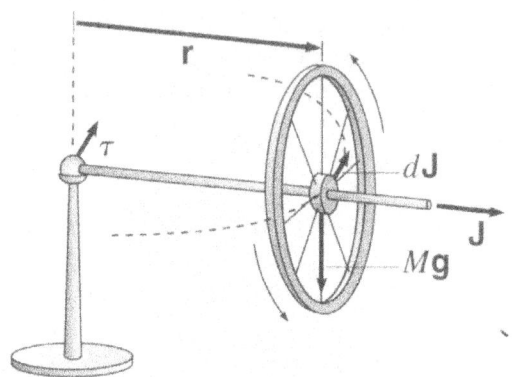

Newton used the top to explain his view, which doesn't make sense. The claim is that while the top, as a gyroscope does not perform work, it does nothing. The question coming to mind then is what the difference would be between the top spinning and the top being motionless and on its side. I use the top to show how the cosmos works and what time is, but I do it with the assurance of Kepler and I use the top by translating what the cosmos indicated to Kepler and not to secure any personal position in science.

Please use your judgement to decide which line of thought is correct, mine showing the line that keeps the spinning top erect or that of Newton showing zero happens when the top spins.

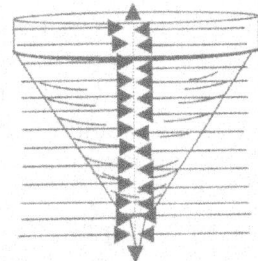

The line that can never start because the line comes from infinity with no starting point eternally

This point is not being in the Universe while it is being in the centre of the spinning top and at the same time it is not there because it has no space and yet it all confirms time that becomes space. In the Universe space is time in history placed in relation to this point. This is the point where the Universe confirms an entire new Universal layout every passing instant by repositioning every possible aspect throughout the entirety to the single point time renews to confirm the Universe.

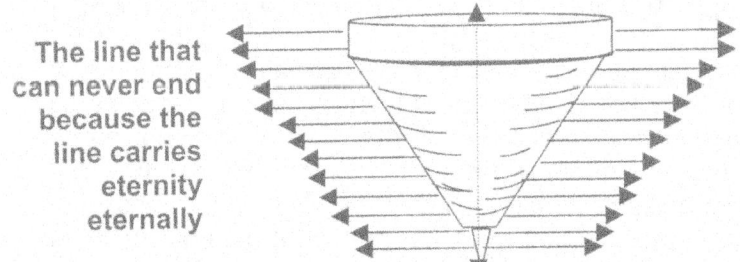

The line that can never end because the line carries eternity eternally

This point allocated on the outer rim of the top is the point where time confirms space that becomes space only as time flows to history forming time as space-time in eternity and from this space-time in time in eternity is also another point where infinity confirms time or singularity which is the same thing, while by movement is getting control on singularity as time inside the atom. This is not zero but this is the Universe in the making.

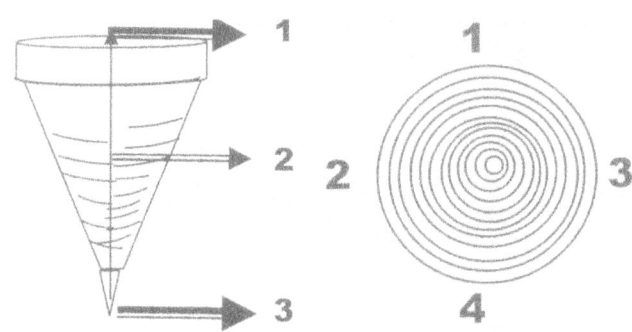

$$\Omega = \frac{1}{J}\frac{dJ}{dT} = \frac{\tau}{J} = \frac{rmg}{J} = \frac{rmg}{I\omega}$$

$$\tau = \frac{dJ}{dt}$$

$$d\theta = \frac{dJ}{J}$$

$$\frac{dJ}{dt} = 0$$

The top being motionless is performing well under the work rate of the top spinning. Newton has all this fabricated nonsense of forces counteracting one another and nullifying work.

The fact charging the issue into being just another ridiculous Newtonian stance of protecting what they try to cover up to give legitimacy to what does not truly exist, is that by establishing independent gravity, the top has the ability to defy the Earth's gravity dominating the top into total submission. Proving the issue of the top doing work or not having work done is in the matter that the top is upright and spinning while the top not spinning is on its side and motionless.

The message coming from the top and the interpretation Newton made about cosmic reality affecting the top in movement is as far off the mark as Newton's Universal Grand Gravitational formulated cosmic concept. What he saw and the truth applying is not even sharing a Universe.

The fact that the spinning top has the effort to keep upright and maintain a stance in independence that defies the mass the Earth enforces on the top alone constitutes to work. The top spinning is denying the Earth to dominate the top while the top is spinning. The top spinning is overcoming the Earth dominating the top into submission. The top is creating independent gravity in defying the mass of the Earth, killing otherwise independent motion by implementing mass in denying independence.

There is a visual difference between the top that is not spinning and is resting on its side and the top that is spinning in a well coherent manner and is very well balanced and exhibits total independence of motion. The fact of motion that defies the mass domination at least proves that work and therefore a presentation of performance is in progress. That completely destroys Newtonian claims with all the other incoherent arguments about one spin direction contravening another spin direction and mass replacing the balance.

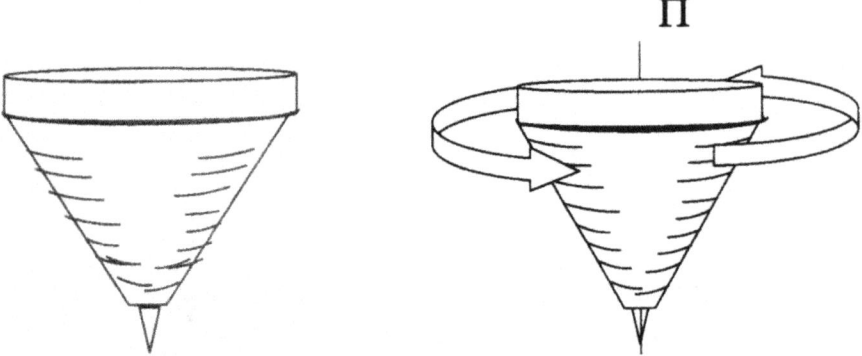

The mass that the Earth inflicts, destroys the perfect balance the top shows and counteracts the performing of the top. The top does not gain gravity from mass but loses gravity from mass because the top spinning is a form of having independent motion or independent mass. The top is counteracting the mass deployment the Earth enforces by keeping a balance between space held and motion providing time.

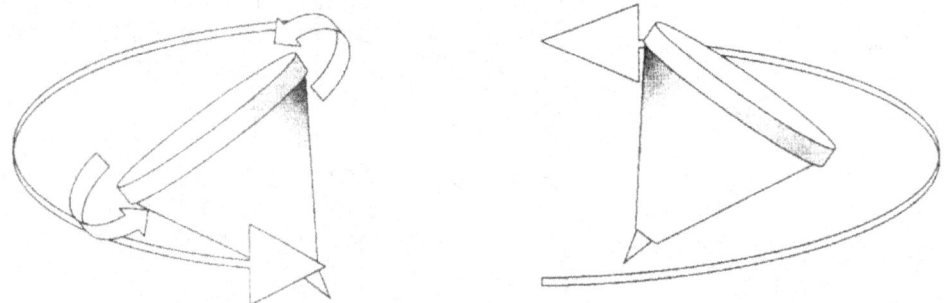

When the top starts to move, the top is in an upright stance. That is part of the motion the top exhibits. The top will spin if the top is rotating and in the rotating of the top, proves independence from mass denying independence. It is not mass putting the top into a balance, but it is the motion the top exerts that defies the control of the mass and that grants the top a poise of balance. When the motion ends, the top drops. When the motion starts to subside, the top goes into a fight for life. When the motion is exuberant, the top tries to break free from all forms of mass by fighting for total independence. The key to the Universe is the fact secured in the discipline that time moves and time can never stand still. Movement is what keeps eternity from collapsing into infinity and ending the movement in time, as we find it happens in the Black Hole.

That independence be reached, we see with the helicopter freeing the body from the Earth. In its effort to break free from the Earth gravity the top shifts the centre line from one side to another side. The effort is an all out effort to break free from the Earth's gravity. The Earth however, draws a line running from outer space towards the centre of the Earth. There are innumerable lines such as that. When the motion of the top is asserting a drive line through the centre of the top, the line is strong enough that it tries to break the Earth line by freeing the line from the holding or grip the Earth exerts.

In helicopters the drive is strong enough to lift the mass into becoming anti-mass. When the drive of the top is strong enough by the motion the top releases, the line sways from one side to the other side in an attempt to find complete individual gravity released from the Earth.

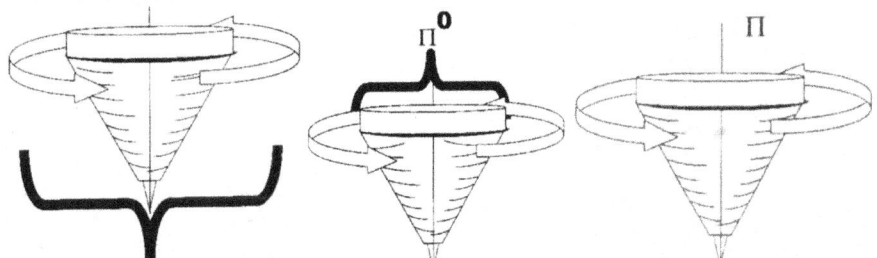

As the momentum reduces and the inertia flattens out, the cyclic manoeuvring of the top ends and the Earth finds a means to take hold of the centreline of the Earth. A struggle assumes and the gravity line the Earth has running down the centre of the top strangles the motion of the top until the top relents in favour of the gravity of the Earth. The gravity of the Earth subdues the motion of the top until the top relinquishes all independence. At the bottom of the motion we again find the top fighting a battle of survival until the top loses all independence in motion or gravity. The top then submits to the gravity of the Earth by accepting the mass the Earth bestows on the top.

When the top is in motion, the top is in a struggle of survival and when the motion is strong enough, the top tries to regain independence by defying the mass the Earth enforces on the top. When the Earth wins the battle for domination, the top loses its independence and with it its gravity or motion. The mass that the Earth inflicts does not produce gravity to the top but robs the top of all the gravity it had.

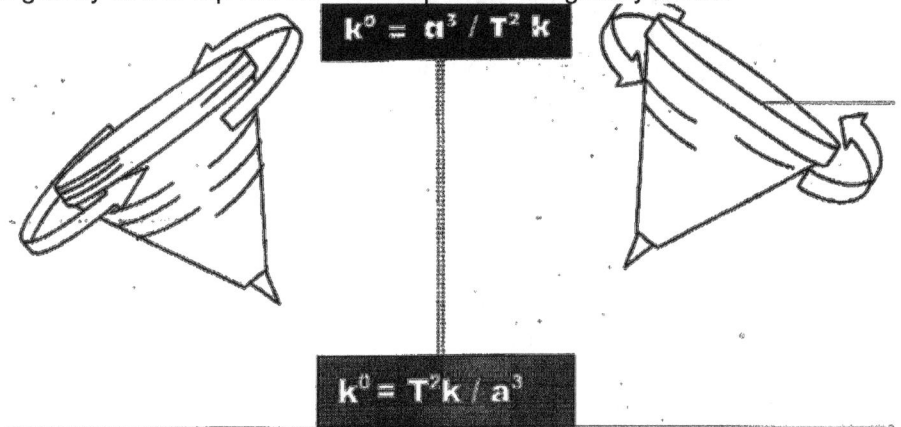

If the argument is brought in that it is the Earth producing gravity which holds the top in mass, then yes I will admit to that but also there is no more top, but in the mind of man. The top is no longer an independent object but has been conquered by the Earth. Because the Earth holds the top at ransom with mass, the top no longer functions cosmically in that it then spins with the Earth as a part of the Earth. The Earth does not give the top gravity by giving the top mass, but robs the top of being independent by engaging the mass increase.

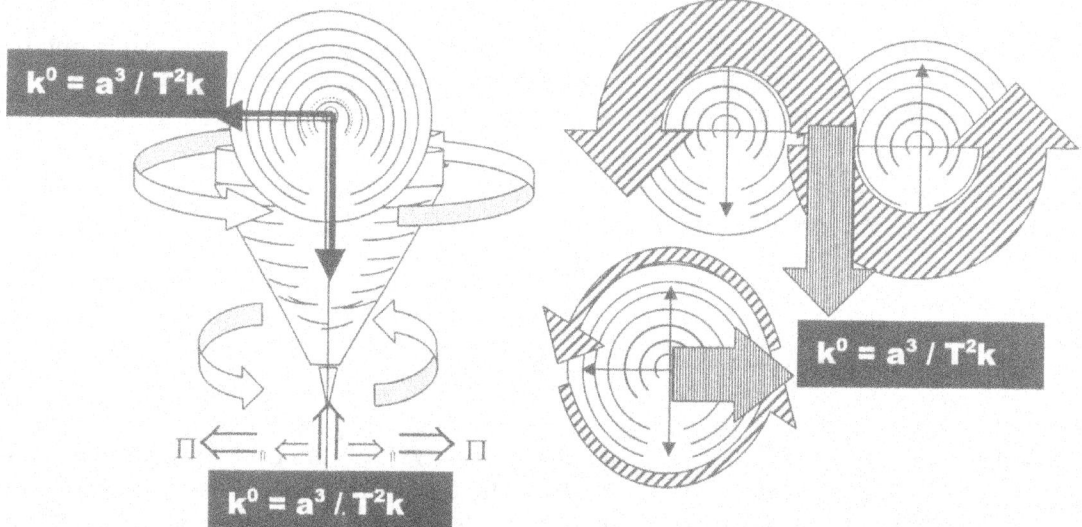

In the rotation certain principles rule events guarding laws. Those principles are as strong as the laws of the Universe in itself just because these principles form the laws guarding the principles that are governing the Universe. The essence forming the laws is that a centre forms a Universe and every aspect is opposing all other sides that are transversely different from all the other possible positions there might be. Every position brings a total new order from what was, to what is and to what will be.

The proof of the line running from outer space, we find in the Kepler's formula. The formula says initially the space $a^3 =$ is equal to the rotation motion T^2 as well as the linear motion k, which is mathematically translated as $a^3 = T^2 k$. Wishing the line away as Newton did when he corrupted the work Kepler produced is complete and utter madness. This is where Newton committed fraud to prove his incoherent arguments in order to validate mass as a factor in the Universe. There is no proof of mass anywhere and this shows where the planets all rotate around the Sun as equals. All planets use similar speeds and the slight variation I explain. Jupiter being 318 times bigger than the Earth is spinning the same speed as what the Earth is spinning and is spinning the same speed as what Pluto is spinning while Pluto is .00025 times the mass of the Earth. If there were any hint of mass, the speeds would have differentiated just as if there were any inclination of mass in the falling process, all objects would have fallen at speeds dictated by mass variations. The planets show very clearly a line running from outer space to the centre of the Sun as the table of the planetary motion $a^3 = T^2 k$, which then forms $k^{-1} = T^2 / a^3$. It is utter fraud to then declare that k has no presence, since the driveline dr is nullified by the connecting line r.

$\dfrac{dJ}{dt} = 0$ Newton, and science, made one enormous blunder, from this stance because zero was encountered, then one of the factors had to be zero in order to mathematically produce zero by dividing.

$\dfrac{dJ}{0} = dt$ or $\dfrac{0}{dt} = dJ$. This disputes mathematics. The dividing of two factors such as DJ / dt represents, can produce any number from eternity to infinity, but it excludes only one; it cannot be zero. It just can't be what Newton said that having $\dfrac{dJ}{dt} = 0$ will constitute to a circle turning and he goes on to say $\Pi \times r^2$ = CIRCLE.

Newton suggested that the factor **k** in Kepler's formula disappears or becomes zero. In that way he misleadingly suggested that T^2/ a^3 but then never gave a factor to which the division will be equal. In truth the factor represents **k** and the correct value of **k** is very well documented! One should think that persons (such as the Newtonians are) that are so highly educated in the mathematical art should not be fooled that easily when some other person comes and breaks their rules that concerns mathematics! But no one can ever be as blind as the one that refuses to see what there is to see. They obscured the truth wilfully and without repenting or showing the least remorse.

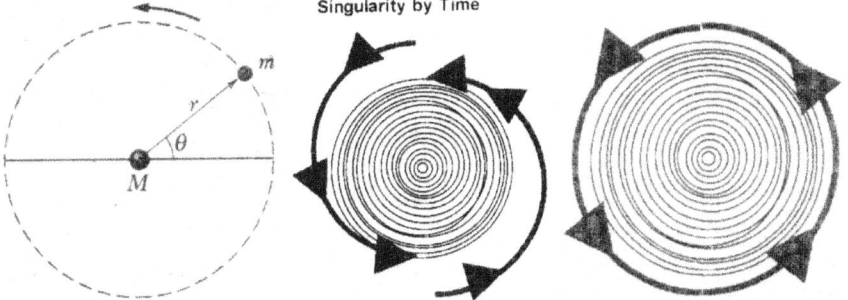

Singularity by Time

Kepler said the space a^3 is equal to the rotational T^2 motion in a linear **k** stance. When spinning, the top secures independence for the space a^3 that the top gains by rotating T^2 and moving in a line **k**. When the top is not spinning strong enough, the Earth grabs control of the line **k** that the top uses to spin because the motion T^2 **k** validates the space = a^3 that the top claims through the motion T^2 **k**. By placing the rotation T^2 in relation to the space a^3, we find there is a negative or a decreasing motion of the space. One cannot have = T^2 / a^3 by disclaiming that the factor that a^3 / T^2 represents k^{-1}. By declaiming the presence of the factor **k** going negative^{-1} is to try and lie to oneself. If space a^3 reduces T^2 by division and it stands mathematically pronounced as that, then to say that, it is not true and that **k** in all forms is **0** is incoherent jabber of the mindless and mathematically incompetent. <u>*I am not about to ponder on this issue since I elaborate this issue throughout the book.*</u>

It is this negative charge of space as a liquid that keeps the top upright and balanced. By contracting the air around the top, we find the top to be encircled by air and the air sustains the balance of the top.

Einstein got the speed of light and time as the same thing while I say the top spinning shows time parted from the rest of the Earth; Time is not in the light factor but in the movement of light. He figured that light travels as time and not in time. Let's reflect on the scenario: light leaves the point of origin as it travels for one year. The light should be one year away from the point of origination. But light always travels in all directions.

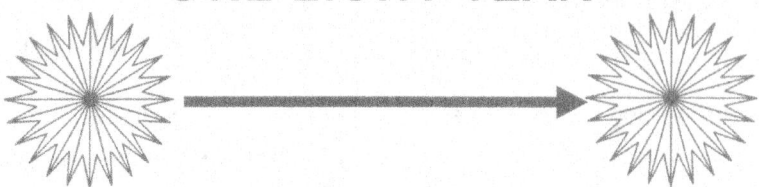

After travelling for one year in directly opposing directions both light photons, not withstanding the direction, will be at a distance of one light year away from the source it came. This is so simple even I can grasp the statement except for the fact that if the Universe were vacuum, the light would not relate. Every as well as any object splitting and then going into opposing directions at equal velocity will be at an equal distance at some point in future time.

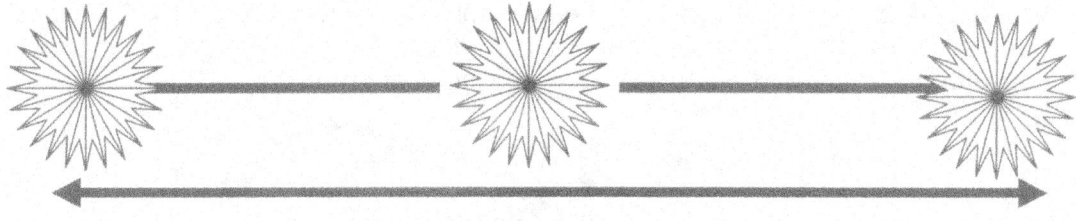

In considering that there is light in opposing travel there should also be light in half opposing travel to complete the circle of the stars' emitting of light. That brings about a circle in four quarter being equal. This is rather as simplistic as science may become. The most astonishing part is that the two points of light will also be **one** light year apart!

Notwithstanding the star from where the light originated that are between the two directions of light, both points holding light are one year apart from each other.

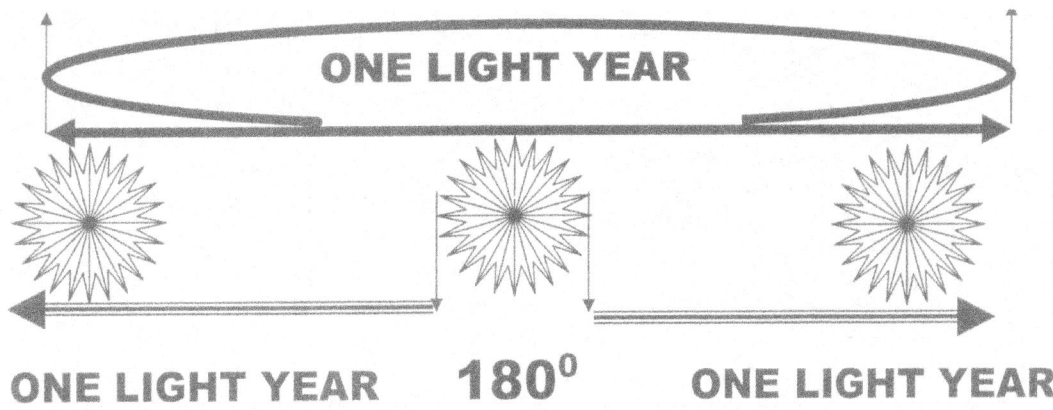

In the same sense while the two points hold a position that is one light year apart and in a total opposing direction, the two points are at the same time also one year apart from the star of origin. The reason for this is not because light is time as Einstein suggested, but the fact is that light is in a dimension that only fringes on the three dimensions of space-time but moves in the dimension of form. A line and a triangle and a half circle

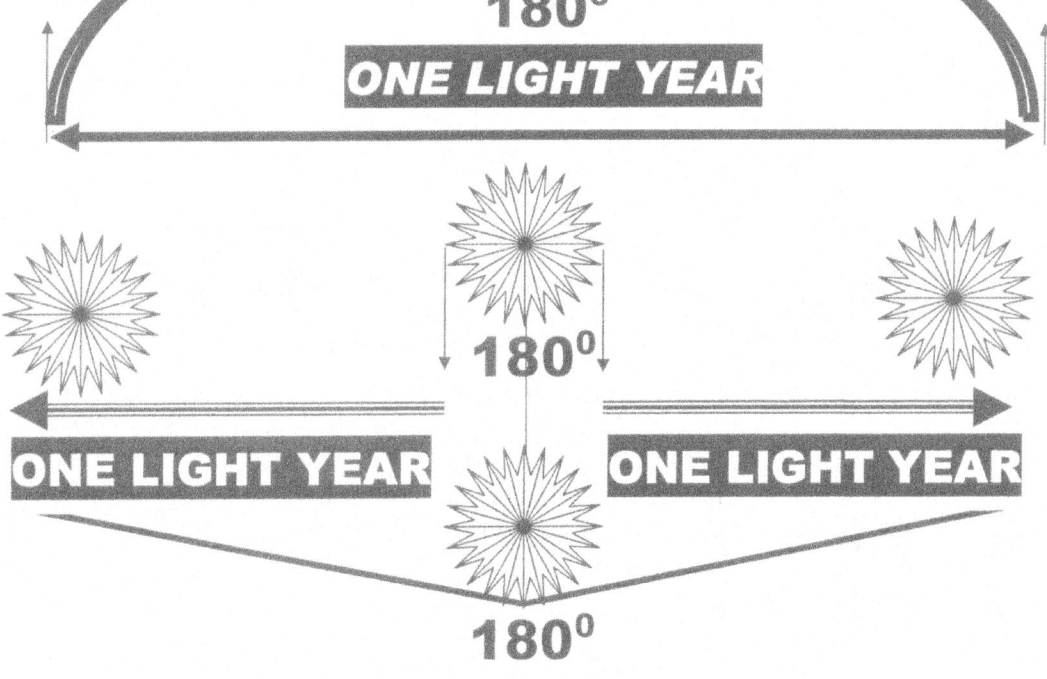

are equal at 180° notwithstanding form differences and in that is singularity.
The only way that can be is when Pythagoras becomes a part of the triangle and singularity effectively becomes one. There then are three sides (3) in relation to the

Triangle represents a^3

Straight line represents k

Half-circle represents T^2

square of singularity Π^2. The factor of gravity is singularity in the square Π^2. That means if singularity then becomes $\Pi^0 = 1$, the speed of light would have another factor in the dimension of space being Π^3 where $\Pi^0 = 1$ becomes in the 3D dimension $\Pi^{1=1=1=3}$ = and $\Pi^0 = 1$. That places light at a second relevancy of 3^3.

The Universe can have no edge and have no end because time will not allow such an end. Because the space a^3 is equal to the distance k the space rotates T^2 in the space will become the product of the rotation by the distance becoming a factor of one. It is the way time operates and that we see in the pendulum. Although the pendulum is proving the opposite of travel by expanding it is the very same principle. Time is motion and motion is converting space back to time.

This is not that foreign with the pendulum.

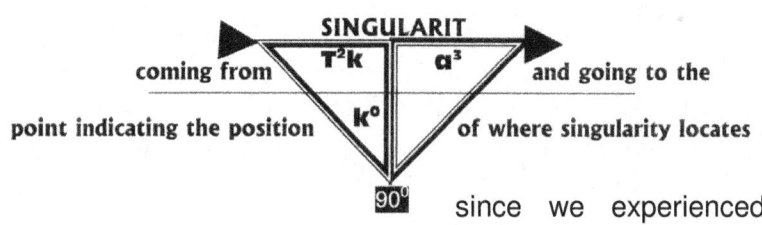

since we experienced

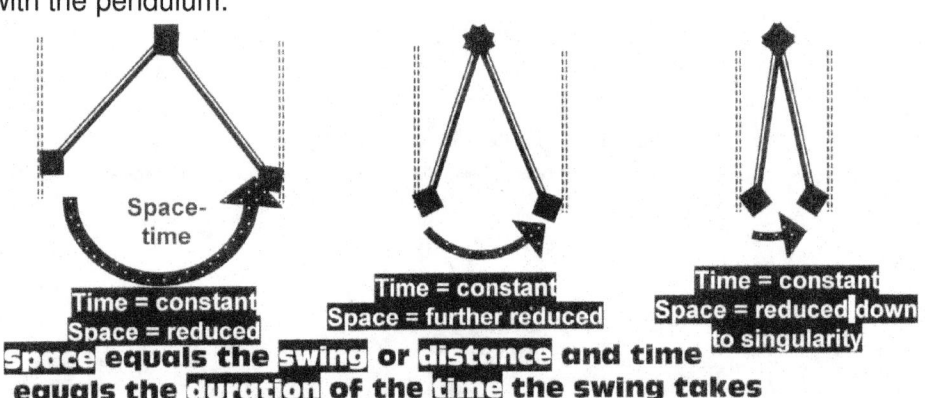

Space equals the swing or distance and time equals the duration of the time the swing takes

Because of space I am not going to give the reasons why the triangle and the half circle are equal to the straight line, but since it is an accepted mathematical statement, I do not have to prove the authenticity, although I do prove the reason why in another book. The pendulum however, proves that time uses all three aspects of direction equally being the triangle, the half circle and the straight line. We know that the space a^3 is always equal to the time $T^2 k$, which is the motion through space of time. This is the statement Kepler made in his assessment about his findings on matters regarding the Universe. We also know that at some point movement becomes gravity and forming space results in the characteristic to put time in relation of motion relating to a specific centre, to put one point of the two factors regarding the time aspect on par. Time involves space and time involves motion of occupied space through unoccupied space. Every action going through motion is using the rhythm of time to move. Time is motion and the way the object moves is the calibration the object has with time in synchronisation. Time has to apply the three mathematical founding forms on an equal basis. Time has to involve motion that puts the line (**k**) in relation with the half circle **T**^{2} that involves the triangle **a**^{3}.

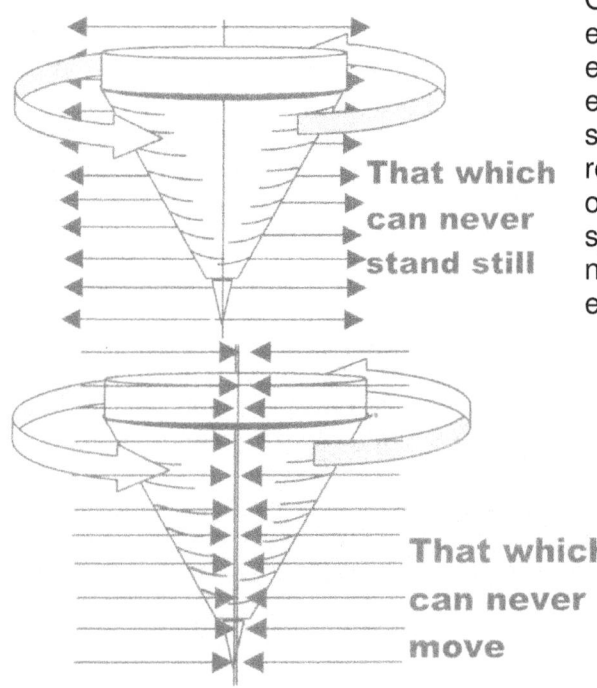

On the outer edge of the top we have eternity that always moves through expanding and because of cosmic expanding the movement can never stop. That space is time gone into relevance or history and is the repetition of time that was before. Since that space in perpetual motion, which can never end, it therefore gives time the eternal factor of time never ending.

On the inside runs a line that doesn't run but is there all the same. The line is there because the line can never be there and while not being there the line is always there and being there in no position the line in its absence is controlling everything that is there or ever can be there.

Any object in rotation will have a middle point, a very specific centre point that does not spin. That point once again hypothetical but none the less must be standing still because every line running from that point in opposing directions are also in opposing directional spin to each other.

From such a point, every other point will be opposing any other point not pointing in the direction to which the first point is pointing, direction it holds. No where the point leads, specific direction will be rotating because at that or wherever, it will be it spins but in the direction outwards.

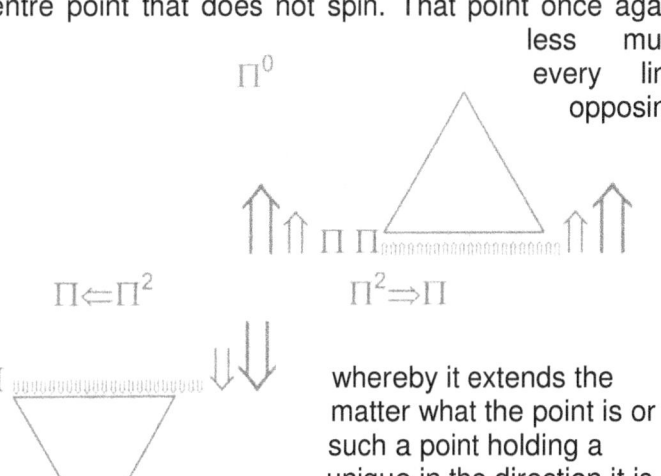

whereby it extends the matter what the point is or such a point holding a unique in the direction it is any other specific point directing not in the direction flowing from the centre point

Motion not only receives from singularity but also contributes to singularity by projecting a governing singularity that controls the motion that controls singularity. Any point will be it opposing itself within the rotating of 180° changing every aspect of its previous flowing characteristics it previously had or will once again have in 180° from there. While in rotation from the point of an outside observer, all may seem static and never changing but to the object in spin every next second will be a diverting from every aspect it was in every second passing, and the direction it held in relation to the direction it held the previous millisecond will totally be incompatible with the direction it holds the very next millisecond of rotation. That proves no point can be static or constant, although it may seem that way to outsiders. The straight line will always convert into the circle as relevancies change and the one adapts to become the other since there comes a point where both are the same. To leave the Earth is forming a straight line that converts to a circle. The circle will be on one side of the Universe while the straight line will be on the other side of the Universe, while the entire Universe are both sides of the single Universe being equal.

Everyone believes in Newton's mass that is pulling mass by destroying the radius which is filled with the gravitational constant in such a way, that the Coanda effect was never before ever been considered as being part of cosmology or being responsible for forming gravity. Fortunately the "everyone" I refer to does not include comets because comets are the exception, because comets fail to collide with the Sun.

However, I can explain in some way, but it doesn't involve Newton's views or Newton's vision of cosmology in any way…

Every vivid indication we so far received of the astronomy photographs portraying outer space, disputes a shrinking Universe concept. It is a fact that the Moon and the Earth are increasing the distance between the two and this is on official record. One should see how the distance increases as the radius grows between the Earth and the Sun from a view science has on the Moon. The Hubble Constant is not an indicator indicating about the growing happening to far away Universes gone into history. All space is growing everywhere and is contemporary and growing at this minute. Only Newtonians who have no concept of time or of space will harbour the notion that the Hubble constant is happening to every Universe except their Universe because their Universe runs by Newton on mass contracting! How slow witted can one get… Space is growing all over and everywhere, it is growing wherever man may conduct studies, even where we now are. Again we have this philosophy that is so typical Newtonian. If there is evidence of something not fitting Newton then simply ignore it, lie about it and forget it because then no one will talk about it and it will eventually disappear into thin mist. That is the way Newtonians conduct all their science and it is going on since the time of Newton. If they don't like what is out there they simply ignore it and it will go away. They like this force idea and they stick with it, whether there is proof about it or not because proof doesn't matter at all…

$$F = \frac{M_1 M}{r^2} G$$

Since the end of the middle ages a force called gravity was identified, but further than that science did not take it. What is gravity, besides being a force? What forces the force? I introduce a cosmic theory that turns the missing questions to answers.

Let us for one second return to the science we all know.

There is an undefined phenomenon in the cosmos, never mentioned (in public) because it obscures reality but is proven in using this foundation of science, the basic formula used to prove all science.

Let's put the mathematical formula into a practical context.

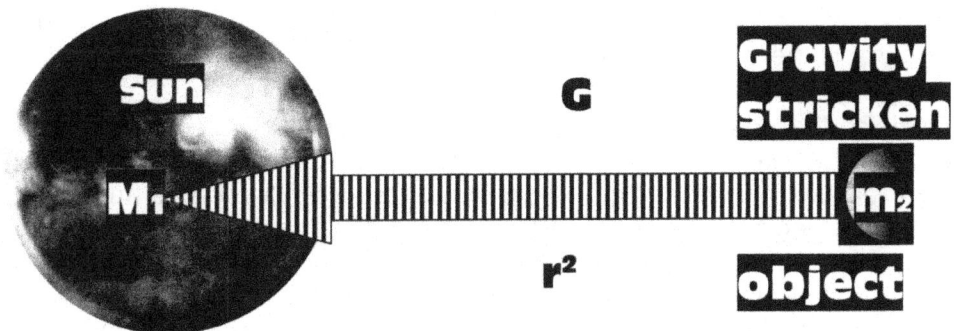

By reducing r, such reducing of r would bring about the same result as enlarging the mass factor of the cosmic objects i.e. the Sun and the planets. It is a very drastic implication that will cause much more than just seasons changing. It must bring about that gravity changes throughout the year…yet the radius does constantly change. Therefore, it is evident that Kepler's name was used when science introduced a formula as follows:

E- e sin E = M

That proves everything but that mass is responsible for gravity. In relation to the next few arguments it is very critical to understand the arguments I present because on the soundness that these arguments represent in the arguments that I make, I am in dispute with the arguments that Newton makes.

$$F = \frac{M_1 M}{r^2} G$$

The entire philosophy of the science of physics rests on the arguments that Newton makes. It is physics that base everything on the fact that mass produces gravity and therefore by the force that mass provides as gravity, the entirety of all physics are founded. I do not dispute mass as a factor in physics but what I do dispute is the way Newton presented the fact that mass has any value between cosmic bodies.

The formula as Newton presented it, is just not baking the beans any more and must be re-examined because there is a lot of contention in this statement.

$$F = \frac{M_1 M}{r^2} G$$

Please follow my line of thought and scrutinise all that I say. That which I touch has resting on it the entire philosophy of astrophysics as well as physics.

What I dispute, is that it is mass that is in control of the cosmos and that mass provides the sticky substance called gravity. Also do I dispute that gravity is a contracting force by which the entire Universe is in collapse. However, I begin by disputing the idea that it is mass that is producing the gravity that supposedly produces contraction that is there to pull planets towards the Sun.

Newton insisted that mass is the influence under which gravity becomes a force. He promoted the idea that gravity is a force instituted by the measure of mass because the mass unleashes the force of gravity on other unsuspecting objects and then begin to pull the unsuspecting objects in, in the same manner as anglers will pull in fish. Once the mass gets hold of mass, the mass reels in the mass from both ends that carries mass and according to Newton everything in the Universe carries mass. We know the strategy behind gravity is to first get hold of a body containing mass. Then it is natural to presume that everything in sight is pulling on everything in sight and the bigger the mass is, the stronger the pulling is. Jupiter is pulling much harder than what Mercury is pulling, because Mercury is much less massive than Jupiter, which is more massive.

Another thing we cannot miss about gravity pulling objects around by the measure of the mass is that gravity pulls mass towards the centre of the other mass conducting the pulling by gravity. We therefore know that all objects are heading towards any object's centre and the pulling of is going directly to the centre of the other object instigated by the pulling of the force, which is produced by the mass the body has. I ask the reader politely to control my facts since so many academics told me with much sympathy in their attitude that I just don't have the insight to understand Newton.

$$F = \frac{M_1 M}{r^2} G$$

Somehow it is suggested that Newton is far too difficult to be understood by a person with my meagre intellectual capacity because God somehow denied me to inherit a strong brainpower with a capacity to crack the limits of the

An Open Letter On Gravity Part 2   Page 466

Universe in a Newtonian fashion. I ask you please to check that I do follow these extremely complicated facts correctly. Those academics I confronted with the issue in hand are of the opinion that it is a God given fact that I am too simple in mind to understand Newton and when taken into account that they are capable of understanding Newton due to the capacity they have in using the brilliance of thinking with mind clarity, they can hardly be wrong about anything.

Every time I confronted an Academic in the physics department, I was told in a very polite and sympathetic manner but also in a very unmistakably fatherly and firm way when they share their opinion with a lesser being such as I, that to their opinion and in line with what they can see about my intellectual status, I am a lesser-blessed individual. There are some people that are intellectually advanced and such persons have a born ability to understand Newton. Then there are those with much less potential and with much reduced thinking ability and in the case of those or should I say we, there is not enough grey matter in our skulls filling the vacuum between our ears to follow Newton. We are those with simple minds unlabeled to understand Newton! I therefore have to accept I was born with much less capabilities and then with that much reduced mental capacity I had to accept my fate which is that I shall never have the ability to understand Newton. With my meagre intellect I do try my best but where it comes to the obvious, then it becomes hard to follow Newton when it is Newton that apparently lost track of his senses. Be as it may, please inform me where I lose track of reality and where Newton loses track with reality. They (our Super Professors in Physics) say I have no ability to grasp how mass entices gravity by mass that is producing gravity as a force forcing gravity to pull over a distance to the measure of r.

They must be of the opinion that I do not understand mass. Well I am not sure that they understand mass because there are as many definitions explaining mass as there are opinions about what mass is. It also could be that they

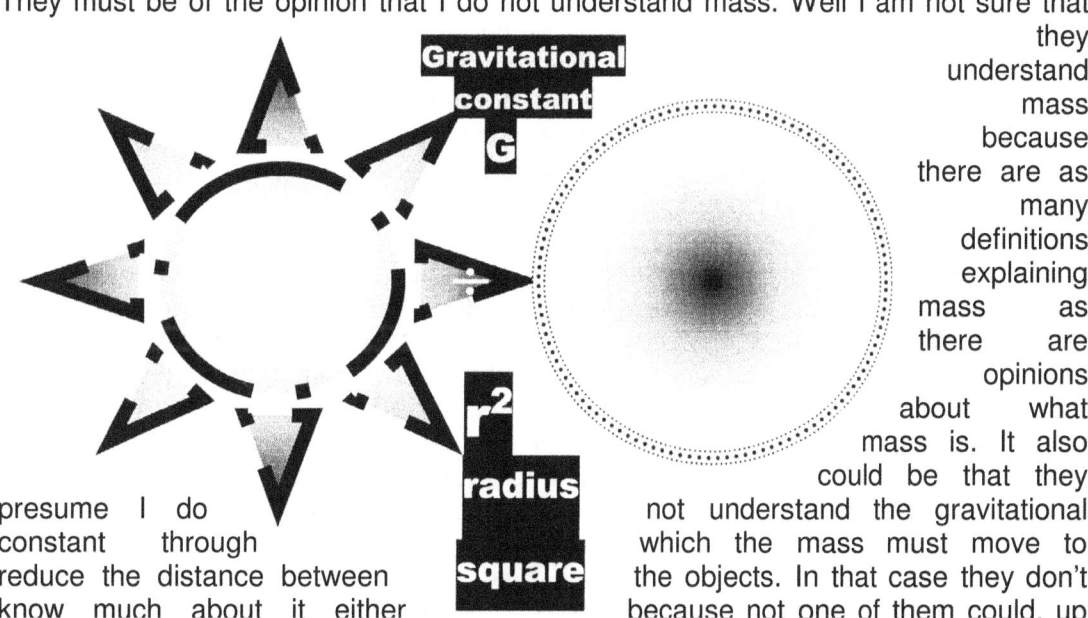

presume I do not understand the gravitational constant through which the mass must move to reduce the distance between the objects. In that case they don't know much about it either because not one of them could, up to now, show me where the centre of the Universe is and where it is that that gravitational constant is pulling all the mass.

To have gravity pulling what there is to be pulled, there also has to be a centre to which all are pulling all there is to pull. It could be that they suspect I don't know what a force is…I do know that…a force is that which scares children at night when the wind howls and which they burned witches for because they said witches and ghost are or have forces. Then again they might suspect I don't know what a distance is that separates objects and there too I think I know what a distance is that separates objects. Other than what I just mentioned, I truly don't understand what they say I don't understand about Newton. That means they are correct and that my mind is so weak I don't even understand what I don't understand about the complex issues they seem to understand. I first wish to investigate what it is about the formula that is so complex that I have no ability to realise the implication of such a dynamic formula and on top of that I wish to share my inability to understand this complexity with you reading this because then you might see how inadequate my understanding is!

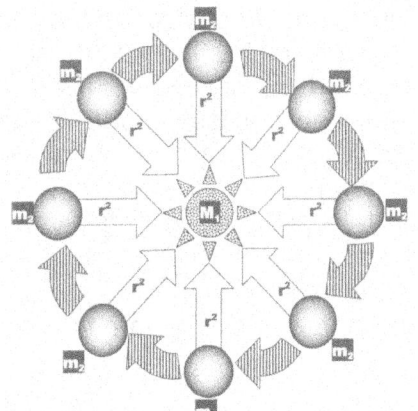

This is what I do understand but apparently it is not enough to understand what I think I do understand. Newton maintained that there is a force that puts relevance or a connection of some degree of pulling power between the amount of mass and the strength of the force produced over the distance. If the distance increases, the influence of the mass coming across the increased distance will decrease. When the distance decreases the influence coming across the decreased distance will increase as the distance decreases. In that the force of gravity that the mass of the comet and the Sun produces, will increase by the square of the distance that is diminishing due to the shrinking of the distance parting the objects. I suppose there must be more because that which I understand and I just shared with you are not enough to prove that I understand Newton!

Mass produces gravity in accordance with the radius distance between the two bodies. This is where my mind gets too weak to understand Newton. Looking at all orbiting objects there are always a wide part and a narrower part in the orbit where the planet in orbit seems to be closer on the one side (E- e sin E – M) or on the orbiting structure side further away from the centre, which was the position that the Sun claims. This does not quite fit the picture of a constant and never changing mass that all structures supposedly have. If mass is responsible for gravity, why is there this flexing wobble in the radius. The lot should be rather constant because the mass is constant. But that is my weak mind as that is what I don't understand about what Newton understands…and with my weak mind I can get no one that understands that. We know that the Earth has no mass increasing and decreasing and neither does the Sun have mass adding and then removing of mass. In that case when considering the practical mathematical implication, the radius has to be at a constant with no mass fluctuating on either side of the factors.

We know from personal experience living on the place all our lives the Earth doesn't change mass as the year progresses. From that as well as the evidence we know about all other things in the Universe with mass, that the mass of all things are a constant and doesn't fluctuate. So the gravity that the mass of the Sun and the mass of the planets would generate should allow for a pretty round circle to be as a result of q pretty constant and accurate r to be in place that will keep the planets circling evenly. That is not the case because we know there is a fluctuating especially in the case of the comet orbit. We then have the task to find what would encourage the deviation. It seems to be unexplainable that there is this fluctuating of the radius notwithstanding the mass factors that produces the gravity never being able to change. Newton had no explanation and according to Newtonian mentality because if Newton did not explain it nobody should think about it. If there is mass then we all are satisfied. This applies to the formula notwithstanding the manner in which we look at the formula.

$$F \alpha \frac{M_1 M_2}{r^2}$$

$$F \alpha \frac{M_1 M_2}{r^2}$$

$$F \propto \frac{M_1 M_2}{r^2}$$

Notwithstanding which formula is used, the outcome remains the same. The relevancies that the divide brings, changes the applicable strength of the force in, whichever approach one applies. Since the lot doesn't change, there is no excuse for Newton's deliberate oversight and in that sense we carry on with the official formula in place.

Or on the other hand

$$F = \frac{G M_1 M_2}{r^2}$$

. It is my understanding when considering the implications about the relevancy there is between mass and distance parting mass when used as it is in the Newton formula, that if the mass is at a constant then the radius too must remain at a constant.

When the object having mass asserts the gravity, the mass employed and the radius parting the objects are large, the objects in mass asserting gravity over such a large distance must be small. If the distance is large then in that case the distance will reduce the force that the mass can produce to a trickle. The force there is between Mercury and the Sun must therefore be ten times weaker than the force that there is

$$F = \frac{M_1 M}{r^2} G$$

between Pluto and the Sun. It is not only that Pluto and Mercury has about the same mass and therefore they should peddle around the Sun equally. Pluto is 39.44 x AU or (5900139992.8 km) from the Sun and that is 101886 times further than Mercury 0.387 x AU or (57909 km). One astronomical unit is the distance from the Sun to the Earth and that is measured in AU = 1 or in km 149 597 870 km. That means the force of gravity must allow Mercury to go 101886 faster than Pluto's being 101886 slower than Mercury. That is not happening. I surely don't understand Newton but that is also surely not because I am too stupid to understand Newton.

The gravity one must find between Pluto and the Sun is then considerably less than the gravity there is between Mercury and the Sun. Mercury is so close that one might form an opinion that the gravity Mercury generates must be in the vicinity of one of the outer gas planets. The gravity Mercury has with Mercury being so close would have Mercury spin at a rate that only Jupiter can manage since Jupiter has the enormous mass in its favour and compared to the sizable reduced radius that Mercury enjoys, the mathematical consequence would lead to the two to be approximately the same in the force of gravity. No other planet should come near to the speed that Mercury and Jupiter have when going into orbit around the Sun. Jupiter has the enormous mass in its favour giving its momentum around the Sun a boost and Mercury is so close that the mass it has is many times bigger than the mass it should have just because it is so close to the Sun. Then Pluto must be at a snail pace and hardly moving in consideration of everything the gravity applying to Pluto has to endure because it is a hundred times further from the Sun than Mercury is while it has the same mass as that which Mercury has. Sorry I forgot that all planets move at an equal pace. I wonder if it is because they share the same mass or are they possibly at the same distance while in orbit around the Sun because there is no mass indication in any of them speeding up or being slower than the other. They all orbit at the very same and equal pace.

Now we get to comets and their bad behaviour. When taking this distance effect further as it affects something as small as a comet, the gravity increase on the comet has to be devastating to the comet. The force that can hardly move the comet out where the comet is hiding from the Sun must be billions upon billions of times more when compared to when the comet is closely approaching the Sun. With all that in mind we now draw our attention to the comet and the way gravity pulls the comet.

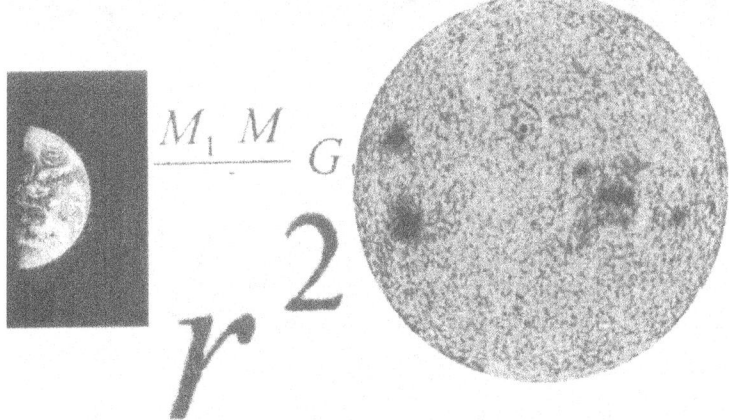

When the Sun gets hold of the comet, the comet is miles away and the miles stretch for miles without end. The comet is where the Sun can't shine. Think of the considerable small mass that the comet has and the enormous distance there is at first when the Sun gets its gravity onto the comet and starts to drag the comet closer. The distance being that large must seem to make the mass incredibly small.

Yet, with the massive mass the Sun has in its favour it not only gets hold of the comet but it drags the comet all the way through all the space and ever closer to the Sun.

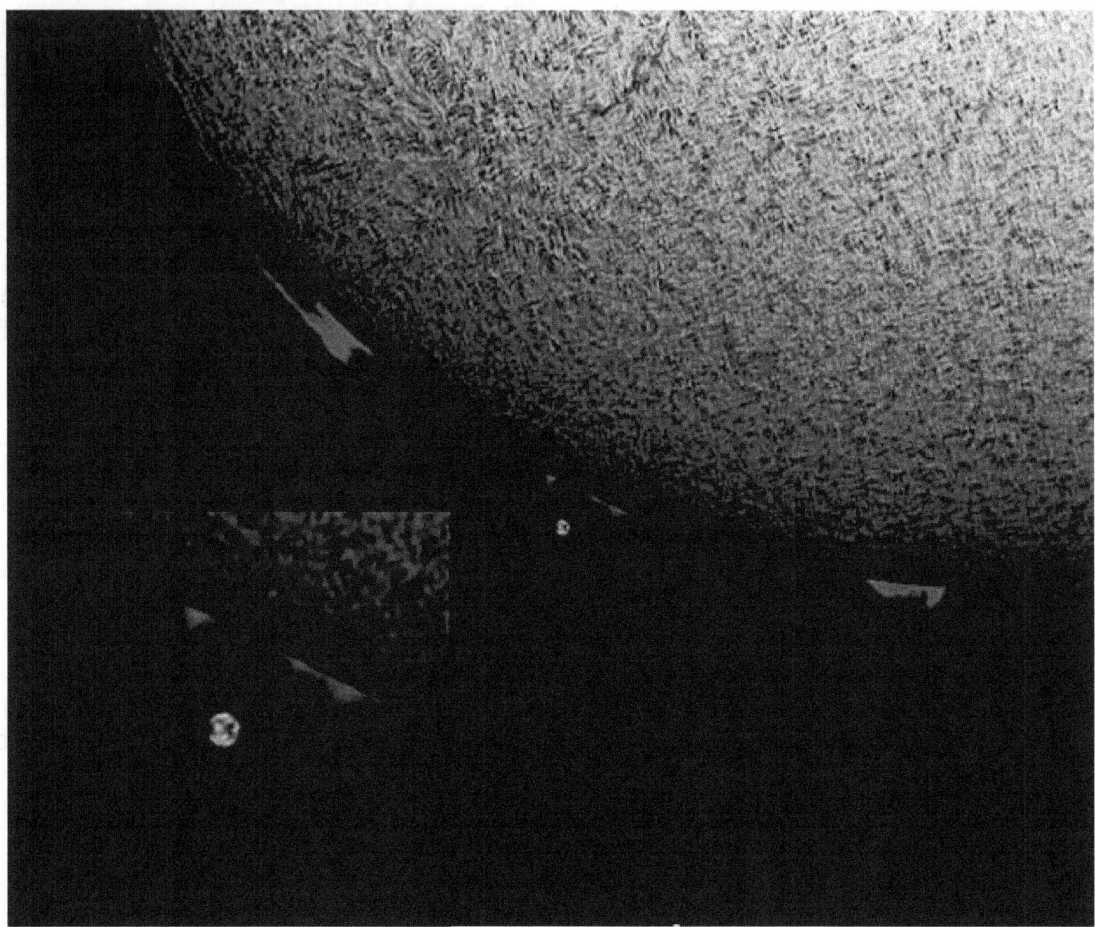

As I said we now come to the difficult part where very few people understand Newton and apparently according to those in the know how, I am one of those being incapable of making sense of Newton. Let's try again and see if I can manage to get it correct this time. The Sun has mass. The mass covers the distance between the comet and the Sun by the mass that produces gravity and the gravity is pulling the comet. Then the comet also has mass and that mass produces gravity that pulls the Sun over the distance between them. Because the distance is coming from both ways, the distance is by the square where it is including the gravitational constant and this lot is in a product to one another that gang up to destroy the distance parting this lot.

There is normally a diverting from the centre or the rotation axis, which the Sun forms by the orbiting planet, but this is much more so applying in the case of the comet ($E - e \sin E = M$). Considering the weakness of the force that initially reaches the comet as it drags the comet from the depth of total darkness where the comet hides in places the Sun does not shine, it is remarkable that the comet does apply the gravity that will pull it on route to the Sun. It just comes to prove how genially correct Newton is! He saw the mass that creates the gravity and the little comet stood no chance hiding from this domineering Sun with all the mass it can display

From the coldest and the deepest of space where. not even the Newtonian imagination can reach, the Sun finds the gravity strength to draw the comet through the deep of space. One would imagine that if it was or is not for the shear size of the massive mass of the Sun, the Sun would not be able to generate a force so strong it could get hold of such a tiny piece, but it does accomplish the gravity to bring the comet from where it is lurking in the depth of space and drag it towards the centre of the Sun.

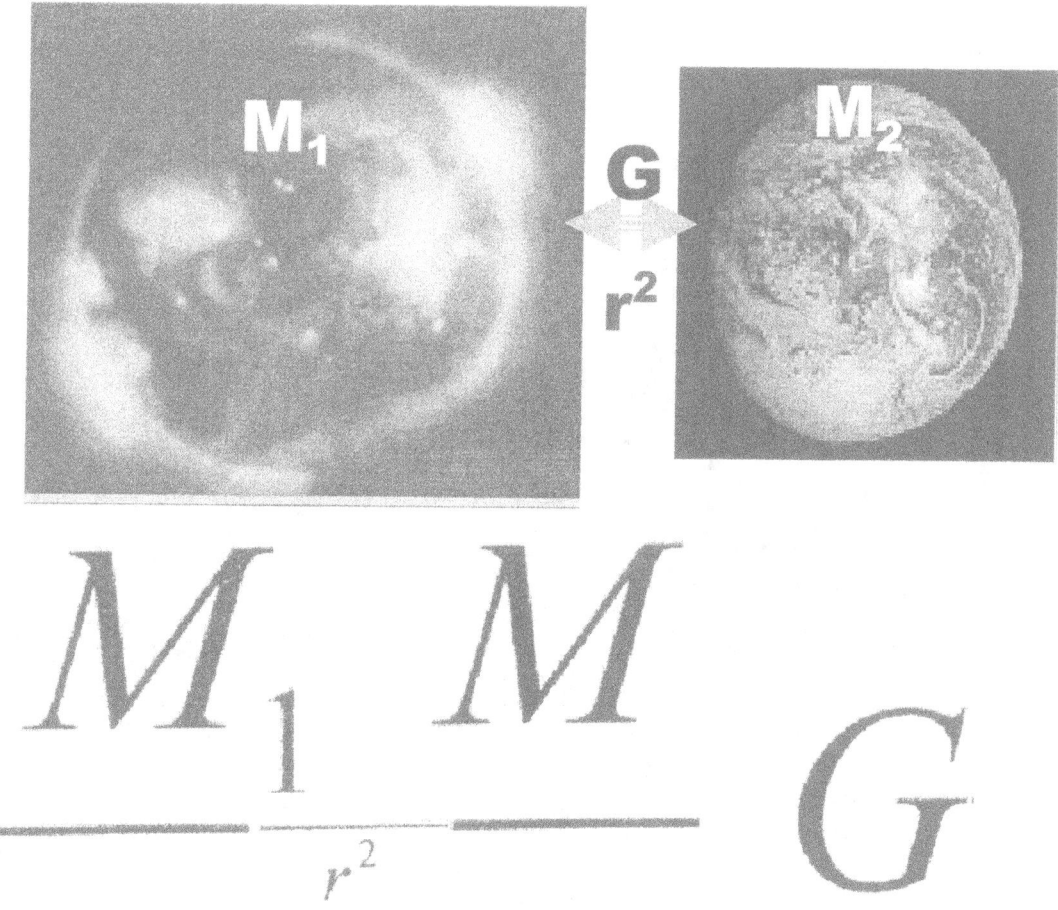

$F = G \dfrac{M_1 M_2}{r^2}$ where F is the force between the bodies.

M_1 = The mass of the one body

M_2 = The mass of the second body

G = The Gravitational Constant

r = The distance between the two bodies

When the radius is big, the comet has little going in the way of what is available in the force department and it is very surprising that the force finds the ability to muster the gravity to take charge of the mass of the comet as to drag it by the "scruff of its neck" so to speak, all the way to where the Sun is anxiously awaiting on the comet's arrival with much anticipation, I might presume. But with the distance that far, the extended distance renders the force that the mass with such small ability can muster, to being quite meaningless. This is what I am trying to say: Divide so many billions of kilometres in the form of the astronomical units into precise what the mass on both ends can deliver and a very small force will apply over such a far distance. Then see how little influence the force truly can assert. In the face of all this evidence pointing directly to how the comet's gravity is incapacitated by such an awesome distance, it is very surprising that the force can move the comet at all.

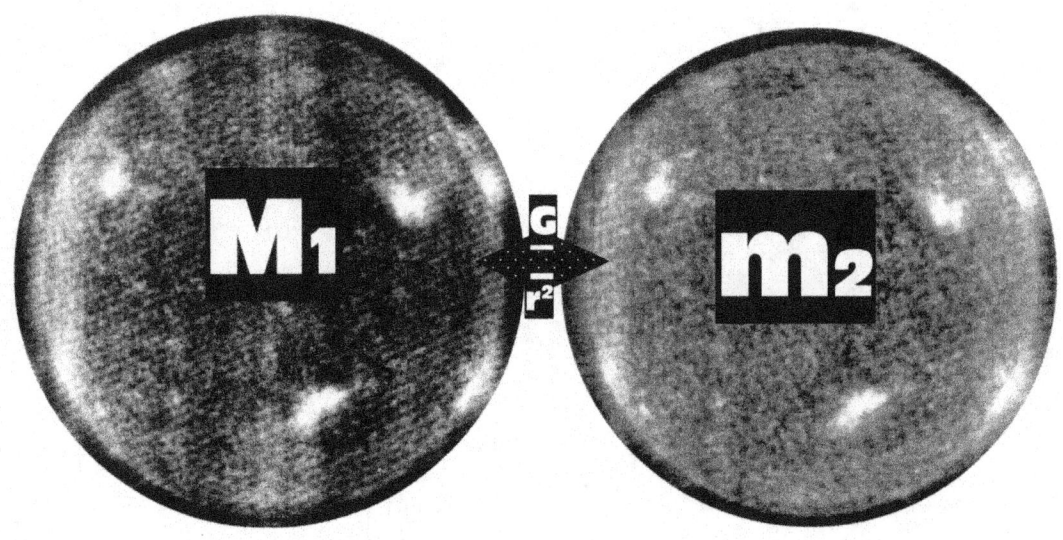

However, all this still notwithstanding, it is evident that there is an inexplicable hiding of an exceptional tenacious force in the mass that lures unsuspecting smaller objects to the web just like spiders do. However, in the case of the mass the web is called gravity.

But with the resilience that only the Sun can muster as it is the only object that has the mass in ability to form the gravity that can extend all the way even to where the comet is lurking, such dragging would have been hopelessly inadequate. But the gravity that

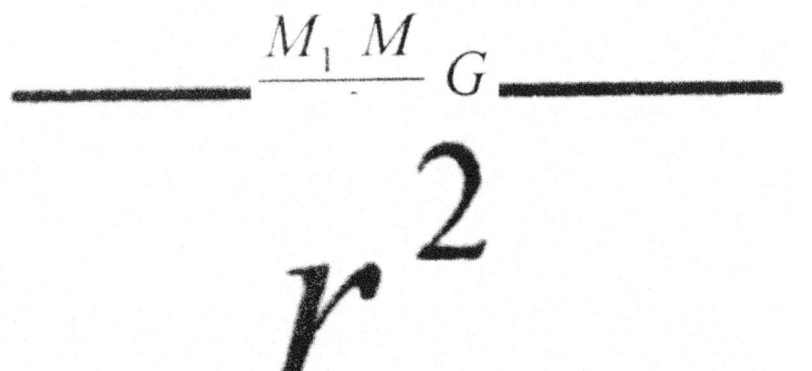

$$\frac{M_1 M \, G}{r^2}$$

the mighty Sun manages with that enormous mass, the comet has little chance in defending its position. The comet is heading towards the centre of the Sun and towards its final doom, since with such gravity coming from such a mass the comet has no chance to defend its position. If any one ever invented an unfair and totally bias fight then it is this fight between the small comet and the enormous Sun.

Alas, it seems that the bookies once again had a say in the outcome of the fight because it seems they rigged the outcome or had some intervention about the outcome…well it is either that or the unthinkable has happened, Newton is mistaken…Newton is incorrect. The comet strayed from its pulling and the Sun lost its grip on the pulling. The mass of the Sun proved insufficient to grip a small little comet by force of gravity and pull it to the centre of the Sun. The comet aims at spot marked X and misses the Sun's centre where all the gravity is concentrated. The mass of the Sun had the comet when the mass was small, due to the enormous distance there was between the two but as the mass grew in potency with the aid of the shrinking distance and with it the force of gravity that the mass produced, becoming ever increasing in the presence of the declining of the radius parting the two objects in mass, the force was unable to direct the path of the oncoming comet to the centre of gravity within the Sun. There is an unexplained diverting of direction, which Newton's brilliance never foresaw… that is if Newton's brilliance did foresee anything at all…and we'll get to that in a while.

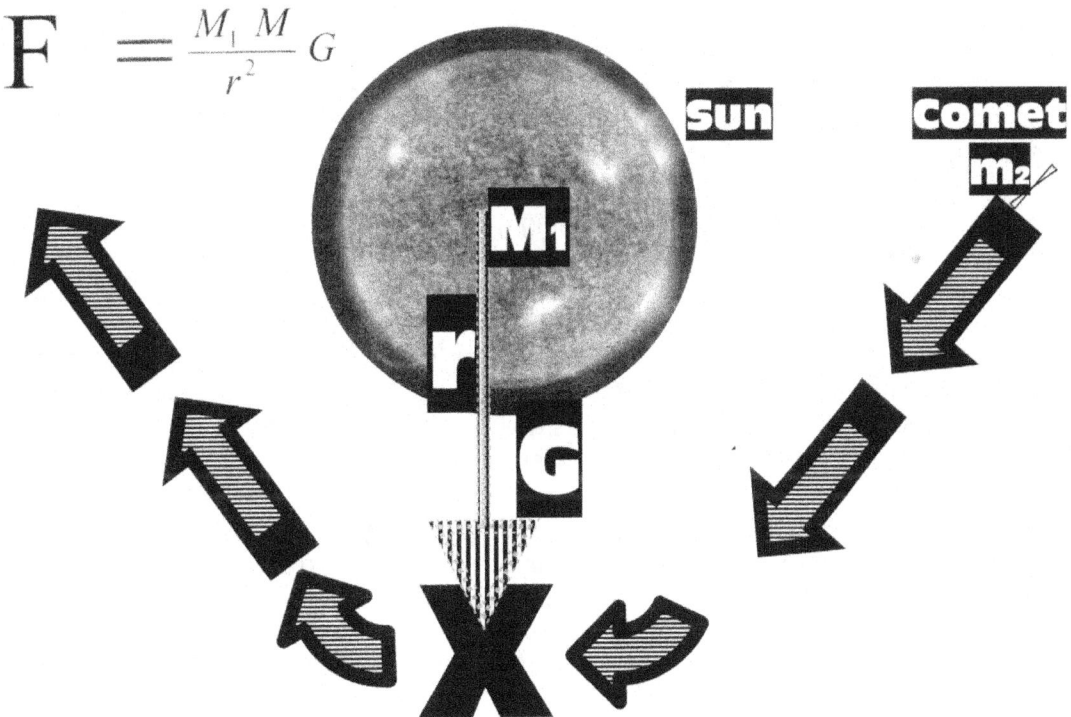

How did the comet manage to escape the gravity wave pulling power and not head directly to the centre of the Sun? What part of Newton's formula makes provision for such detouring of the comet to escape the clutches of the Sun?

$$F = G \frac{M_1 M_2}{r^2}$$

The relevancy of the big radius by the square reduces the influence that the mass supposedly brings about. The further the objects are from each other, the smaller will the gravity be that the mass can inflict. A long distance has a small mass that produces weak gravity.

It started when the radius was big which made the mass to be so small that it produced an insignificant force in gravity. That pulled the comet out of the shadows and into the Sunlight. The force was strong enough to grip the comet and pull the comet towards the centre of the Sun where the strongest gravity mounts an attack. When the force of gravity grew, the influence the force made started to tarnish because it allowed the comet to aim for spot X, which is way outside the centre of the Sun. Somehow with an ever increasing gravity, the comet got brave and fighting fit. In the presence of overwhelming mathematical evidence to prove the contrary and in spite of mathematical laws and principles applying, nevertheless the comet prevailed and got the better of the Sun.

It missed the centre of an ever-increasing gravity force notwithstanding. In spite of the size of the Sun increasing that should make the aim that the comet produce all the more sharper, still in spite of that, the comet deviates from its aim. This is not the end…it is the beginning of Newton's problems. Up to this point it seems as if Newton was a profit sent by God to foretell us other far less important mortals about the future and about the future of the comet. The Sun had its mass providing a force called gravity and the force called gravity was going to take the comet to its eternal grave, as gravity will bring the comet its final demise.

We find the comet fulfilling the Newtonian prophecy. We find the comet behaviour vindicating all human belief in the mighty genius. We find the comet proclaiming the genius being beyond question or doubt. We find why the science of astrophysics adheres to Newton in the smallest detail. There now is all the proof that Newton saw what no other could see and Newton read what God used when God used the Formula to invent the Universe. There can be no doubt…except for one small blemish in the whole thesis. The comet avoided the centre where the gravity must be the strongest. How the comet did it must be written off, as just bad behaviour and not following orders like good subjects should do. It can't be anything serious because Newton makes no remarks about it.

The comet becomes very spiteful and tries to elude the inevitable outcome Newton forecasted so many years ago. It aims for a target next to the Sun and in that misses the centre of the Sun because it circles around the Sun...and for four centuries all the main brains on Earth never saw this flaw and still insist that Newton never puts a foot wrong...that is rubbish.

You are addicted to Newton. You do not understand Newton because if you did understand Newton as far as cosmology goes, then you would reject Newton. You were force-fed by your superiors on Newton at the time and during the time when you were a student and being force-fed, you either had to grow addicted to Newton or die from Newton. Being where you now are, it means you are not academically dead but mentally mutilated which is as good as being brain dead. That means you took the second option by becoming addicted to Newton.

You saw comets come and you saw comets go. You did not question the reason why the comet went because of your addiction to Newton. You accept Hubble's expanding, yet you go about looking for a critical density in the hope you do not have to abandon your Newtonian addiction. You know the Big Bang cannot be possible while Newton's mass annihilates the radius between masses because then the Big Bang had to be the Big Crunch and from where we are there is only one enormous out explosion that had no Newtonian implosion. The enormous force that were supposed to contract the small radius that was in place during the Big Bang never imploded on the mass by nullifying the radius through the tremendously large mass and small radius being present at the time. All the facts modern day science accepts do not stroke with Newton. Is it because subconsciously everyone is silently admitting about Newtonian incorrectness?

Now I come and I tell you your addiction is killing you. As all addicts do, you are not willing to kill your addiction although you do realize that your addiction to Newton is senseless, stupid and altogether wrong. The second choice facing all addicts including you is to hate the messenger that comes to take the drug from you. You will hate me, as the messenger. You will be willing to kill me being the messenger of evil because you see me as the evil that wished to part you from your addiction, which is Newton. You now find the bizarre feeling to put your anger, distrust and vengeance on me and on the book you are reading. I have seen many academics stop reading at the part you now have arrived at. Brave yourself and throw out your fears by reading the rest. It will eventually cure you. This is what truly happens...

If you are not addicted and can find sense in arguments, then how could anyone with a clear mind defend the rubbish Newton suggested? The addiction in Newtonian Philosophy incapacitated the physic academic's insight and vision while it is clouding their better judgment. How else can one explain the ardent fixation on the ridiculous by persons that should supposedly be overflowing with common sense and clarity in understanding? Their ability to defend Newton speaks volumes of how effective the brainwashing is that is part of the education system in physics.

The mass of the Sun brought the comet in range and to the point where the strength of the force of gravity of the Sun is most and up to where the force the mass provides is at it's greatest and seems to be unconquerable. With the radius tarnished to being almost representing a factor of one, the force of the gravity becomes eternal, since the small radius will charge the mass into greatness, which will unleash a force that will drag the Universe around if it could get hold of the Universe at such a close range.

That does not happen because at the point where the force that the mass must produce is at its point of ultimate victory, we find the comet avoiding defeat. The comet not only avoids the centre of the Sun and final destruction, but it wins the fight completely. In the jaws of absolute defeat when the comet should have been clinging to the ropes and living to survive the last desperate moments, the fight swings unexpectedly in favour of the loser.

The comet rushes past the Sun as if the comet is unaware of Newton's predictions. The comet behaves in a manner that leaves Newton in doubt. That cannot be tolerated. Newton cannot be wrong because Newton is always correct. The comet cannot speed off into the distance where the dark is much and the light is little. The comet cannot move around the Sun and then slip past the Sun because we know Newton said the mass of the Sun is pulling the comet as it is pulling everything there is.

The thought alone of Newton possibly being incorrect is more than what the mind can bear and the principle alone is outrageous. Still, it seems that the man that has never been proven wrong proved himself wrong by predicting the demise of a small comet crashing into a big Sun and where the mass of the Sun produces a force of gravity that pulls the comet towards the Sun and then into the Sun, that is not happening. That which pulled the comet towards the Sun is now pushing the comet away from the Sun. If mass was pulling, then what is now pushing, because the silly little comet is in defiance of the great Newton and is rushing into the blackness again where the Sun hardly shines? If it is mass pulling then what is pushing?

Newton's vision gave us all the answers to the point where the fishing of the Sun starts but then as things really get serious, Newton slips away like the comet and leaves us all without an explanation. Newton fails to provide answers and that leaves the academics looking silly. It is at this point that I don't understand Newton because I am too stupid and uninformed to understand Newton. Newton now gets beyond what I can comprehend and that is true.

If the reducing of the radius depended on the radius becoming systematically smaller, as the case will be with a string winding around a finger while the object is spinning around the finger, one would have to find that there is a continuous reducing of the radius as every circle comes about. I have had this argument put to me in the past, but it does not hold water. With the radius already that small, the circle should then reduce on a deliberate and a continuous basis, and moreover keep on reducing until such time as a collision is finally inevitable. This is not the case and again rubbishes even that argument. There is no string shortening attached between the objects pulling one another by the force of gravity simply because there just is no string in the form of gravity pulling by force.

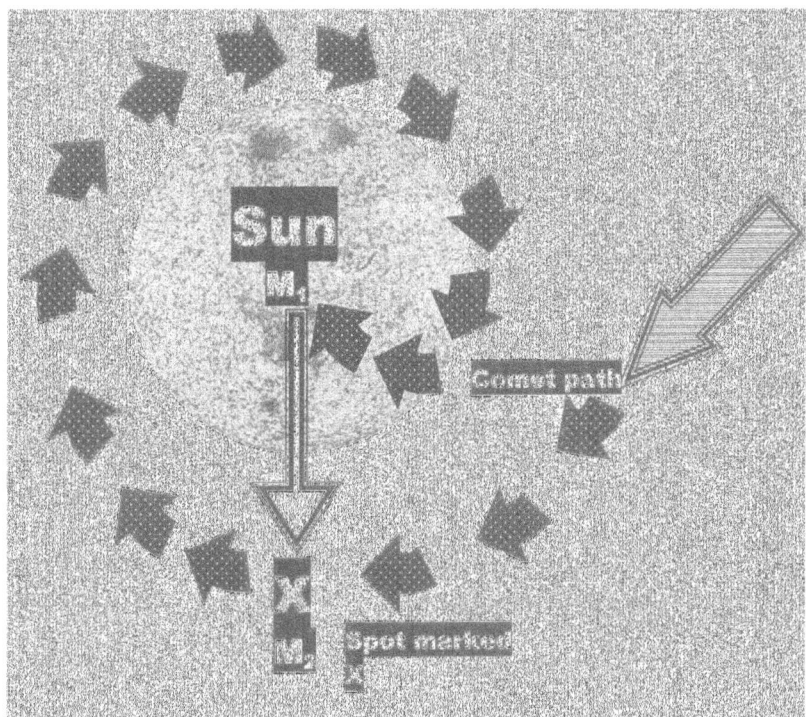

At this point I have to share my stupidity with you, the reader. I have to declare my shortcomings and indicate to you the reasons why the Academics have my mental ability under suspect. If it was mass that did the pulling, I can't find anything that does the pushing. If it is mass that produces the gravity that produces the force that devours the gravitational constant, what is pushing the comet away and into the dark unseen and unknown?

If mass was pulling, Newton clearly forgot to mention what is pushing the comet away from the Sun. First the comet does the unspeakable by avoiding the Sun and then if that was not the limit in unstable behaviour, the comet starts to move away from the Sun. What happened to all the mass that was so eagerly doing all the pulling? Where did the mass go that had the comet coming through all the dark night sky? Where is the comet heading? Might it be where the mass had the comet coming from? What drives the comet away if mass drives the comet towards? It is those questions that I cannot answer and are responsible for making me too inferior to understand Newton and all the complexities attached to Newton.

One may even suspect that if the mass was insufficient to provide a force of gravity strong enough not to pull the comet into the Sun, the comet would at least begin to circle in a reducing fashion around the Sun until the comet falls into the Sun. One may argue that the comet may come at a speed where the Sun is too little to stop the velocity of the comet but that argument doesn't actually make sense.

$$F = \frac{M_1 M}{r^2} G$$

The speed will ultimately not change the direction from the centre to a point where it totally misses all apparent targets and coming at speed will benefit the Sun as well as the comet in aiding the effort of giving more of the gravity that the mass provides. I know such an argument is outrageous considering the mass of the Sun and the speed of the comet, but hey, at this point Newton needs all the help he can get…and he needs help.

But without getting silly, all arguments we might think about in order to save Newton's reputation is bordering on madness. The comet is small. The comet is providing a force as best it can to enable the Sun to accomplish what the Sun set out to do from the start. The comet is pulling as hard as it can and in that, the radius reduces by the square. It is hardly as if the comet is trying to prevent the seemingly inevitable destruction it is heading towards. The comet by mass is enforcing gravity as much as it can to self-destruct. This is in aid of the Sun's efforts to destroy the comet. While this is all going on, it is at the same time not happening. That which is pulling is now pushing because the comet is escaping into the yonder.

If the comet's mass does assert gravity onto the Sun by measure of mass while the Sun does the same right back to the comet using the same grounds and mass principles in doing so, it must be conceivable and detectable that all the planets rotate around the Sun at different speeds, seeing they all have different mass bringing about different forces of gravity. The rate of orbit must vary considerably as Jupiter must spin around the Sun billions of times faster than the little comet we just now observed does rotate around the Sun. That I already proved is what is not happening.

I stood accused by many academics I crossed paths with in the past that I am not familiar with Newton and because of my poor academic background that I am not capable of understanding Newton. That is not the case and I have to be very adamant about that. I would accept such accusations if Newton's science explained all of science. That is hardly the case because I can state four very prominent cosmic principles that no one can explain by applying Newton's claims. The Roche-Lobe, the Titius Bode principal, the Lagrangian five-point position and the Coanda gravity contraction is what Newton's gravity formula cannot explain at all. Neither can the Big Bang theory be the starting point if cosmology insists on using the application of Newton. I am aware of the Critical density theory and black matter, but those arguments have not found proof in the slightest way and in truth serves as an escape corridor because science is at the end of its tether with the phenomena contradicting Newton. I admit that I am lost at finding a starting point introducing the book because the issue remains comprehensive when dealing with issues of cosmic proportions. I shall explain the four unrecognised phenomena I use in proving my statements. With my introduction of the phenomena, which I named the four cosmic pillars, you will find it obvious why science does not accept them even if it is documented throughout the Universe and is quite commonly found. It totally annihilates Newton's formula of $F = G(M_1M_2)/r^2$.

In a following article we are about to investigate such a scenario and then when it comes to light how much faster Jupiter spins in orbit than does Mercury on the very inside orbit and as Pluto on the very outside orbit we then can start to vindicate Newton and find the resolve about the comet behaving very awkwardly and out of step with the rest. With the massive mass Jupiter has in enforcing gravity, there can be no more arguments about how the mass brings about gravity to have Jupiter move at 318 times faster than the Earth around the Sun because Jupiter is 318 times more massive than is the Earth. Then surely all scepticism must be nipped. That is also not happening and also that I already proved. The closer any two cosmic objects come, the stronger the force should be, with eventually no force in the universe being able to keep them apart. This is just not happening!!!

There is no indication of truth about a contracting solar system as Newton proposed

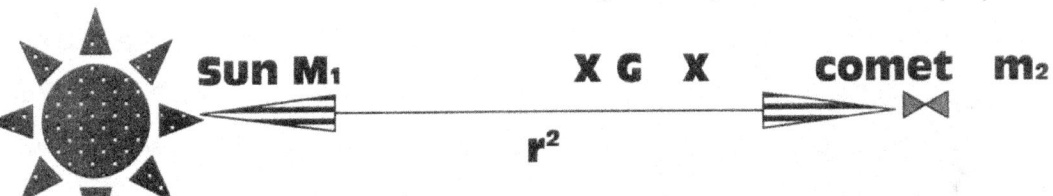

and as Newton's followers promote a contracting Universe while when seen from the Hubble perspective where he introduced the Hubble Constant…and not from any other evidence seen through the Hubble Telescope, such evidence we see through he lens of the telescope that which we see, is just not matching Newton and the reality is that it is contradicting Newton notwithstanding the fraud such as we find in the critical density issue.

As explained, there is some discrepancies about calculating the force of gravity, because gravity would apply as nicely as it does if it was the perfect balance, a balance exists in space of equal measure bringing about equal seasonal time. The biggest discrepancy and a practical denouncing of the official version of the comet's flight around the Sun come from the actions committed by the comet.

The Sun gets a grip on the comet by mass inflicting gravity and as it gets hold of the comet, it drags the comet through the solar system straight ahead to the Sun just as Newton predicted the Sun with all its mass producing gravity will do.

As Newton had said, the gravity that the Sun and the comet mass induce will pull the comet to the Sun. This at first appeared to be true. As we all know the comet moves to the centre of the Sun just as Newton predicted but with a slight complication and a change in the venue, the comet no longer aims directly at the centre of the Sun but aims at a target outside the limits of the space that the Sun occupies. It follows a line aiming towards a point to the side of where the Sun is located. Again this is the part my intellect is apparently insufficient to follow Newton because Newton in his formula never mentioned this apparent deviation from his prophecies. It seems the Academics are correct in saying that I don't understand Newton but they never try to explain this part that I don't understand about Newton. The comet aims at some point gravity does not hold and misses the point where gravity secures the strongest force, which is right in the centre of the Sun.

Should one give Newton the benefit of the doubt and disregard this missing of the centre of the Sun, the position that the comet should aim at, then one would reason that it could be that the gravity was not generated strong enough to locate the centre at such a great distance as the comet was at first. It could be that the reducing of the radius comes in instalments of a few circles. His formula does not indicate these presumptions that I make but hey, let's try and give the guy a chance. The comet might just take a cycle or two to wind down as the radius reduces and one should wait and see if it is not that which Newton meant when he said the mass by the mass is dismissing the radius between the objects.

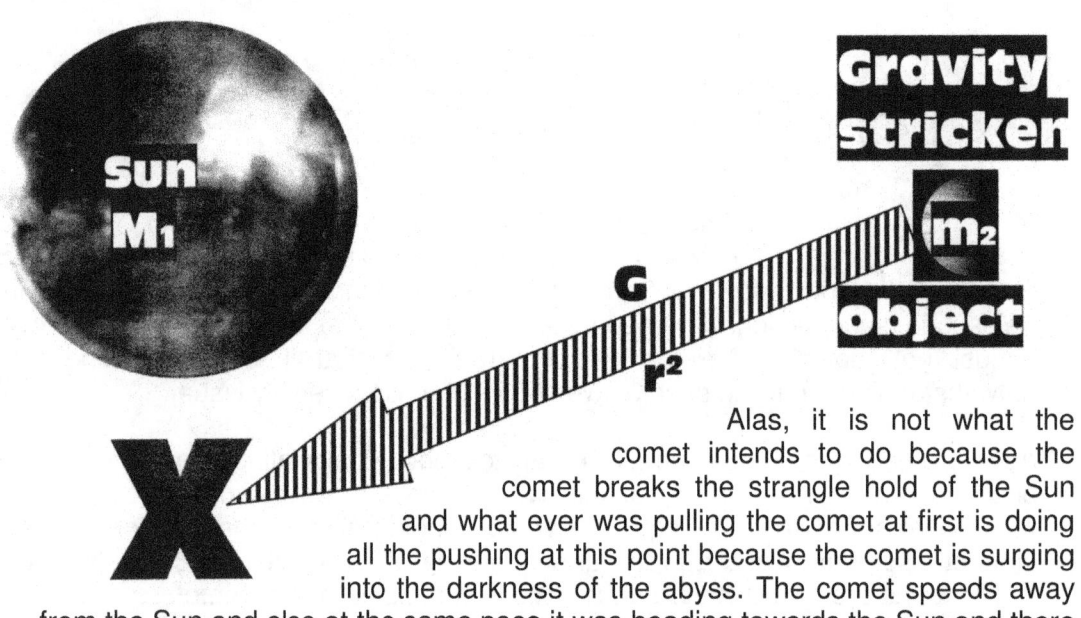

Alas, it is not what the comet intends to do because the comet breaks the strangle hold of the Sun and what ever was pulling the comet at first is doing all the pushing at this point because the comet is surging into the darkness of the abyss. The comet speeds away from the Sun and also at the same pace it was heading towards the Sun and there is no altering to the speed in any way. The Sun seems very much unaware of the comet behaving in opposition to what the great Newton predicted with his formula. Again I admit I don't understand Newton but I suspect the intelligence of those that do understand Newton because they might for the same measure believe the story of little Red Riding hood and other forces fairy tales offer. When we enter the world of forces what is to be expected?

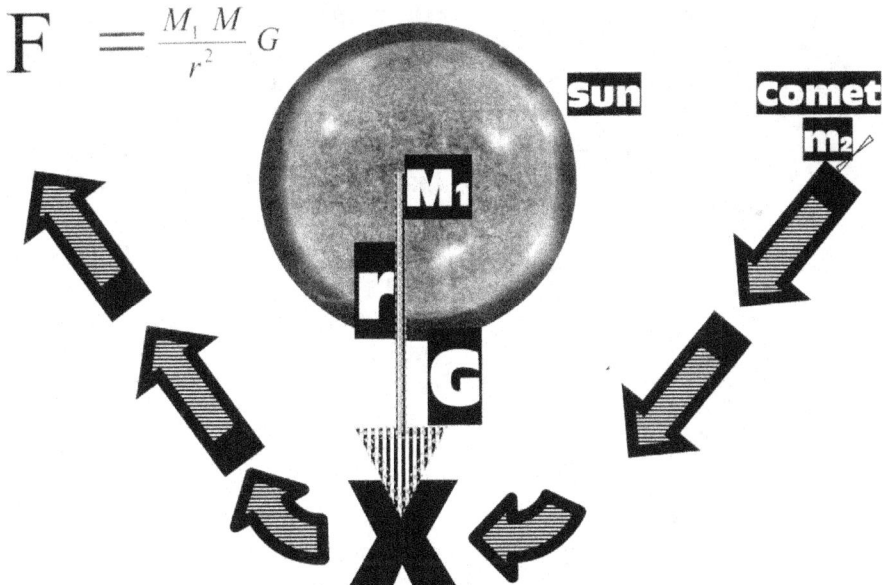

Newtonians improved on Newton's force by creating so many more forces legally without the Church interfering in the process of creating forces that the witches went on strike. They don't work any more! The poor magicians and witches no longer has any authority when it comes to forces because from the one side the Church roasted then alive for having or allegedly having forces and from the other side physicists stole them blind by taking away all the legal forces flying around as forces. To some life is not fair!

I evaluated the four cosmic principles found widely but never before explained because with mass one cannot explain the four principles and Newtonians can only explain when they cheat by using mass.

From these cosmic phenomena I produce a path of cosmic development, preceding the Big Bang. The problem that comes from this is that I take the reader from a point and lead the reader through the explanation of the existing principles, pointing out how they are flawed and introducing my explanations and proof and substantiate my argument. This is a path one has to follow. There is no point where one can drop in, or out and in again and maintain the golden thread of understanding. To conclude, only this: As I bring proof of existing evidence in cosmology about phenomena and of which science acknowledges the existence but science is failing to understand or explain the correctness thereof.

How can the Universe expand? Well, it cannot, and that is yet another illusion the Newtonians create through misunderstanding. What is in it is in it and it cannot grow, as much as it cannot shrink. It cannot expand and it cannot demise. It is only a consistence of changing relevancies, where the relevancy flows away from one part of eternity or singularity (space) to another part of eternity or singularity (time).

...And then in the face of an ever-expanding Universe where all is drifting apart at a rate they hold a formula for. Notwithstanding, I also at the same time have it on the best authority that this cosmic mess with this likeness is going to become a future star. This blob of expanded gas with no form or no structure being part of an ever-expanding Universe that started as the Big Bang is going to contract and become a dense star with a bright shining light. How can a cloud of hydrogen contract into a star within the realms of an expanding cosmos...and then the Newtonian clowns insist they are in consistency with reality?

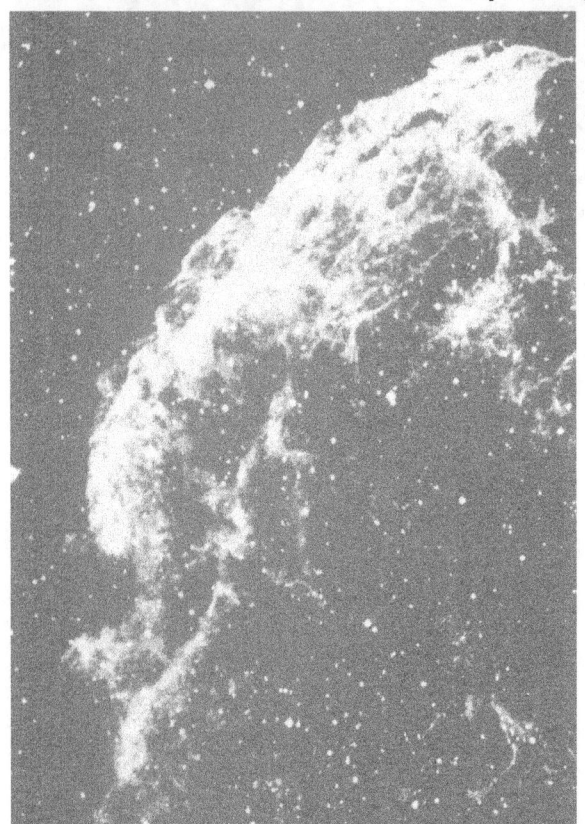

Having two objects in some relation where the one body is small and the other body is large, we find that the smaller body (any small object) can be falling down to the larger (which can also be the Earth) by implication of size holding mass. According to Newton's formula every one of the objects should have its own value of mass that is supplying gravity with the aid of a Newtonian invention thought up as the gravitons. Then in comparison the Earth must have the most gravitons. By implementation of its gravitons that is in relation with the mass, the mass putting the gravitons at work has in the case of the falling objects, insignificant and an unrelated value when compared with that which is influencing the larger object and more so where we use the Earth as the example. However, the two lesser and falling objects are in their own individual dual to see who reaches the Earth first because it is their personal gravitons that are inducing their individual and totally different mass, which is producing the gravity to fall. Let's compare an iron ball in matching size and compare such a fall to a wooden ball falling the same distance under the same conditions. It stands to reason that the iron ball's gravitons should give it a superior advantage just because of superior numbers of gravitons working. This comes about because the two objects are in a position where they compare in relation to one another and share a common second factor, which is the Earth.

In relation to the Earth, the gravitons of the two balls do not come into consideration, but this does not play a part, since the Earth is a common factor when only the descent of the two falling objects are compared to each other by performance measuring.

The balls, however, are put in a situation where they stand in relation to each other and the graviton of each ball is in a contest to see which one will produce the highest mass. When compared to one another, the gravitons should give the heavier ball a sizable advantage in its falling effort, since the mass which is higher in the iron ball is pulling stronger on the mass of the Earth than the wooden ball with less gravitons and therefore less mass is pulling on the earth. But Galileo said there is no heavy or light, big or small, since all objects sharing similar conditions fall equally.

Galileo was the first to indicate space-time with his pendulum but all academics persisted in failing to notice that Galileo was the first to prove gravity is simply equal motion to all. But as everyone from Newton on, was bent on their mathematical games, they all and including Newton didn't recognise and failed to see this aspect. It is worse when considering that Newton supposedly is the master on motion.

This is where all academics show their criminal intent and corroborating to conspire with the intention to defraud the public as a group. If Newton's mass did play a role, objects of lesser mass has to fall slower than objects with more mass and objects being massive has to fall faster. If it does not, then mass plays no part. They fall equal and thus mass do not play any role. When bringing this to the attention of the academics, it is not only that the academics I approached in the past can't explain this; they get angry with me and deny this fact being a factor in evidence at all.

Their tactics they then unleash by trying to pretend I am the fool that cannot comprehend the full picture of what Newton implies, does not scare me as it did not scare me in twenty-eight years, because I can see the criminal liars they are by covering Newton and following Newton's conspiring fraud.

The fact is that it seems they are the proven fools unable to see the full picture and the extent incorporating all the facts of what really and truly applies.

It is either Newton or Galileo because they are on two very different plains. They never said the same thing and never can say the same thing. This is how the academics also miss the four phenomena, in which I found a manner and a means to explain, but I shall get to that later. If it is mass that applies, then the heavier object will fall faster than if mass has no influence. They fall equal at all times up to the point they hit the Earth. If mass does apply but has no influence, how then does mass apply? When they stop their motion and being part of the Earth's motion, only then do they receive the mass they show from then on. This is because something is pushing them from space towards the Earth and pulling them from inside the Earth. That with which they fall is then pushing them because the falling object then became part of the Earth and is no longer part of that which is falling.

Newton stated that $\frac{M_1 \times M_2}{r^2} G$ = Force and the mass in each case remains the same as well as the gravitational constant. This means that all stars must collide, which makes the phenomenon called the Roche limit highly illegal to have presence in the cosmos, should the cosmos try to uphold Newton's laws. The Official Policy Protectors never try to explain the relation between Newton's laws and the phenomena we find in the cosmos. Take the binary star system forming that brings proof about the principle we know as the Roche limit. The binary stars are systems where two stars spin around each other and never collide. These stars are many times over the size of our Sun. When one applies the same Newtonian formula as given above, these massive giants must crash into each other, destroying themselves in the process. The enormous mystery is not in the apparent misbehaviour of these giants, but the fact that this is known to science since the previous century. Relate the binary once again to the comet / Sun relation and there is a distinct similarity.

There is an existing example one can use to show as well as to prove that when some structures in the cosmos has a matching size and they do come into conflict by coming too close to each other, there is a process where the larger turns the lesser to liquid. In the space that develops between the two as they spin around a joining axis somewhere between the two stars in confrontation, they have occupied space sharing. By something reminding of electrical power, the lesser one of the structures are turned to heat in space by the other and larger structure.

The larger object is not pulling the lesser object closer as Newton suggested, but the stars confirm the very opposite by the larger star that literally dissolves the lesser star before consuming the liquid that form as the lesser star overheats. The turning to liquid of the lesser star overheating is deforming the lesser as it is turning as much of the lesser object's interior into raw heat than the superior star then accumulates and absorbs the heat to benefit the superior structure as to sustain the growth of the more prominent object. This has gone unexplained for centuries because it does not fit into or qualify as supporting the view of mass forming gravity that Newton envisaged.

If the structure proves too large, the superior structure turns the lesser compatriot into heat. Then being heat, it will apply gravity and admit such heat into the ranks of its atmosphere, but not before the superior star turns the solids into plasma and the plasma into photon fragments, which then is good enough to become heat that is condensed to the value such as light has. This process where this devouring happens has a known name for centuries and while the known name was given no one tried to marry this process off onto Newton's suggested formula, merely because the marriage ends in a divorce before the meeting can take place. There is forever the talk about stars presumably going into collisions, such as Newton's formula would suggest. In spite of this we are all aware of binary stars that spin one another into the oblivious but they never collide. The one tries to put the other in a state of liquid heat, and then dissolves it by placing the overheating star into the bigger star's atmosphere. By liquefying the lesser star, the larger star is capturing the lesser star and cannibalising it. The head on collision as Newton suggests is not evident throughout the entire Universe. When there is no obvious dominant one in the binary, the two stars will spin one another into pulsars where there is no clear dominator, but evidence and proof about how or where they collide has never been found. If I understand gravity correctly the spinning should produce some collision of some kind as Newton's $F = \dfrac{M_1 \times M_2}{r^2} G$ = Force will let to believe when and if it had any merit of truth about it.

For literally hundreds of years science taught the world about mass pulling mass closer and the mass that pulls mass with all the gravity it can muster will eventually and finally have a Universe coming together and form one lump of you know what…

$$F \; \alpha \; \dfrac{M_1 M_2}{r^2}$$

This formula embodies Newtonian physics on all levels and represents Newtonian physics in all proportions. The belief in this formula is so highly regarded that all Newtonians would rather believe that the Bible can be wrong, than would they believe Newton's formula can be wrong, and this also includes even those concerning themselves as the most ardent Reborn Christians. They, all the Newtonians from whichever denomination or level of development would rather think of God that can be wrong but Newton holding this formula just can't be wrong! Thus I say to all Newtonians and this remark does not exclude anyone in particular: If you are a Newtonian you are oblivious to what incorrectness there ever can be in relation to what Newton said. Although there was one man that did prove Newton wrong but for proving that part that Newton is wrong this man which I am referring to will be the chosen one that shall never receive the Nobel Prize.

Newton claimed that the Sun is contracting the Earth while the Earth is contracting the Moon while the Earth is in return contracting the Sun with the Moon contracting the Earth and with this contraction going on in all directions the lot will eventually re-unite. In this way the cosmos came about and in this way stars form. The one dust particle pulls the other dust particle until the lot forms a lump of material that becomes a crumb and from that a star comes about. This all happens under the magical influence of mass establishing the presence of gravity and gravity pulls stars to form from dust.

I have been accused in the past by academics in the field of physics to be too mentally impaired to understand Newton and that is true. Please allow me to explain yet another part of Newton that I cannot understand.

The cosmos formed from dust that through some magical intervention of gravity compressed into solids that forms the stars we now have. Big stars blew out and the dust that was big stars became hydrogen that formed the little stars we now have, except that they have the opinion that our Sun is almost as big as any star can get. The official opinion is that the stars form when dust collects by the initiative of gravity between them and from that the stars compress the material to form units. There are rings of material that gather under the auspices of gravity working the magic because with the aid of mass the gravity compiles units and the units become strong enough to form crumbs and the crumbs become pebbles where the pebbles become sand storms that still further compress to be become rocks that form gas and end up as stars. What a lot of Newtonian rubbish that is.

So your response will now be so what is new? This is the part that is above my inferior intellect: How did they get there. Remember they were rings of dust that by the magic of gravity collected to become a solid structure. We now have one solid structure claiming space at 58×10^6 km and another solid sphere claiming some space at 108×10^6 km and then one solid sphere at 149×10^6 km. We see a small ring with a small structure filling a small space going around a big area in a wide circle.

Teaching students that it is mass pulling the planets around the Sun while the evidence that is supporting this matter is very skimpy and dodgy, such teachings is a way of committing an act of folly by which then perpetrators are committing fraud. With so much evidence that is lacking and yet academics still insisting that students accept the fact that Newtonian presumption on the matter is correct, is brainwashing.

It is outrageous to force ideas onto students by disciplining thoughts and controlling the minds of the students into believing that the mass is in charge of the gravity, which is pulling matter onto matter while all obvious evidence so far lacked any proof. A presumption remains a presumption until it is proven. Providing uncountable and undeniable facts never substantiated Newton. He made a presumption in the case of his apple falling, which remains a presumption until now. In such an event the presumption remains a presumption until facts prove the presumption accurate. Then only does the presumption become fact.

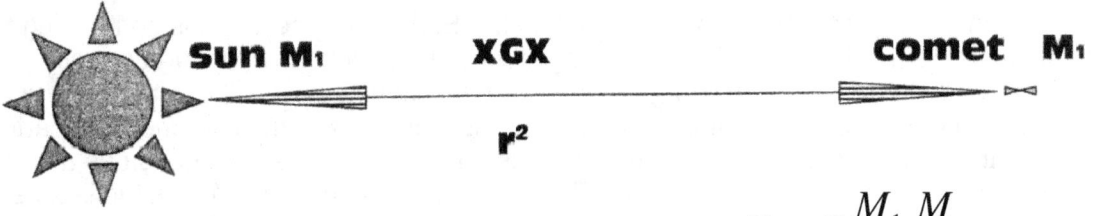

At the start the first sketch explains $F = G \dfrac{M_1 M}{r^2}$. The rest is an unknown and putting facts aside and fiction in place. When bringing in intellect while putting stupidity aside, then not even Newton can fill in the blank spots. It is all make believe and it becomes so clear that the academics in physics are keeping the minds of students busy with the biggest cover up the world has ever witnessed. Your local academic is not a wise old man having all the answers he wishes to share with you...instead he is a shrewd criminal that wants to deceit you with lies in order to carry on with the biggest cover up ever produced.

From wherever the comet is coming, its coming straight to the centre of the Sun

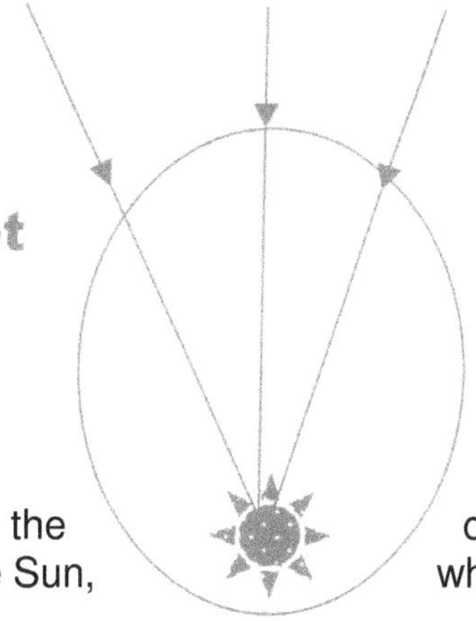

The comet performs all the other manoeuvres as the sketches indicate except returning to the Sun, and it is all the manoeuvres that the comet does, except running into the Sun, which Newton's formula totally ignores. Yet on this very principle every aspect of modern physics is based. If the formula forms the basis of all physics used by science, then it is the basics, which are around for hundreds of years that suddenly are trash and simply does not perform as it is supposed to. If there is out there one professor in physics that will tell me the profession didn't know about the comet not falling into the Sun then I can show you one liar as I cannot that equal poker playing game of while lying about what their hands have. Because of space-time differentiation within the star, the time lapse in the atom becomes a separate value to the surrounding space.

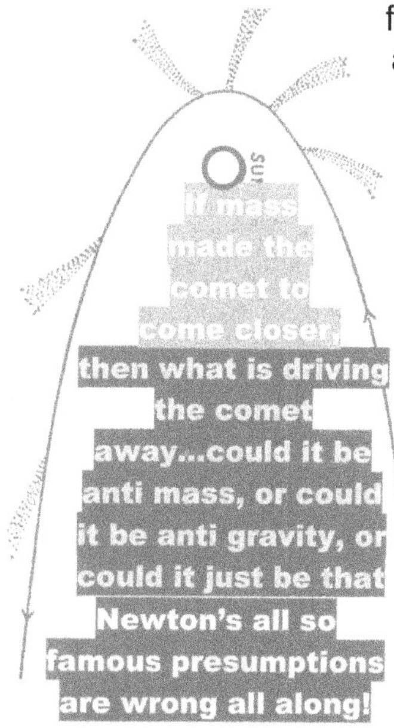

If mass made the comet to come closer, then what is driving the comet away...could it be anti mass, or could it be anti gravity, or could it just be that Newton's all so famous presumptions are wrong all along!

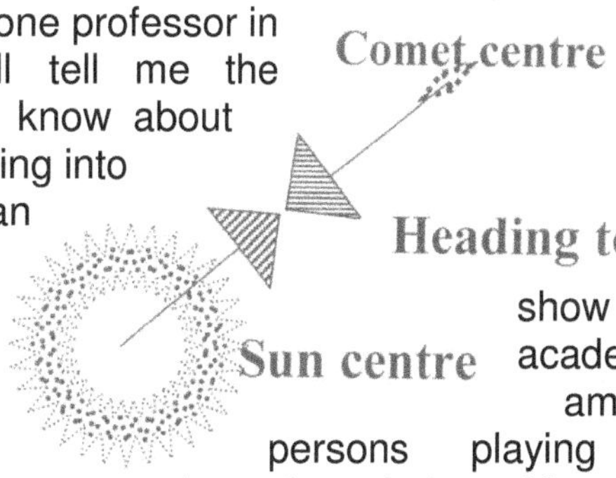

Comet centre

Heading to Sun centre

show you academic's amongst persons playing the deception being his equal

As the time value increases within the atoms that are about to fuse, **the intensity of time demolishes the space component. With time between the two atoms becoming eternal, all space is nullified. With no space, the atoms share time** and therefore, in **future space-time becomes a single unit** and a new element is produced with complete new space-time related values. This is put very simplistic, but IN THESE BOOKS the whole process of fusion is explained in detail.

Every one, including even everyone having the mentality of a motor mechanic such as I, or those being of the lower cast in the social order such as I, (because I am just a motor mechanic), knows that Newton's formula is preposterous. If mass is doing the puling, then what is doing the pushing when the comet makes its great escape?

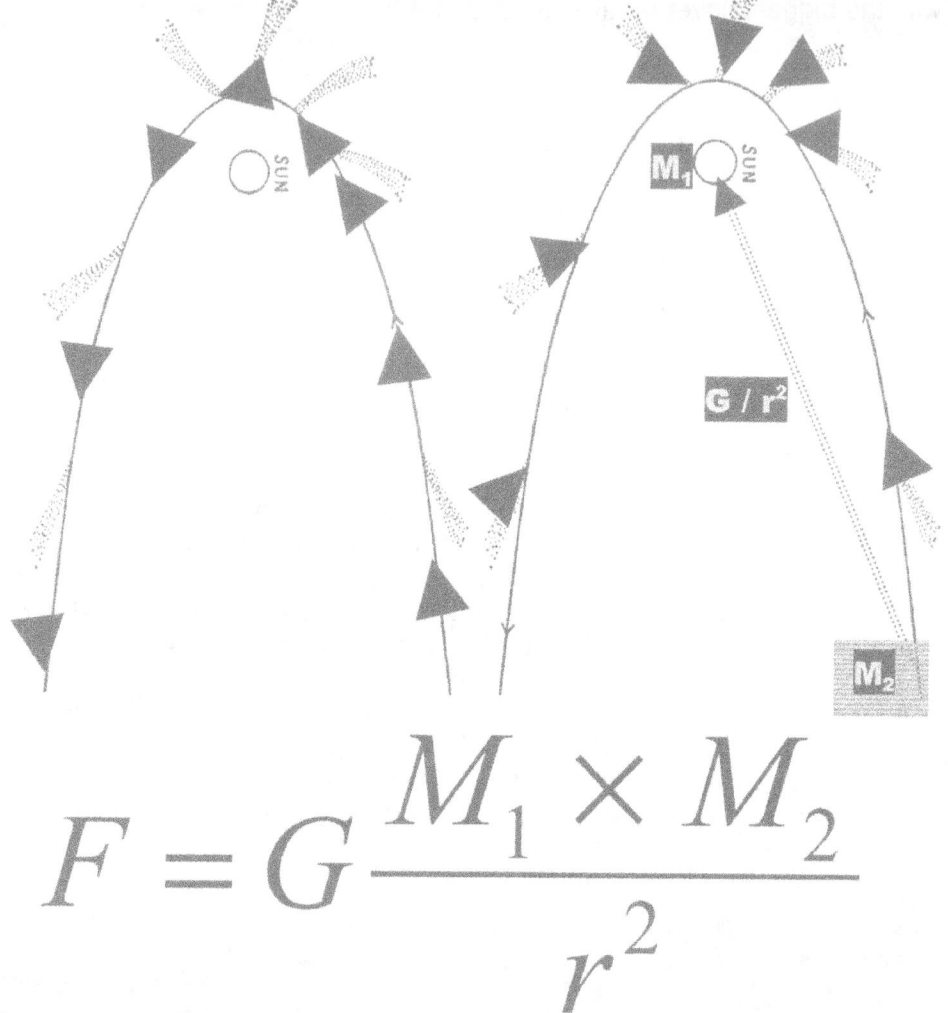

$$F = G\frac{M_1 \times M_2}{r^2}$$

There is no mass that can pull, as there is no gravity doing the pulling just because there are no forces flying around as much as there are no "free electrons" waiting to be used by someone looking for "free electrons" as spare parts. What everyone is very sure about, is that the comet should be heading towards the centre of the Sun if it is heading anywhere and by missing it means it is not heading anywhere near the centre of the Sun.

If gravity was what Newton said gravity is, then we know that the comet should be coming for the centre of the Sun no matter where the comet is coming from. It is this surety we humans have about gravity that convinced us to believe in Newton's concoction, he deliberately mislead everyone to believe. It is this notion of gravity or motion coming to the centre that helped the bearers of Newton's corruption that we also now know as Newtonian High Priests used to keep us fooled and mislead.

$$F = G \frac{M_1 M_2}{r^2}$$ where F is the force between the bodies.

M_1 = The mass of the one body, in this case the Sun

M_2 = The mass of the second body, in this case the comet

G = The Gravitational Constant that fills the nothing that fills the radius between mass one and mass two

r = The distance between the two bodies

By now it should be evident that this absurd formula carries no composure any more and we have to find the truth. This is the pivot of the argument I have about conspiracy, treachery, misconduct, corruption, neglect and purposely defrauding every new student into mismanaging the minds of so may students through hundreds of years. In three hundred and fifty years and hundreds of intellectuals not one ever saw that the comet is doing the great escape! Is there truly not one that could see in the space of three hundred and fifty years that if the comet is coming by mass pulling, then what is having the comet departing?

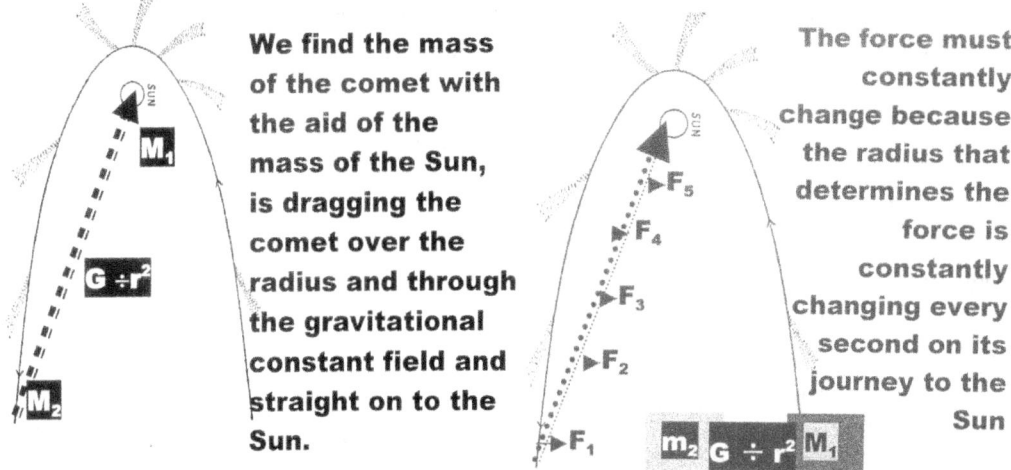

We find the mass of the comet with the aid of the mass of the Sun, is dragging the comet over the radius and through the gravitational constant field and straight on to the Sun.

The force must constantly change because the radius that determines the force is constantly changing every second on its journey to the Sun

It is said that the mass of both objects find the ability to generate enough gravitational constant destruction to destroy the radius that is parting the mass on either ends until all the gravitational constant is eaten up. On one occasion they say the where the gravitational constant is a void holding nothing but then they say this void holding nothing has a value of 6.672×10^{-11} N m/s²/ kg². Only a Newtonian can give a void holding nothing a value of 6.672×10^{-11} N m/s²/ kg² and believe their views is correct.
When I confront academics on this issue, they end the interview as I am told that Newton is not for everyone to understand. Only a selected number are, as privileged to see Newton and others must accept this fact not understanding Newton, in their stride. So I suppose they say I must accept I am too stupid to understand Newton.

An Open Letter On Gravity Part 2

If the comet arrives by method of where the one mass is pulling the other mass along to a point where the small mass will end up inside the large mass by reducing and eventually destroying all the gravitational constant that is filling the radius...

Ever heard about the King standing with his magic clothes that only the wise could see and all others that were fools were not wise enough to see the magic clothes. Everyone in the Kingdome, including the King, wasn't going to be a fool in the eyes of others, so everyone admired the magic clothes. Everyone was talking about the Kings new magnificent wardrobe and the splendour of his suits because with the clothes being magic, only the wise were able to see the clothes.

Everyone was admiring until someone in the form of a child saw that the King was walking naked in front of everyone. When she cried out loud that the King was naked there was a Kingdome filled with fools that were too stupid to realise how stupid they were. This is one fool Newton is not fooling. Maybe, because I am not so smart as to be fooled, or maybe the others were suckers stupid enough not to wish to look stupid facing that they couldn't see where someone saw mass in a place where no mass can be put to task in order to participate in forcing gravity. Maybe some were more stupid than they would admit.

If there is anybody out there that thinks that a person is going to convince this motor mechanic that those Newtonian Physics Priests that sports the best developed minds and the most intelligent brains humans can offer, they that form the definition of intellect, did not see this miscalculation Newton made, then that person that is about to try and convince me, is going to have one hell of an argument before I am convinced.

Then what is the method the comet employs, that the comet uses to depart and leave the Sun on a voyage where it again fills the gravitational constant by increasing the radius between the sun and the comet? Did Newton forget to inform any one about this or did Newton simply never realize his error he made.

Newton never thought of the question as to what happens when the comet returns into the blackness of outer space and what then drives it on its way going away. If mass was pulling then what is pushing when the pulling ends?

I have to believe those who are the cream of Academic society when they tell me it took almost four hundred years for not one Newtonian to realise the comet arriving as it was coming closer and the comet departing as it is speeding off is driven by mass pulling mass. If this is not corruption then what qualifies as fraud and deviation? I am expected to believe that not one Newtonian ever saw this abnormality as it went four hundred years without some person detecting this? If there is anyone prepared to convince me that in all that time that went by from the time Newton first came to the conclusion with the idea that mass pulls mass and by such pulling it destroys the radius, there was not one Newtonian that realised the comets departing is annihilating the comet's arriving by the pulling of mass, then I am the corrupt one.

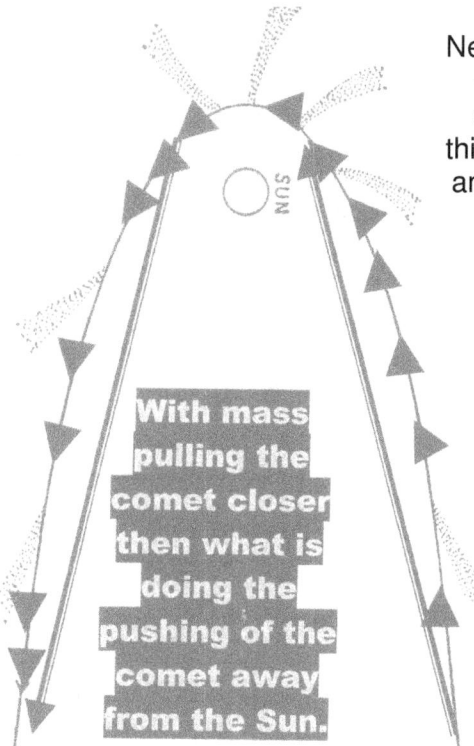

I can't see that anyone ever can believe that there is no conspiracy by physicists to conspire to corrupt and defraud on a scale not yet equalled in all the time since Adam was corrupted and was defrauded by Eve. I am not prepared to have anyone with a sane mind and an honest heart tell me that a person knowing this and still ignoring this and to top the lot, is still believing Newton never made a mistake is a honest person. How can anyone ever believe another word they utter? There is not even misconduct and an attempt to defraud that can even try and match this corruption. Is there one academic that will under oath declare that in the ranks of physics worldwide not one knew about this contradiction? Then those in charge of students teach them Newton's failures without blinking an eye! Is there anyone that is prepared to pick an argument after this as to establish the validity of $F = G \dfrac{M_1 M}{r^2}$? I had some arguments in my lifetime but that will be a battle he just will not be able to win notwithstanding, even if that person was Albert Einstein in person. Those Highly Educated Brains-walking-on-legs called Newtonian High priests of physics had over three hundred and fifty years to see there is one hell of an error going on in Newton's suppositions and it just don't add up. If one tries to convince me they were stupid in awe by the magnitude Newton had to the degree they never realised Newton was at error, then my name is hound, and my name is not hound. It is so obvious from the way the Newtonian High Priests called Physics Academics protected Newton's fraud, that it is clear that they obviously were collaborators in Newton's crime. They knew for centuries Newton committed blatant fraud but they all conspired to hide the fraud. No one will convince me that no one in physics realised there was a mess and no one will convince me that the Academics were not committing blatant fraud. If not, why did they all try for so long to cover up and when I came and showed the problem as well as the solution, no one was interested in my story. I was ignored notwithstanding the evidence of malpractice that Newton committed!

The rotating of the comet and the rotating of the top is identical cosmic principles that all work by the same rules applying in the cosmos throughout. The rotation of any object has certain specific demands and motion that applies time in time through time, which is one of the cosmic principles we now call gravity. Gravity works on motion or speed differentiation and mass has no ties with the concept at all.

$$F = G \frac{M_1 M_2}{r^2}$$

The more we investigate, the more Newton's Universal gravitational concept of forces pulling by the way of mass becomes less plausible, and if you don't agree, then you are a Newton junky that is corrupted to the foundation by your Newton addiction. This formula fits just as unlikely with the likes of the way Newtonians propagate how the start came about.

Newtonians suggest that normal gravity started at 10^{-43} seconds. Newtonians also suggests that at the time, the Universe was the size of a neutron or somewhere in that vicinity in size. GUT or the grand unified theory proves that at a point the entire Universe became light. It is a defining position in time. It also is rather ridiculous to suggest that the Universe started at that point since that point in truth is neither here nor there.

The Universe exploded into substance from which matter came but never states or suggests any position before such an event. According to mathematical evidence there is substantial proof that the Big Bang began and GUT or the grand unified theory produced the attempt to describe the strong and weak nuclear forces and electromagnetism in one single mathematical theory. Somewhere before 10^{-12} seconds of counting the Universe cooled to about 10^{15} K the electromagnetic and the weak interactions acted as one single physical force. Science recons that unification may come about at temperatures of 10^{27} K, which was the temperature of the day at 10^{-38} seconds after the Big Bang. This statement echoes my viewpoint but one has to look carefully for that to surface. The strong and weak forces are only a reference to the atom's bonding ability and the ability to unite the flow of time.

An Open Letter On Gravity Part 2

For many hundreds of years every person in physics were aware of this flaw but did nothing about it. I call that criminal and I call that deception because they took money from others to betray the innocent, the young and the trusting vulnerable, by spreading untruths to those still with unblemished innocence being pure at heart and forcing young minds to believe what they well know is not remotely true. That is an act of a criminal mind notwithstanding what motivated the person to commit the crime. Defining a criminal is to have a person that is prepared to place himself or herself in the centre of the Universe as to deprive all others of truth and possession in order to further the needs and wealth of the criminal while the criminal acts by never thinking about the rights of the victim or the harm done by the criminal's action in the matter. Spreading untruths and forcing payment for such actions is most certainly criminal! You students in physics go on and confront your Professors about the truth. Insist on the truth for once and all. Mass has nothing to do with gravity but to prevent gravity from being in place. If it is mass pulling the comet as Newtonians declare, then they better inform you, being the students paying their salaries and in that forming the reason why their wallets are filled every month, what is pushing the comet away. The comet is leaving as fast as the comet was coming. Confront them because you have the right to insist not to be lied to about Newton and his shambles and criminal fraud. Go on and ask them to explain these Newtonian inconsistencies and see how they try to carry on with the criminal deception and covering of the truth that has been going on since Newton first thought up this scam. This is but a small part of a big picture uncovering the scam Newton came up with and which all the academics are knowingly still participating in… and read on for I am about to inform you of more criminality the academics came up with and with which they force feed you. You will see fraud as you never saw fraud before.

All the Academics to whom I refer are also those, whom I had dealings with, where I concluded that they have this attitude that they consider themselves (every one in their group of distinguished academics) as the only ones amongst us that are blessed with thoughts. They have this idea that they are the most privileged group and that they are the only persons that are being blessed with insight into facts that matter and truth. Being in a class of their own puts them in the bracket as those that are exceptionally gifted with brainpower. They are the soul group to control thought. In this attitude they also see all of us being the others and the rest as the ones that merely ponder on instinct. It is this most privileged group with mind and thought that see they fit enough to be able to charge me as being a retard that is in a state of stupidity. To their most admirable view, they see me, as suffering from a condition being with some disposition that all others would consider as having a lamebrain. This condition is a mental state where facts move past you at a rate that the person, having the lamebrain is unable to follow or process facts and where such a person can cope only with parts of the events as flashes while allowing much of the detail to pass. In other words I am not understanding proceedings or not following the entire compliment of events as all the facts are discussed and therefore not being able to gather a picture that will portray the entire library or compliment of an accumulation of facts. Therefore, I wish to share with others the facts, as I understand them, because Academics are convinced that I can't understand Newton's gravity and they are correct.

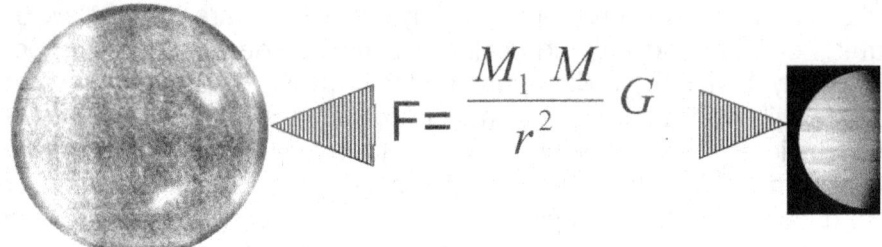

This formula embodies Newtonian physics on all levels and represents Newtonian physics in all proportions. The Bible can be wrong. God can be wrong but Newton holding this formula just can't be wrong. If you are a Newtonian you are oblivious to what incorrectness there ever can be in relation to what Newton said. Although there was one man that did prove Newton wrong and for that he shall never receive the Nobel Prize.

Newton claimed that we (all of us forming a Universe as a group) are in a shrinking mode where everything is rapidly growing smaller as it is driven towards one another on account of the mass it forms. It states that irrespective of whatever distance particles are apart, the particles will get hold on one another's mass and in accordance with the mass, it receives a force that drives the one towards to the other. Remember, the distance has no limit since that mass will focus on the mass and the driving will bring the lot together. From such pulling that they call gravitational pull, dust grouped together a long time ago as dust clouds and pressed this lot into a structure as dense as the Earth and the inner planets are. Please allow me to explain yet another part of Newton that I cannot understand.

I sit with these nagging questions about how things started the Newtonian way. I am not sure if it is I being incredibly stupid, or are there others with similar suspicions about the gravity fairies becoming a little lazy. It was about between 15 and 13.5×10^9 (hey, when I say this I am only echoing Newtonians) that the lot actually started. At first things were rather hectic. Things happened and there was a lot of action with happening going about.

Big Bang to 5×10^9

Timeline

Apparently at first there was this explosion. To have an explosion there has to be an inside. The inside has to be going outside to be in an explosion. The inside was filled with whatever went exploding. For an explosion there has to be an outside as well, giving the "exploding something" somewhere to go to. If nothing was on the outside, what was the nothing, because if there was nothing on the outside where did the something on the inside start and where did the something on the inside stop to allow the nothing on the outside to start? Where did nothing on the outside stop, because if nothing had a start, then that nothing must have an end, since nothing has a start? What made the something exploding stop and had the nothing to which the something that was exploding go while the something was exploding? What made up a border where one can say this is where something stops and nothing starts? Again nothing filled what was apparently then the same nothing that also was vacating so that something could fill the vacating nothing while something was exploding. Is that Newtonian or what…

In the first 5×10^9 years there were the grand repulsion, the Hot Big Bang, the quark era, the weak force ended and the proton era started. Then came the electron era, the gamma-ray era, the neutrinos ran loose after which helium started. The problem that this brings is with the expanding, the "pressure reduced". What then made the heavy atoms form if it started with helium and the helium should have reduced a lot of the pressure, since the helium takes up a lot of the pressure? There is a lot of helium to form. That raises more questions except if you are a Newtonian. Newtonians never ask questions because they simply size the object and award mass. When they have the mass no more questions are asked because all other questions are unnecessary, they know the mass. Here are a few feeble questions Newtonians do not seem to care to ask because Newtonians consider the answers (I suppose) feeble.

At one micro second there was matter and there was anti-matter. If matter is heat as we see with the atom bomb disintegrating into heat, and the naming being a thermonuclear explosion, which inherently refers to heat then what would anti-matter be when it is supposed to be the opposite of heat? Going anti means it must be cold which it then went disintegrating into cold. No Newtonian seems to care to give anti-matter a sensible explanation except to say the matter ate up all the anti-matter and then there was only matter left.

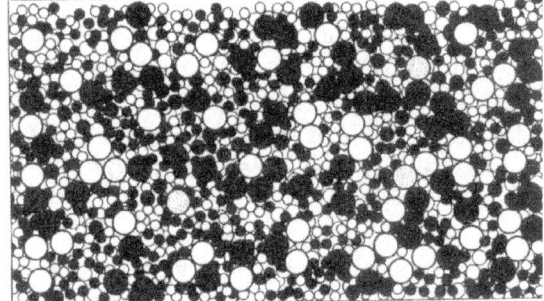

No one cares to say where the antimatter went because it is not important in accordance to Newtonian principles. If matter ate up antimatter then antimatter has no mass and without mass, then who cares about antimatter? It has no mass and therefore it has nothing of substantial importance. Where did the antimatter go since the antimatter could not leave the Universe?

If the matter ate up all the antimatter why did the matter not double in size as it fattened with all the antimatter in its belly? This is the closest Newtonians ever came to mimicking Little Red Riding Hood and especially the wolf that ate up Grandma and still had enough room left in its belly to go after Little Red Riding Hood. If they can show where the antimatter went and not just discard it into the nothing component, then they would have had some degree of believability. But throwing the antimatter away as in banishing it from the Universe without explaining where the antimatter went is hardly believable. In fact it is so Newtonian even Newton could have thought this one up. Then after the hearty meal, the pressure relented a lot. It had to because half the Universe devoured the other half of the Universe, which left the first half still feeling very hungry. That is the function of mass. It is a hunger one object with mass has to get hold of the object next door that is also with mass and devour it. Therefore the gravity eating mass by mass is a hunger that never stops being hungry. The main thing is the pressure reduced considerably. This raises questions that require answers.

Protons Freeze Out.

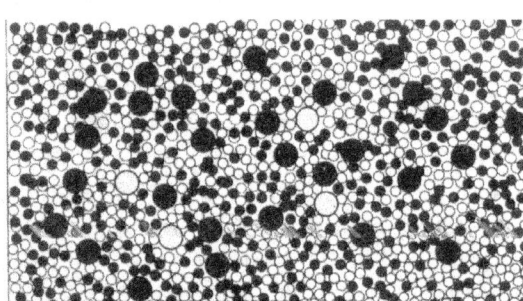

Then after a time lapse of one hundred micro seconds the protons froze out. From nowhere came protons and then instead of nothing there were protons and nothing. What made the protons be there…you guessed it…of course it was Newton's formula. It was the formula God used the Create the Universe. It was $F = \frac{M_1 M}{r^2} G$, which was what was used to create the Universe when God did not create the Universe but the Universe came into place by the magic of mass creating a force that establishes gravity.

Please allow me to explain that which nobody in his right mind can explain because it is a Newtonian principle. When that which was not even yet had mass to pull that which was also not even yet and the two not even being yet but also had mass, started pulling together to form protons with mass, then what else could it be but to be magic. From nowhere came protons by the magic of mass. There is no explaining required as to what formed protons since protons have mass.

I suppose that according to the brilliant Newtonian thinking mind that it is not worth thinking about why protons came about because we know there were nothing and something and something exploded into nothing and in this flash by the absolute magic of mass that produces gravity came the proton. Then who needs a God to create when you have a magic force with the ability to conjure? There was an explosion and then there were lots and lots and lots…of protons. Now there was a Universe filled with the exploding of protons and… also filled with nothing. Then forming the protons stopped as if it was ordered to stop.

Electrons freeze out at one second.

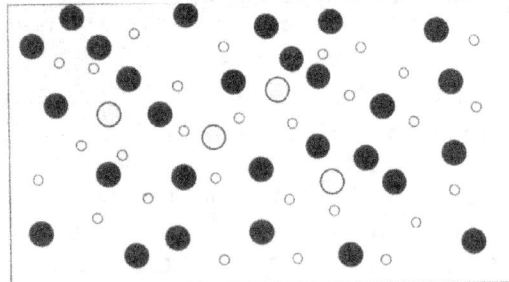

After such a long response as that it took the full period of one second in lapsed time, the electrons came about. Again we have the electrons coming from nowhere and the electron was forced into being in the Universe since they came from the nothing in which it was. From the protons to the electrons it all transpired by the magic intervention of mass pulling mass. What had enough mass to pull enough mass to form electrons? This question was never asked because no one ever answered. There were electrons…no there was nothing and then there were electrons and the electrons came from the region where nothing today still is and also back then where nothing simply was. No one knows what intervened to have the electrons arrive and what formed the mould to have the electrons form in such a perfect duplication. All we know is that there was nothing and then there were electrons coming out of nothing.

The Big Bang made both matter and antimatter in abundance. What made up the matter and antimatter is never answered. From where did the matter and antimatter come from that suddenly popped out of the blue and were there in abundance? That fact no one knows because it surely had no mass. It was matter (what ever matter is) and it was antimatter (what ever antimatter is) but sure as hell it could not have been $F = \frac{M_1 M}{r^2} G$. After the annihilation of most of what was (now was it the matter or was it the antimatter for who will know which apart from what?) but this led to particles forming as particles and other particles formed not as particles because they formed as antiparticles. Have you the imagination to grasp what concept would an antiparticle have? What would consummate to form an antiparticle?

Never was there one Newtonian that offered to describe what the differences are between a particle and an antiparticle except that the one is the anti of the other. A slight imbalance left a residue of matter: then came the heavy protons (lots of mass so not to worry, we have mass) and neutrons (except that neutrons never received mass) and no one knows why they never received their quota of the mass that went around. What brought the residue about is also an unknown fact that has to remain unknown because trying to find any answer will be very against the Newtonian grain of having things done. But all the while the pressure was relieving and the pressure was drastically reducing. Why did the protons turn to atoms and not go back to where it came from because if it was the pressure banging the Universe into shape the bang then was losing pressure and then had much less to bang about. Why did it form any atoms because the pressure was releasing at an alarming rate. The bang was strong enough to push whatever was around into protons and then left the protons to be protons. Why did everything not bang into protons, because we find that electrons banged into shape later on?

Why did whatever remain in the hope of later becoming not protons because there was only nothing and protons going around at that time? Then the bang came around and banged whatever was in need of a bang into electrons. Why did that which then came about later banged it into electrons and why did that which banged into electrons not went the route of forming protons when the proton forming was in place? What made whatever formed electrons not form protons but made it choose to remain proton free and later served as electrons? It is reasonable to think that the biggest bang came first and that biggest bang then banged anything and everything into protons.

Why did anything remain and why did a much lesser bang come and banged that lot into electrons when the electrons at first were able to escape the violence of the first bang that made the rest become protons? At first the bigger initial bang formed the much heavier protons and then with a much more reduced bang (I would imagine) there was enough whatever formed this lot left to form an exact and appropriate number of electrons. Is that not appreciably coincidental? And behind all this sorcery sat the magic of the force that later became known as gravity. It all happened coincidentally.

Coincidentally there was matter and there was antimatter that came from nothing and 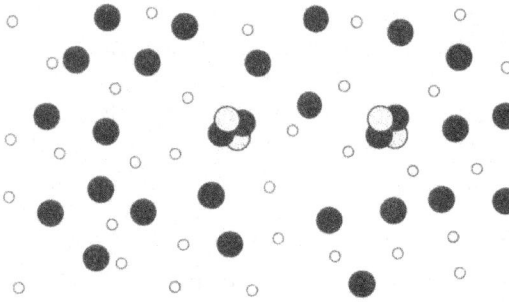 was when joining matter that didn't come from nothing (or that much I suppose) went exploding and since that event it was going nowhere, because there was simply nowhere to go. Then a coincidental bang came and it was banging because it came about being coincidental. One has to believe in fairies to believe this lot came about that precisely while being coincidental. I do not for a single minute dispute what happened as it is explained by the Brainy Bunch...well that is if only someone somewhere can offer an explaining as to where what came from. Coming from the blue as if by magic is very unscientific to my taste.

However, the coincidence seized to stop because then another coincidental and unexplainable thermonuclear explosion converted neutrons plus protons into helium. The lot didn't weld into one lump when the lot was very hot.

They first cooled off and then when the lot was cool enough then only did they weld into many very small lumps we call helium. The helium nuclei did not trap them when things were hot. It trapped them when things cooled off. Now I sit with the question that where does $F = \frac{M_1 M}{r^2} G$ leave us? Al this time r^2 became larger and things got 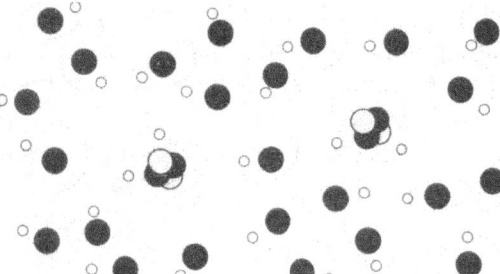 colder and nothing became a lump that was holding one single form. The lot decided to wait and then formed innumerable small spots of whatever. What made the mass able to form small lumps and not the big lumps when the mass had a very small radius to deal with? What triggered this smaller thermonuclear explosion that the earlier larger thermonuclear explosion did not waste by explosion? The earlier explosion should by all norms and standards have wiped out all possible later explosions. After all, the Universe was not a warehouse filled with containers that was left at different locations, which made the containers explode at intervals. This lot was as mixed as nothing ever was mixed that came afterwards. Then hydrogen formed. The blast was not strong enough to push this lot into clumps of Uranium or something even heavier. It pushed the lot into helium. No one is interested how the pressure exerted again to enable the proton to push into groups of up to 93 in the cluster.

Another name the Newtonians use to describe this era is the Hadron Era. If it is true that the proton has mass and if it is true that the proton pulls the proton by mass as we find Newton suggested, then with the protons all being equal in mass, then how did $F = \frac{M_1 M}{r^2} G$ not pull every atom into clusters of billions of protons. Why did it form clusters of two protons and when was there enough pressure afterwards to form those huge many multi clusters of protons? In an explosion such as that with everything going into nothing, one would expect an evenly distributed explosion raging for a while and then going soft. After it raged there should not be any material left to go pop and pop again after the first all consuming bang.

Then after the gas expanded, that gas eventually had to become solids…wait there is something wrong and it is more incorrect than just being Newtonian…how did gas become solids if the gas did not compress into solids during the very first bang? After all the decompressing that followed bang after bang, there was force left to push gas into forming solids. How did the explosions that came afterwards have the force to push protons into clusters coming down all the way from 93 per cluster to 1 proton in the cluster? What made this differentiation take place? Why did this not plainly happen during the first all compressing explosion that occurred. If it was $F = \frac{M_1 M}{r^2} G$, then what made some groups pull harder to form clusters coming down from 93 protons per cluster all the way to one and left most other cluster at one in the cluster? Why did some have the force with it and other have the force acting against it? Then we had a number of events going completely against the grain of $F = \frac{M_1 M}{r^2} G$, but during all this lot Newton is surprisingly never mentioned when this part is explained.

Big Bang to 10 X 10⁹

At 12.5 X 10⁹ years we had galaxies coming about.
At 12.2 X 10⁹ years we had the oldest quasar coming about.
At 11 X 10⁹ years we had the element-making boom with elements forming. Never is the process explained by using much or only a vivid elaboration of the intricate detail during the process of how the elements formed as mass pulled mass but also before mass was not pulling as it should have been pulling with a much-reduced radius by the square in place.
At 10 X 10⁹ years, which is more or less one third of the time used we had the solar elements coming about.

The first part was frantic. The second part was hectic. The third part was to say the least boring. During the first part we had all this forming of whatever could form. Then the second part started off with the forming of innumerable galactica and quasars and ended with the Sun and its planets forming at round about 5 – 4.5 X10⁹ years ago.

Big Bang to 14 X 10⁹

This last third of the timeline, the Earth came about with the other eight sister planets. Sorry, as I recollect they were already here because at 4.5X 10⁹ years they all suddenly and inexplicitly jumped to life. Why did everything that was going to happen, happen during the first and second parts and the Earth went spinning around the Sun with nothing better to do during the third part of the Universal timeline? During the second part a Super star structure formed out of nothing coming from nothing and then blew up into nothing. That was the material that laid the grounds to form the solar system. A star formed that was many times bigger than the Sun. That star blew up and again became dust and gas. That went on and took the best part of the second part of a three-part forming Universe. Then during this same era the dust once more gathered and from that the Sun formed at 5 X 10⁹ and soon after the planets all formed at the same time being 4.5 X 10⁹ years. Then nothing happened all day and it went on since then. During this last part there just was no action because even getting the solar system in place took up most of the second era and that left the third era with the Sun and the planets already formed from the muck and leftovers that remained as the scrap that was cast away when the first Big Sun went bust.

The cosmos formed from dust that through some magical intervention of gravity compressed into solids that form the stars we now have. Big stars blew out and the dust that was big stars became hydrogen that formed the little stars we now have, except that they have the opinion that our Sun is almost as big as any star can get. The official opinion is that the stars form when dust collects by the initiative of gravity between them and from that the stars compress the material to form units. There are rings of material that gather under the auspices of gravity working the magic because with the aid of mass the gravity compiles units and the units become strong enough to form crumbs and the crumbs become pebbles where the pebbles become sand storms that still further compress to be become rocks that form gas and end up as stars. What a lot of Newtonian rubbish that is.

The way that gravity works is much more complicated and the description fitting gravity is much more comprehensive than demoting everything humans don't understand to the likes of gravity. The entire picture is very complex and I am sure that not one Newtonian even thought how involved the issue is to move any object from one location to another location. In its simplistic Newtonian description it would be to shift the lot from here to there because it has mass. The truth is that it takes such a lot of interaction between cosmic realities that only describing how the function operated took me the writing of one book containing 760 pages to get the whole issue debated. It involves time and it involves matter but Newtonians never once in three hundred years took the time to define what matter is in the context of time and what time as such is. But how can I ever get to the part where I can explain real issues when Newtonians are so brainwashed and goggle eyed about Newton that they won't even read the simplest issues Newtonians can't explain. Newton's formula underwrites the entirety of what Newtonian physics represents. Newtonians saw that God played games as the Romans did with Gladiators. God gave one object mass and so did He give the other mass and then God placed them apart. The objects then had to fight their way to one another where the Universe formed the arena. Then God had the two objects fight it out to the bitter end where the one devoured the other and the winner was not only champion of the other but became also the other. The stronger had the most mass and could pull the hardest and that was exactly as the Gladiators in Rome did.

It bears no surprise to see the people five hundred years ago going around and showing such a mentality, but really, one should have a better view of the overall picture now that man has developed in mind somewhat. The cosmos is not an arena where fights are fought till the bitter end. There has to be a mathematical plan. The Sun is not a coal stove that is going to run out of firewood at some point and then become dark. That is the mentality of druids to go and advocate such a humanistic view. The cosmos is about eternity repeating eternity within eternity. The most important issue is to see the cosmos for what the cosmos is and not see the cosmos for what man in the cosmos is. What is in the cosmos can go nowhere but remain in the cosmos and because time brings space to the cosmos, time cannot remove space from the cosmos. Time can only reposition and reallocate space in the cosmos.... and then we have Newton and his Newtonian followers...

Put yourself in a big hall filled with a cloud of dust. Then imagine that you are in outer space, in this vacuum filled hall and the hall is outer space, where the cloud of particles are not going to flow in any specific direction, but will hang around suspended for the time of one eternity. Now supply the reasons why you would think that all that dust that is going nowhere and having no direction to go because there is no specific direction to go, then will find motivation from somewhere, choose one spot that is going to be the future centre of the planet it is going to form, and head in that direction. Why will all the dust find inspiration to flow to one specific point? This argument is as much rubbish as the rest of Newton's arguments about cosmology are!

That cloud of dust, to which I am referring, is not the one represented in the picture

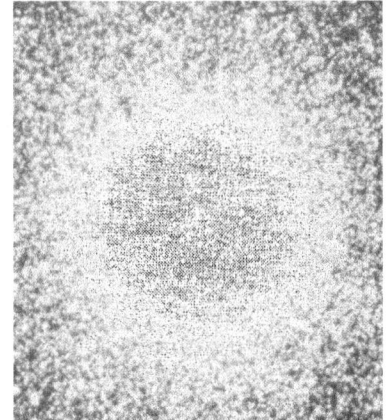

next to this writing, although the cloud of material that the solar system had to come from, according to Newtonians and should they be correct, would not be that much dissimilar in structure and in form than the one represented by this picture.

From this cloud areas must have come that brought divisions as clear as only planets can represent. From these clear defining lines must have formed borders that then became future structures that still hold spherical shapes to this day. What decided where the edges would form when at first all the material was equally disposed and evenly allocated? What determined where every planet would be and according to mass nonetheless, when all the material was spread evenly and placed at random? How did the divisions arrive with everything spread proportionally and evenhanded? Yeah, sure I don't understand Newton but it is not the result of my stupidity!

An Open Letter On Gravity Part 2

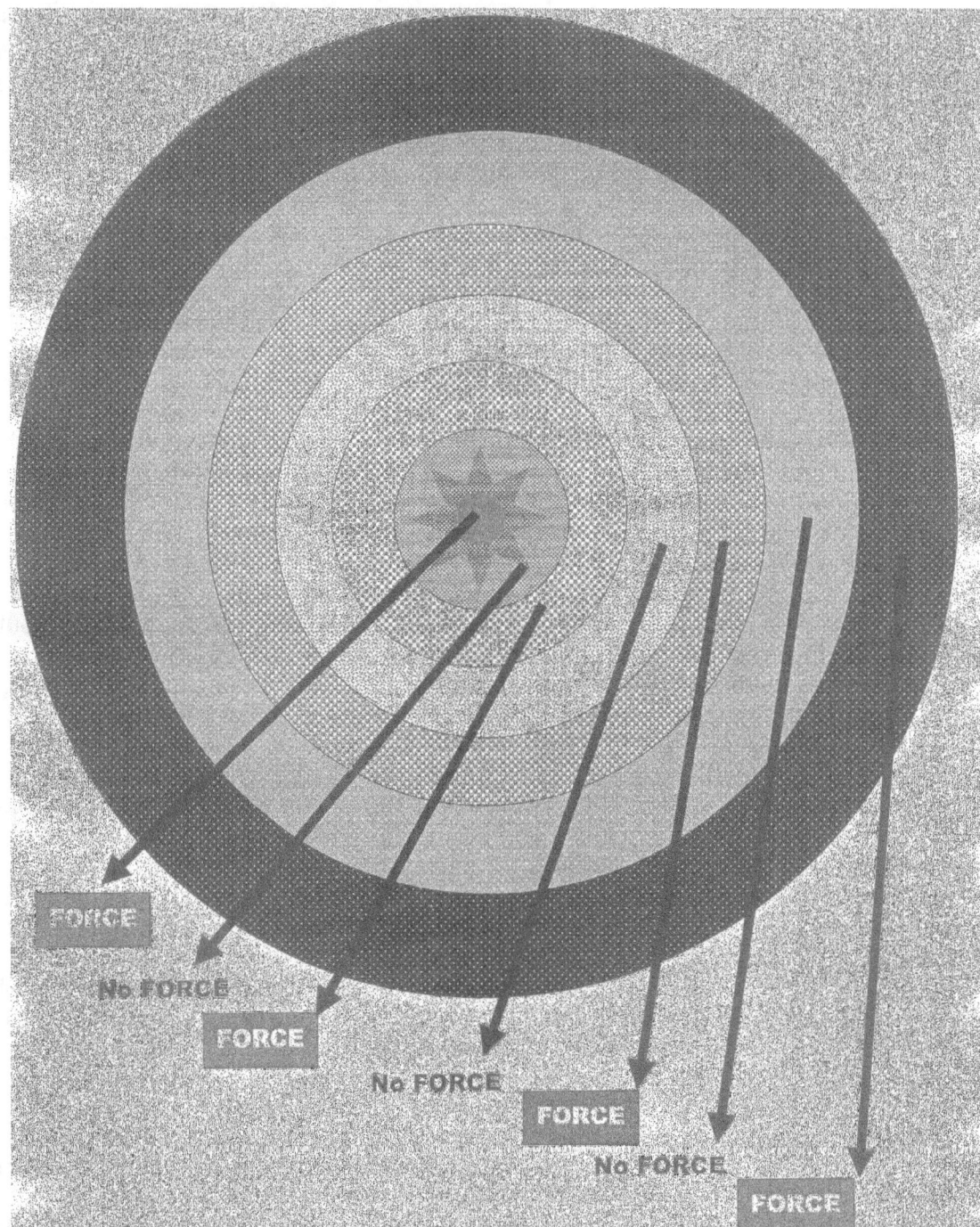

How did the gathering take place, seeing that there were definite collection spots hat became holding points for material and if the collection spot is placed in relation to the area that covers the gravity by which the collecting is done, there is no room on paper to indicate how small the point of gathering is? There must have been certain no – go areas for collection and other areas earmarked for such gathering otherwise a lot of rubbish was in the way of a lot of other rubbish.

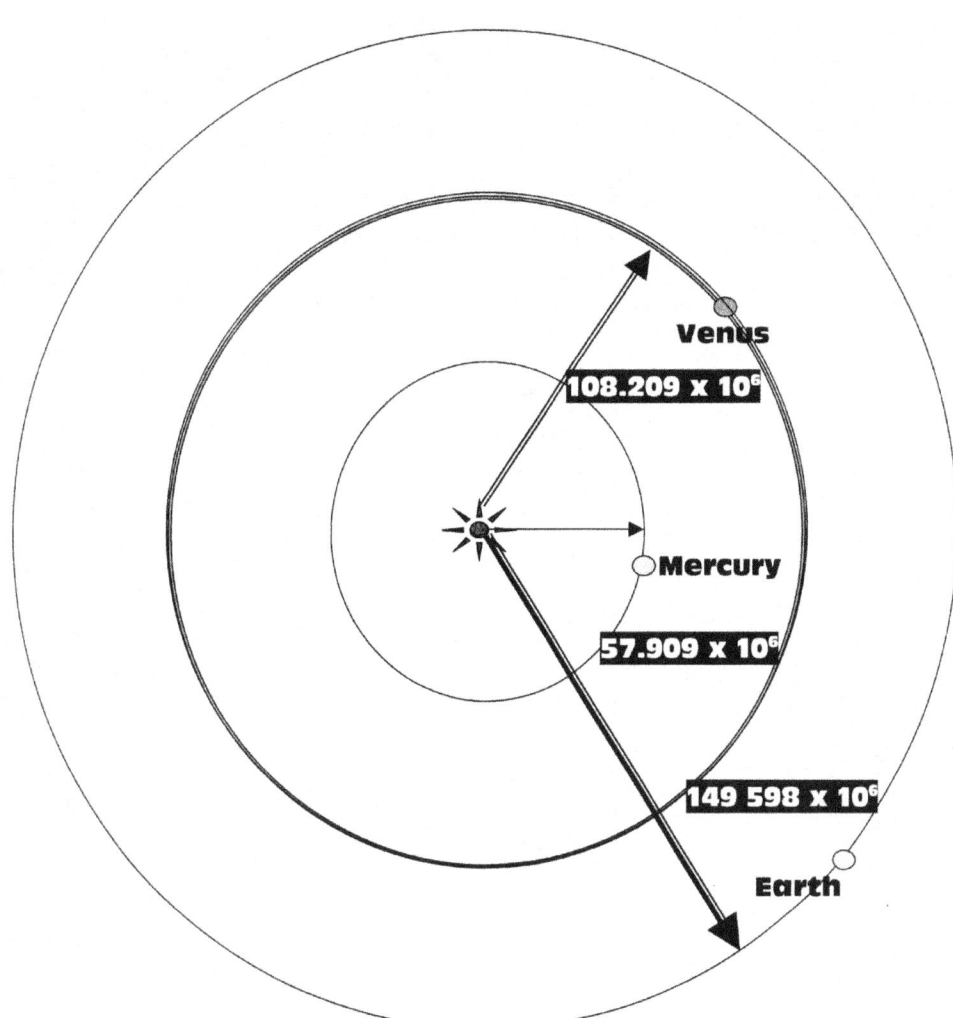

I feel sort of embarrassed when sharing the following stupidity with you, the reader. I include this presentation of how the solar system officially came about by implementing the mass theory as Newtonians present it and now you can share with me in my absolute stupidity. The following will underline the reasons as to why I am too stupid to comprehend Newton's genius and now one and all will see how shameful I should be for being so stupid as to not understand the simplest aspects of Newtonian brilliance.

In the centre there formed the Sun and in this area where the Sun is, is where the smallest area is. It is therefore conceivable that in this area gravity can be the strongest to collect the most dust that was a left over from a previous giant star. This part I still follow to my surprise. The Sun came into the position as the Sun compressed all the dust that later by the magic of gravity turned into hydrogen. Therefore, to this day the Sun is a bowl of gas with much pressure. The Sun had a much smaller area to collect its material and then could manage to collect so much more material in a much-reduced time. No one ever mentioned the time it took to do such collecting and no one ever mentions or speculates on the time when such collecting started. It ended some five thousand billion years ago and the job was done some five hundred billion years before the planets were done with their job. Then the planets finished some four thousand five hundred years ago.

A very funny part of the whole story is that the planets all had the same period during which they ended notwithstanding such enormous mass differences between some and others that had such huge areas to scramble material from. In the end they all worked as a team and had the lot finished in time honouring a date of completion and was on schedule for some grand opening.

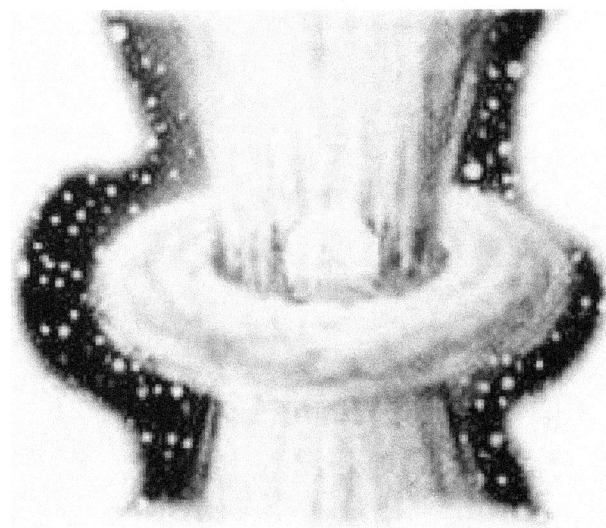

Just think how far the gravity fairies had to run to get Pluto's dust into Pluto. Think of the distance that those gravity fairies had to travel in order to go around the Sun in the huge circle that is required to do so, and then to collect the dust into one container. Not only that, but it collected so much dust that it made the dust into solid rock leaving the gas as solids. What a feat that was and still it took the gravity fairies that were responsible for Pluto just as long as it took the gravity fairies taxed with the task of forming Mercury to do the same job in the same time with billions of cubic meter of less space to cover. It seems that some fairies worked their buts off to get the job done in time and others strolled along as if they had all the time in the world ("solar system" would fit better) to get the job done.

We have the inner planet being Mercury orbiting in some sort of orbit at 57×10^6 km from the Sun.

To the outside of Mercury we find Venus orbiting at 108×10^6 km away from the Sun.

To the outside of Venus we find the Earth orbiting at 149×10^6 km away from the Sun.

To the outside of Earth we find Mars orbiting at 227×10^6 km away from the Sun. With this we can carry on for all the other planets but this much will be sufficient to explain my argument. So, your response will now be so what is new? This is the part that is above my inferior intellect: How did they get there? Remember they were rings of dust that by the magic of gravity collected to become a solid structure. We now have one solid structure claiming space at 58×10^6 km and another solid sphere claiming some space at 108×10^6 km and then one solid sphere at 149×10^6 km. We see a small ring with a small structure filling a small space going around a big area in a wide circle.

An Open Letter On Gravity Part 2

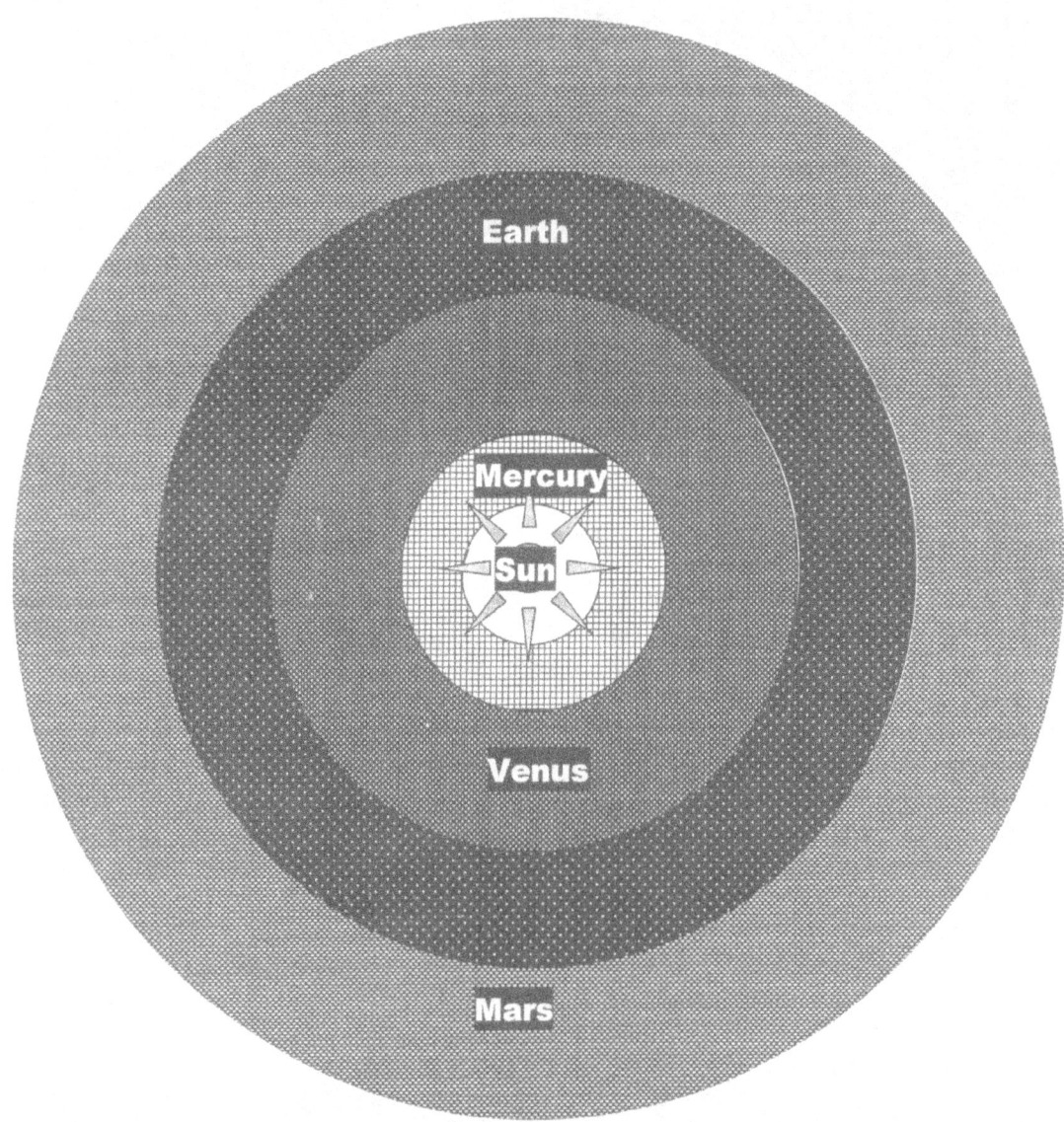

When regarding the official myth, Newtonian brilliance propagates a big circle of dust formed that eventually became the Sun with planets we now enjoy. If you are a scientist that only work with facts and totally reject any mythology, then enjoy the following facts and spot the mythology as you go along. This is a pure Newtonian fact or it is mythology, depending on your perception of what is Newtonian truth and what is scientific mythology. It falls in the class being just as Newtonian in truth as is the idea that it is mass that is provoking gravity to jerk other mass around. All the facts I now present is official as you may find and is taught throughout the world and is propagated as being more proven than the Biblical holy truth. The Newtonian gospel that I am about to represent is preached in every Newtonian classroom by every Newtonian in an official capacity wherever one might locate a Newtonian priest.

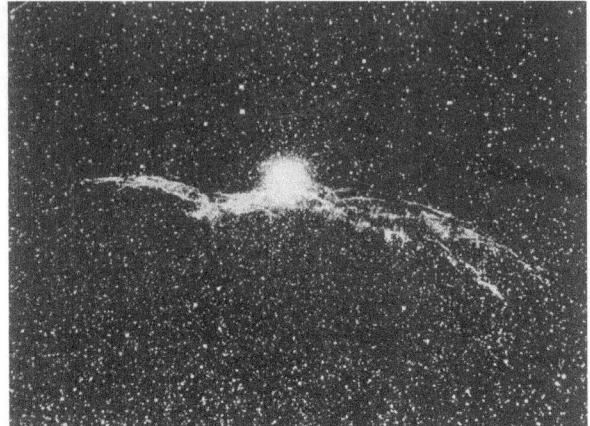

There was a centre ring of dust that accumulated five hundred billion years ago and this ring of dust became the Sun. It took most of all the dust in the centre to form the Sun...well that is if you will take the chance and believe the Newtonians. Then comes the real complicated part that I just haven't got enough brain matter to understand the true complexity of, or so the academics say, Newton's overall picture. Outside the centre ring there were rings forming from the dust that remained after all the dust in the centre settled as the Sun.

What the Sun did not use was the leftovers that the planets used and because the planets were so slow in getting finished they had to be content with what the Sun left them. The Sun had its fill and it was up to the planets to collect what the Sun had no use for. Every planet was in it for its own growth and I suppose no planet helped the other planet in the collecting of dust.

The rings accumulated and the rings progressively took five hundred million years to form planets. Notwithstanding the fact that Jupiter is so many times more than Pluto and Pluto is many times further than Jupiter, the lot came about at the same time about four thousand five hundred years ago. The lot formed the same time with Pluto on the outside taking just as long to assemble from the dust than did Mercury on the inside. The fact that Pluto had a ring billions upon billions of kilometres longer than the rest did not interfere with the completion date and the fact that the gas planets are so much bigger, still had no influence on the date the forming of the planets were completed. Also no Newtonian finds it odd that the smallest of the lot with the least gravity to show, formed the densest with the least gravity to show while the biggest could only manage to compress into gas. The inferior planets with the least mass and the least gravity had the ability to form hard solid structures with iron in the centre and with silicon outside the centre. How did such inferior gravity compress the inferior planet into a hard structure while the massive superior structures that truly sport a lot of gravity could only cope with forming some gas structure? These are nagging questions considering my inferior ability to grasp complex situations Newton explains so vividly. But with my inferior grasping of complex Newtonian issues there are more facts that totally exceed my ability to grasp.

You might share the opinions of the Newtonians in thinking I just do not have the mental requirement to understand that which I am now in this page sharing with you as something I do not understand, but when you get condemning and judgemental about me, then see what you understand about what I do not understand about Newtonian genius.

Gravity might represent magic because it is force related and all that but superstition

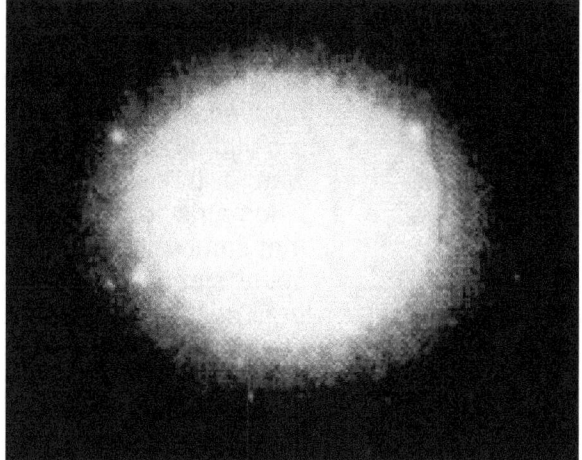

and make believe can take reality that far and then even the biggest improviser of the truth must call halt when the truth is circumvented beyond the limit. Newtonians might have a high regard for Newton and might be able to substitute the truth with some fairy tales but going over board and beyond the realistic then becomes part of the issue.

The picture that comes about in this argument is even exceeding the unlimited world of improvising for the truth as far as Newtonian improvising stretches. Let's refrain to the information that the picture holds. In the centre there is this huge Sun that has the gravity to maintain the solar system even to this day. Around this Sun were dust storms like we cannot envision. The dust was all around and the dust had to be in the form of a sphere since the dust was the remains of a super star that went out of gravity control. The dust floated around the centre Sun and the dust was evenly spread since there is no evidence of any form of an uneven Superstructure blow out.

The Sun had already formed and thus the Sun was in place. The Sun being what the Sun is was as strong in gravity as the Sun presently is. The Sun filled the centre as the

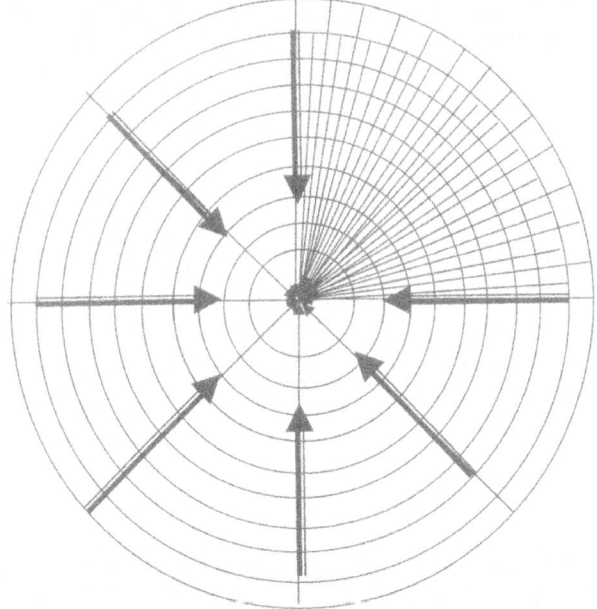

Sun still fills the centre and with all the mass the Sun has, it has the gravity ability to control the solar system as no dictator can control any country.

Every facet of gravity is targeting the centre and that applies today as much as it applied back then. We see that we have $k^{-1} = T^2 / a^3$, which Newton falsified to read as $T^2 = a^3$, but nevertheless it still proves that all gravity centres towards the Sun. How did some of the debris, dust and gas circumvent this gravitational pull and bypassed the Sun to land on the other side of the circle and combine with material on that side of the divide where the Sun then is the centre divide?

An Open Letter On Gravity Part 2   Page 509

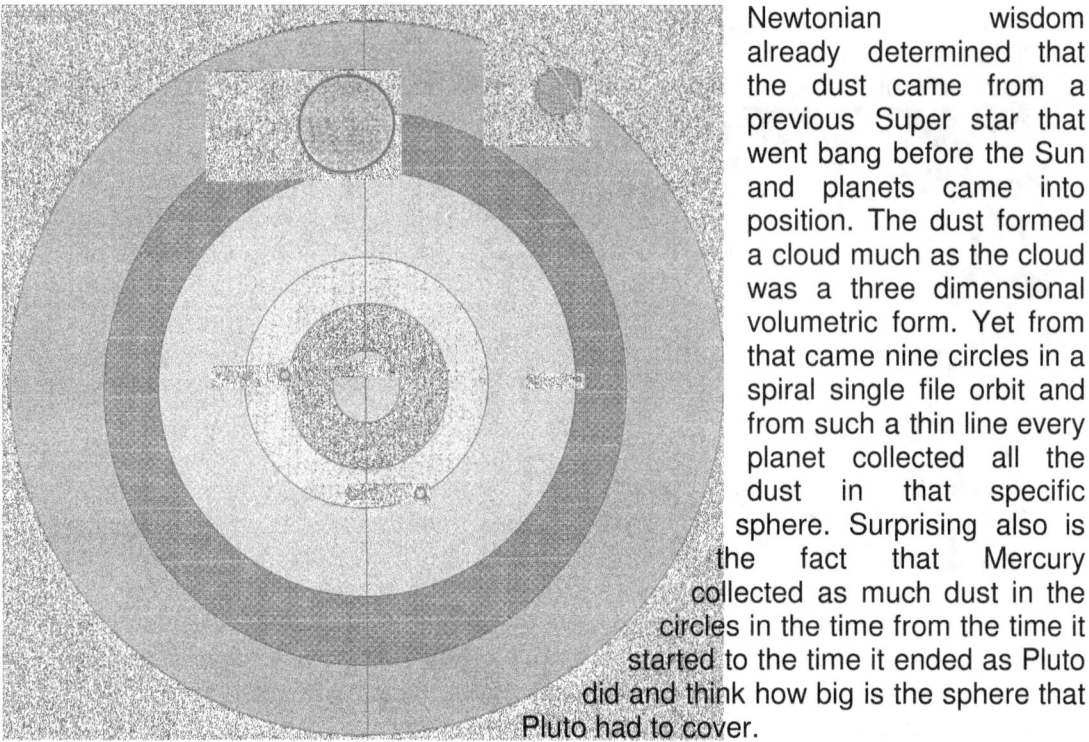

Newtonian wisdom already determined that the dust came from a previous Super star that went bang before the Sun and planets came into position. The dust formed a cloud much as the cloud was a three dimensional volumetric form. Yet from that came nine circles in a spiral single file orbit and from such a thin line every planet collected all the dust in that specific sphere. Surprising also is the fact that Mercury collected as much dust in the circles in the time from the time it started to the time it ended as Pluto did and think how big is the sphere that Pluto had to cover.

When considering the orbit distances and the orbit radii from the Sun centre to the planet centres measured in astronomical units, it requires quite a thought through explanation as to how the planets formed from debris spread across such a massive area. One has to think clearly to reach a conclusion as to how the dust assembled at one point in the specific. In the centre of every structure dust formed from that centre outwards to the debris edge of the planet. The around the was spread in a circle right that the dust orbit and there is no evidence more than other already favoured certain areas piece by piece from areas. The debris was collected How did the dust cross dust to become a solid as we have in planets.

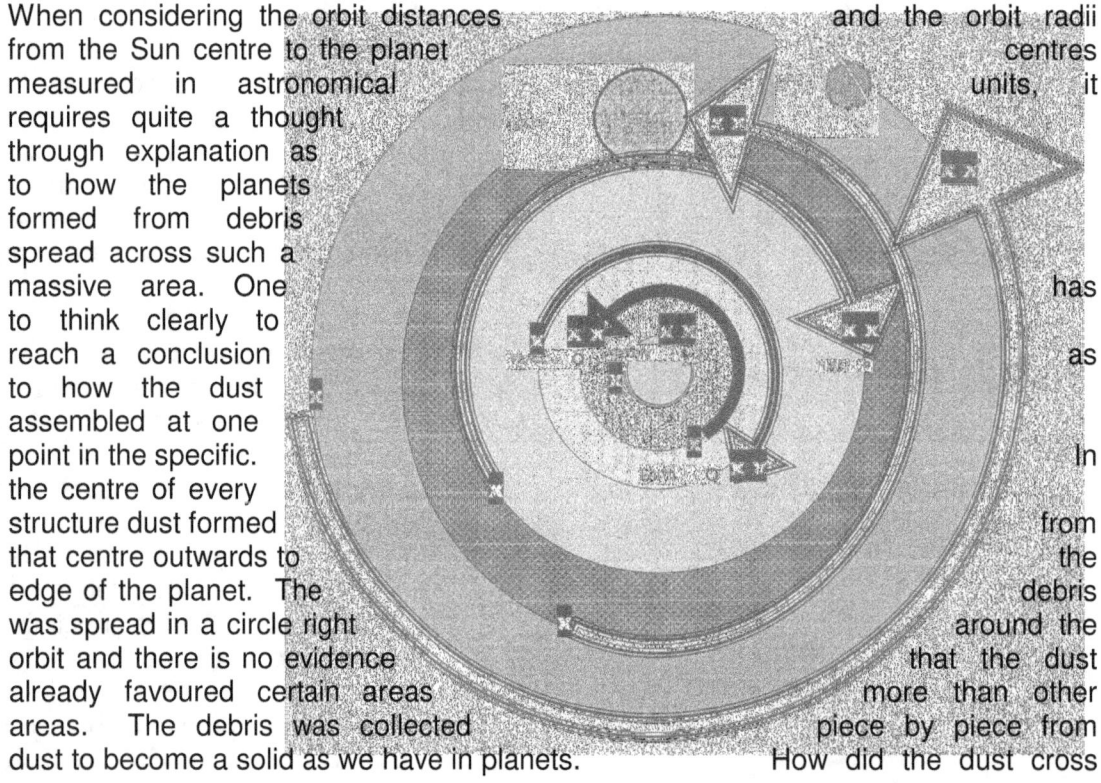

the divide that the Sun's gravity blocked? If the dust moved from one side to another side, how did it get there and not tangled with the gravity of the Sun?

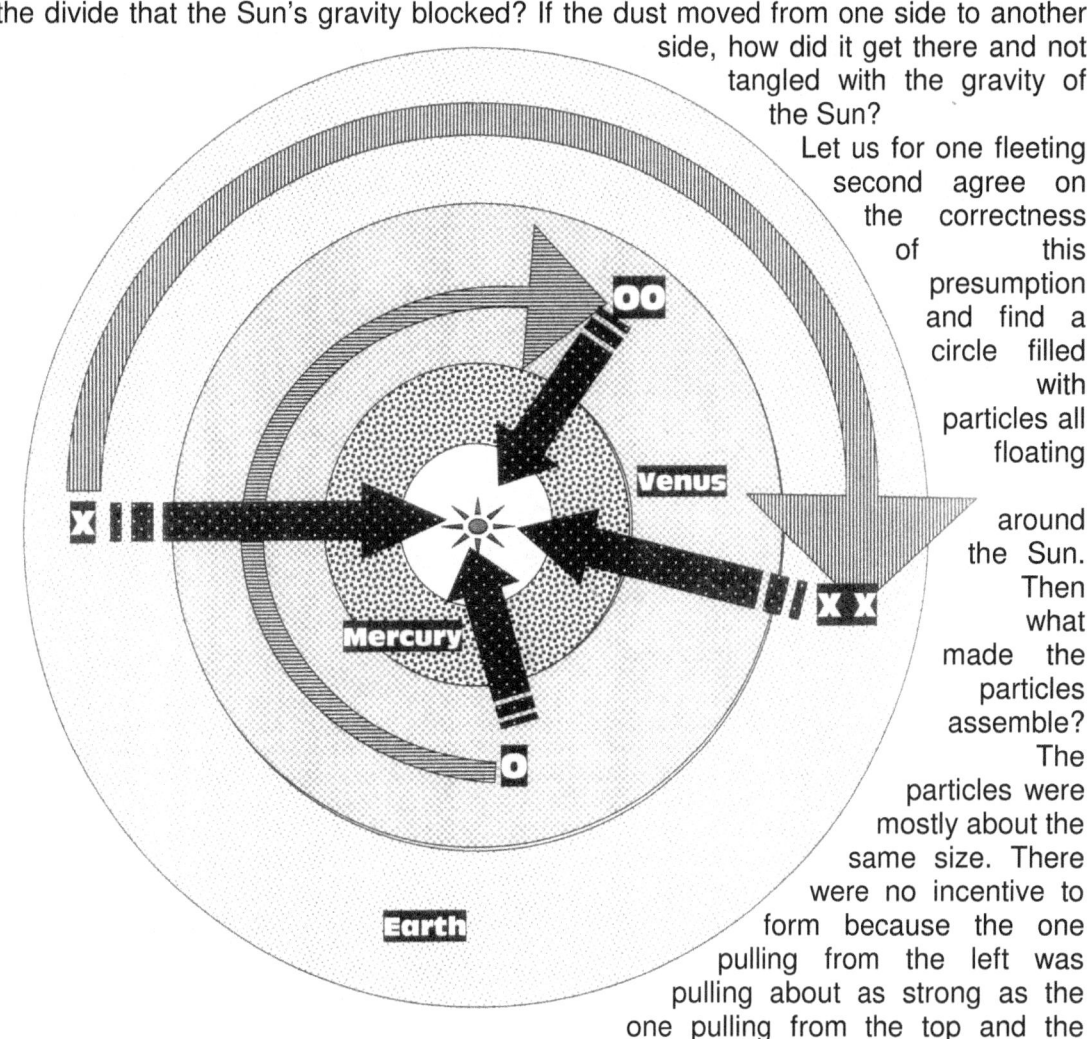

Let us for one fleeting second agree on the correctness of this presumption and find a circle filled with particles all floating around the Sun. Then what made the particles assemble? The particles were mostly about the same size. There were no incentive to form because the one pulling from the left was pulling about as strong as the one pulling from the top and the other pulling from right or from the bottom. If you think this is a riddle then hear my next question. Why did the particles join in one spot?

What made particles being in spot X move all the way around to form a unit with particles XX.

What motivated particles at the spot marked O to move to where the particles were that filled spot OO?

Why would the debris cross the circle to the other side of the Sun and unite with debris on that side? You might say that some moved faster than others did and in unequal speed they cross paths. That is not plausible because then we are stuck with more unresolved issues such as why would some move faster through nothing than others did because where they move there are no winds. So what would drive them to move in any case and why would whatever drives them, and then drive some faster and others slower? Digging deeper gets one dirtier because there are always more unresolved issues bringing frustration.

In the process where the planets gather material, there has to be one centre in the entire of the object that will contract and gather the material that will become the planet. It is not a centre line because the centre line is divided by time duration. The mass is at first very much evenly distributed throughout the entire debris field. Then we have in the centre of the Sun still dictating the orbiting pace of $k^{-1} = T^2 / a^3$ and if that is valid even today then how much more did the Sun's dictation not apply back then? But according to Newtonian myth all the debris located, compacted into planets that became much more dense than what the Sun still is today. X moving to XX became one Unit and we don't have debris still floating around in a circle that was the left over on the assembly line. Why would the dust particles on the one side of the forming structure move around in a circle to join the other particles that were completely on the other side of the forming structure? What convinced a particle with less mass than a grain of sand to move across the circle to the very other side? It had to move to the other side because all the grains finally formed a unit that is today the planet.

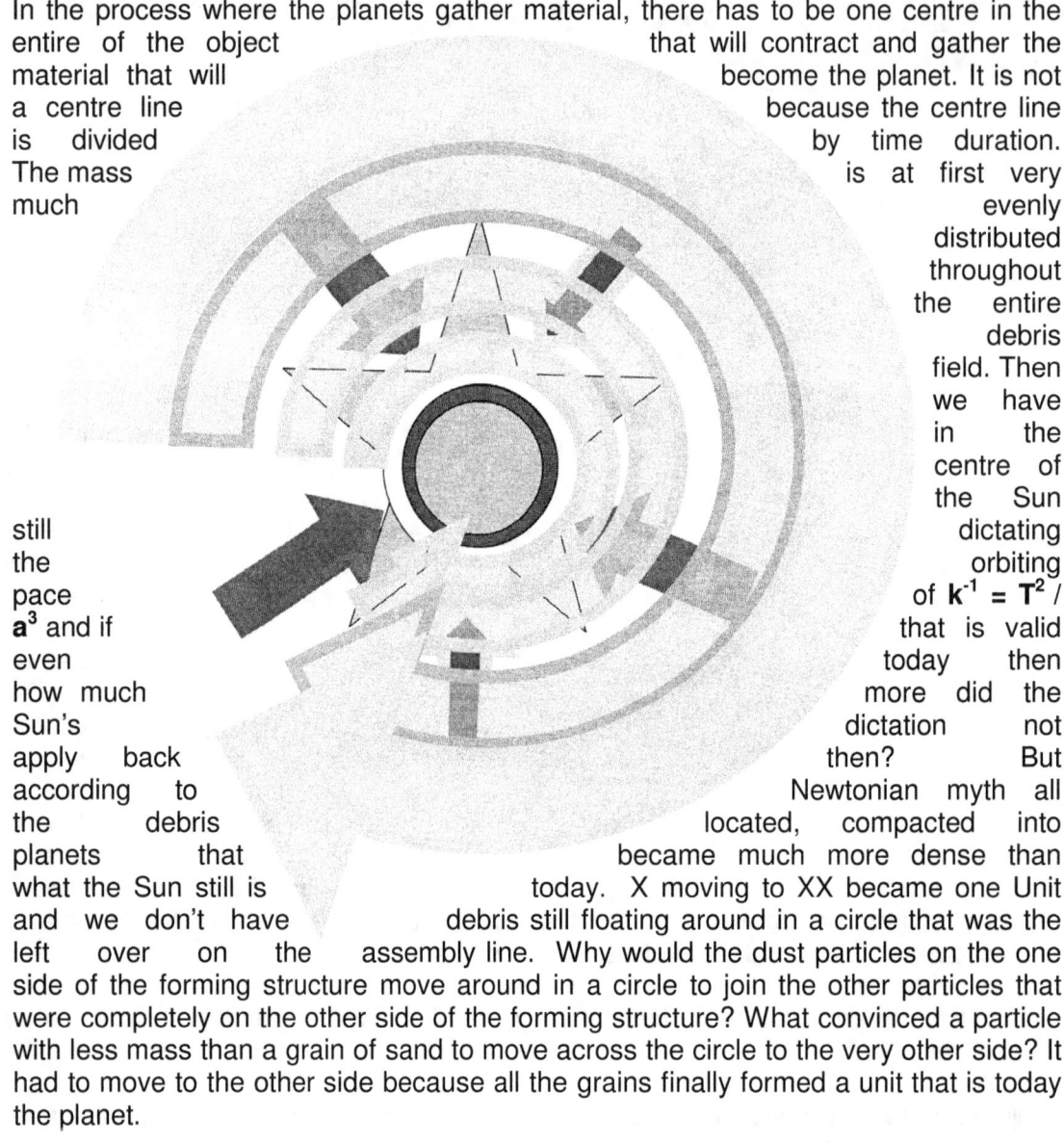

By the Sun filling the centre and having the biggest mass and having this enormous all ready formed mass (the Sun is 5X 10⁹ years old while <u>all</u> the planets are 4.5 X10⁹ years old) the centre's mass in the Sun was splitting the positional formation and the layout before assembling the mass of the debris into two opposing locations no matter where the positions were. From the one location all the debris has to move to another location on the other side of the radius and splitting the diameter into two radii, it places all debris where the debris has to move in a band around the Sun to avoid the gravity that the Sun does inflict on all particles falling into its gravity range.

How did the allocated debris find a way either around the Sun or circumventing the Sun? Please hold in mind that we are talking about a nine-layer sphere filled with star stuffing scattered randomly over an area going from the Sun out all the way to Pluto and even far beyond. The debris formed as a sphere with nine layers (that is if the debris did result in forming nine planets) and if there were not nine layers, why then did nine planets plus an assortment of debris form? I didn't cock-up this tale. I am just the narrator coming up with questions; so don't start questioning my sanity in this affair. This is Newtonian science as bright as it comes. First with the input of some inexplicable reason (and that must be the force that we can only refer to as the magic of gravity) these nine layers drew flat forming a line, which the planets used to move around.

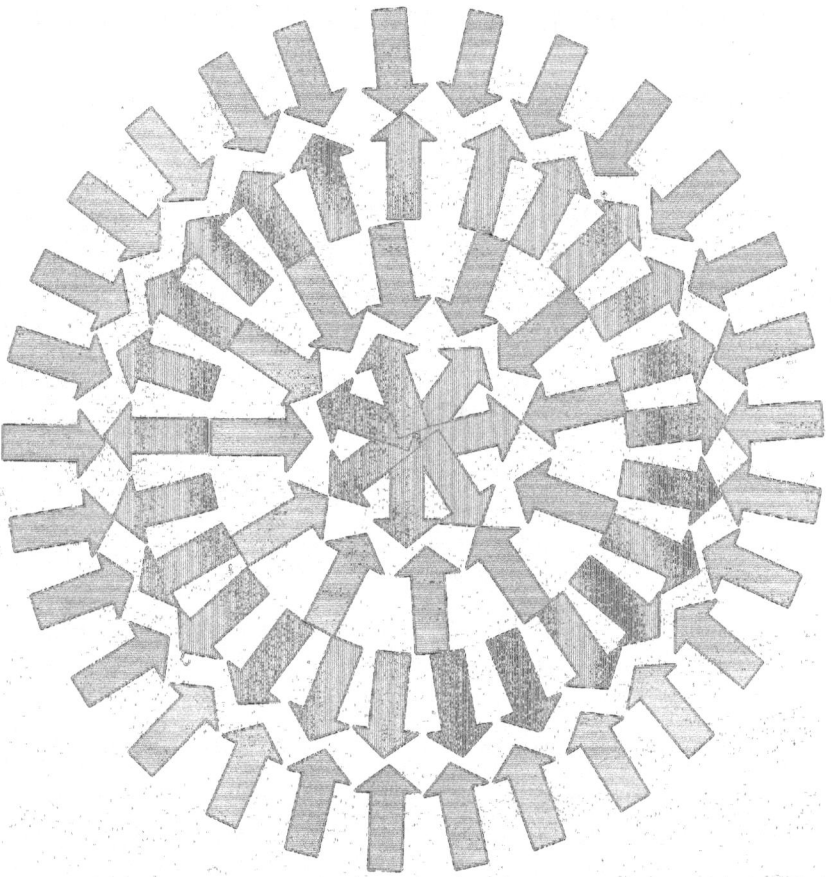

The sphere that held the debris became a circle, as the sphere had to draw flat in order to become the lines that the futuristic planets were going to follow when the magic of gravity assembled all the debris in nine compartments where every compartment became a future planet.

When regarding the official myth Newtonian brilliance propagates a big circle of dust formed that eventually became the Sun with planets we now enjoy. If you are a scientist that only works with facts and totally reject any mythology then enjoy the following facts and spot the mythology as you go along. This is as pure Newtonian Fact or it is mythology depending on your perception of what Newtonian truth is and what scientific mythology is. It falls in the class being just as Newtonian in truth as is the idea that it is mass that is provoking gravity to jerk other mass around. All the facts I now present is official as you may find and is taught throughout the world and is propagated as being more proven than the Biblical holy truth. The Newtonian gospel that I am about to present is preached in every Newtonian classroom by every Newtonian in an official capacity wherever one might locate a Newtonian priest.

There was a centre ring of dust that accumulated five hundred billion years ago and this ring of dust became the Sun. It took most of all the dust in the centre to form the Sun...well that is if you will take the chance and believe the Newtonians. The real complicated part which is which that I just haven't got enough brain matter to understand what the true complexity is thereof, or so the academics say, is as follows: Outside the centre ring there were rings forming from the dust that remained after all the dust in the centre settled as the Sun. The rings accumulated and the rings progressively took five hundred million years to form planets. Notwithstanding the fact that Jupiter has so many times more mass than Pluto and Pluto is many times further than Jupiter. Nevertheless the lot came about at the same time about four thousand five hundred million years ago.

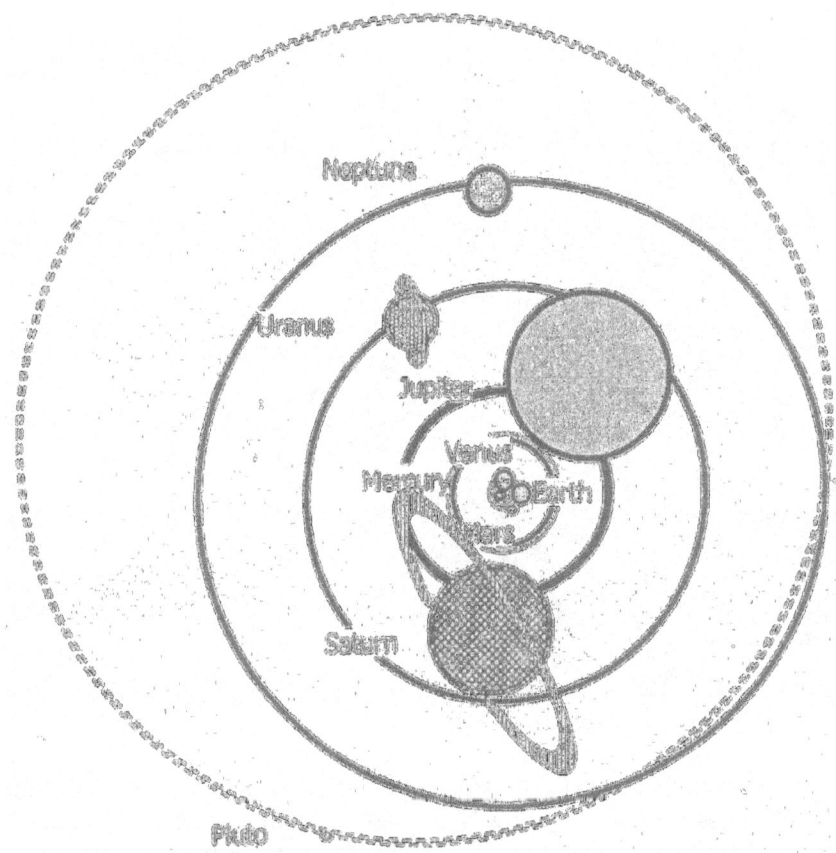

The lot formed the same time with Pluto on the outside taking just as long to assemble from the dust than did Mercury on the inside. The fact that Pluto had a ring billions upon billions upon trillions of kilometres longer and bigger than the rest did not interfere with the completion date and the fact that the gas planets are so much bigger still had no influence on the date they were finished as planets. Also no Newtonian find it odd that the smallest of the lot with the least gravity to show formed the densest with the least mass that were producing gravity to compress the available dust into rock while the biggest could only manage to compress what they assembled into gas.

The inferior planets with the least mass and the least gravity had the ability to form hard solid structures with iron in the centre and with silicon outside the centre. How did such inferior gravity compress the inferior planet into a hard structure while the mass of superior structures that truly sport a lot of gravity could only cope with forming some gas structure? That is some of the Newtonian wisdom that goes above my ability to understand. How did the dust get that hard? These are nagging questions considering my inferior ability to grasp complex situations Newton explains so vividly. But wait because in spite of my inferior grasping of complex Newtonian issues there are still more facts that totally exceed my ability to grasp.

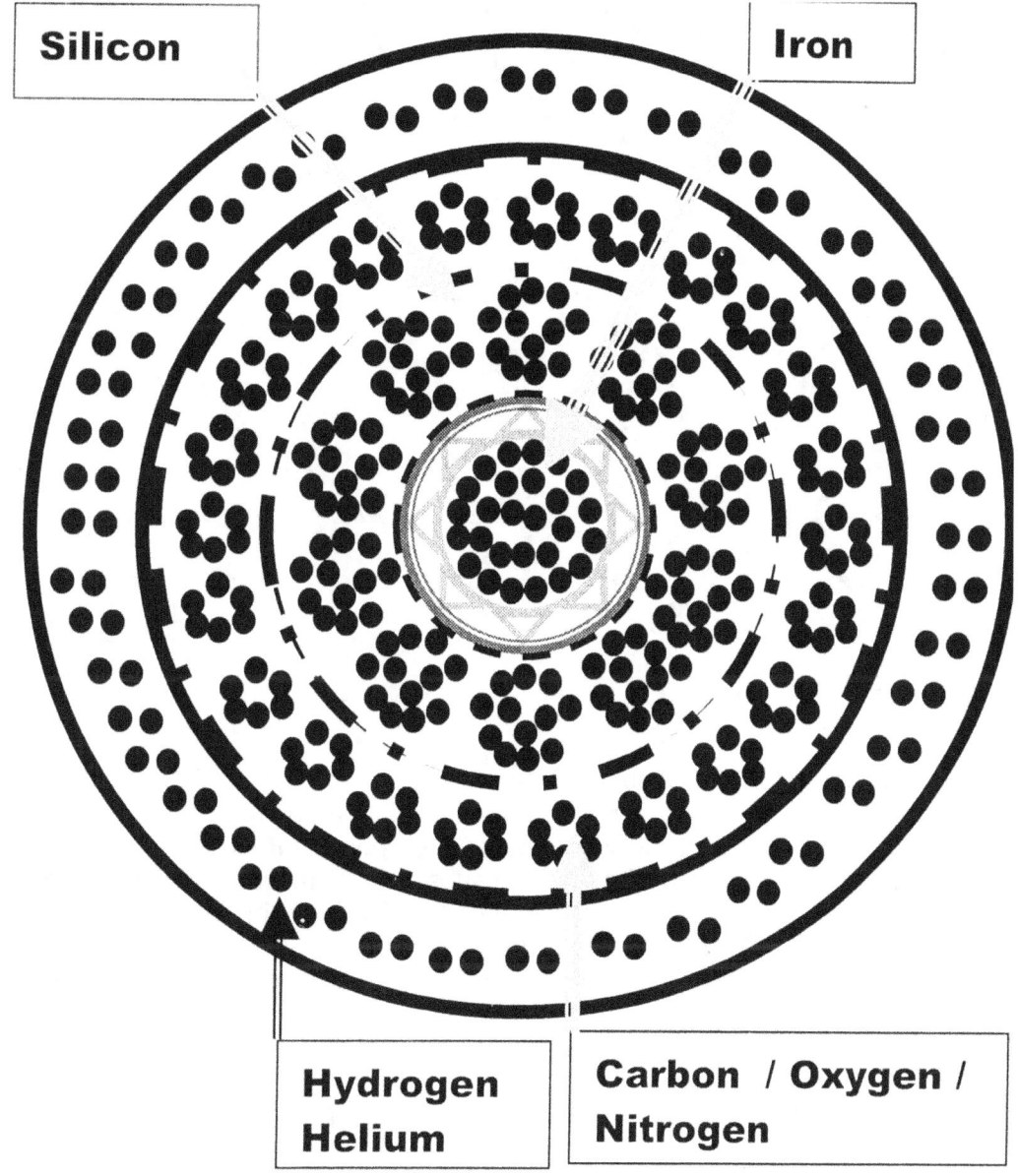

A problem Newtonian myth do not address either is the placing of the most dense material on the inside of the planets and having the lighter elements on the outside in the air. Why would the heavy material be inside while the thrust coming from the spin should thrust the heavy material outwards?

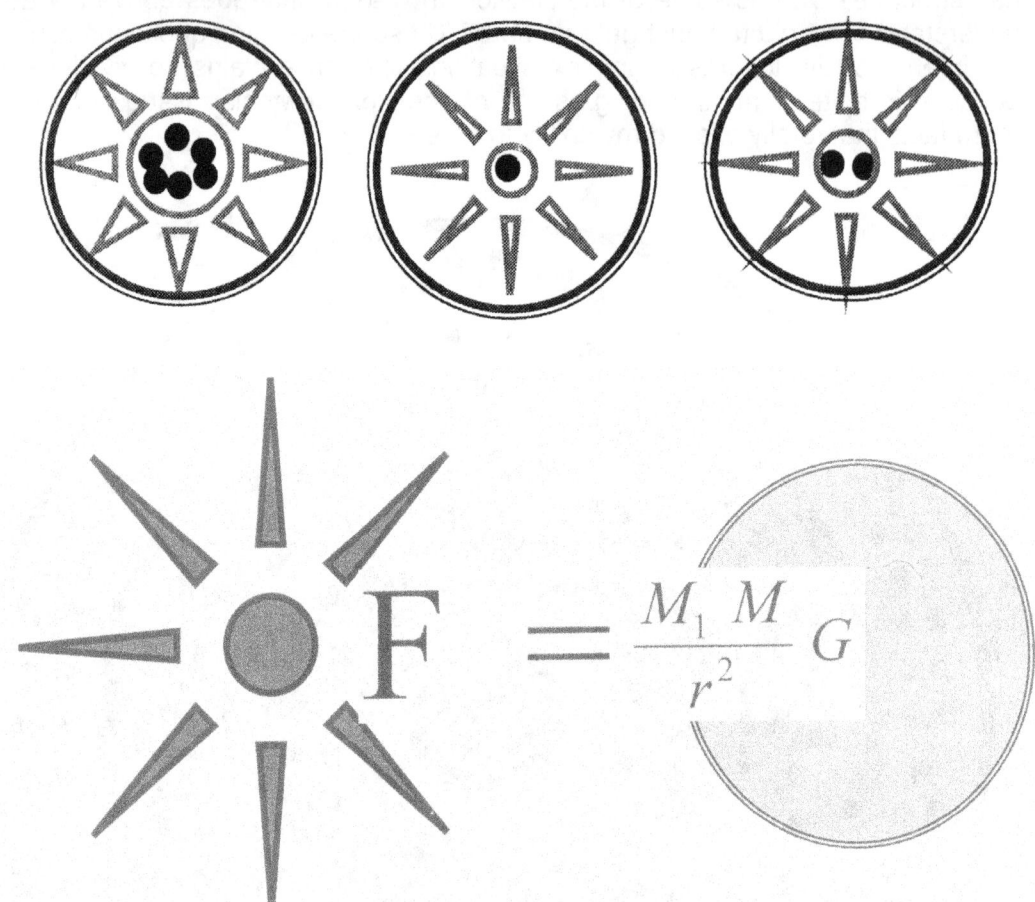

Looking at the full Newton formula, the formula does not suggest in any way that the mass will entice the gravity to pull material around a bend or in a circle formation. The pulling is directed in a straight line between two points offering mass. The gravity directs a flow of a space filled with material substance to move directly from one point towards another point by using the diminishing of a straight line between the two objects in mass. The shortest line between any two points is a straight line, and since the gravity will induce a maximum pulling, it will move to destroy the line as easily as possible.

$$F = \frac{M_1 M}{r^2} G$$

The gravity has the purpose to combine the mass that the radius parts. To do that, the gravity will employ a means with the most results to remove the radius by the square as soon as possible. What can it be about Newton that I don't understand because of my weak insight into Newton's brilliance? What about the formula Newton provided is it that I can't follow? Do you realize that the academics see me so incompetent that they never tried to show me what it is in or about the Newton formula that I am too stupid to see. Not one academic thus far showed me what I am unable to see…I have a notion it is because they are unable to see what I am able to see…

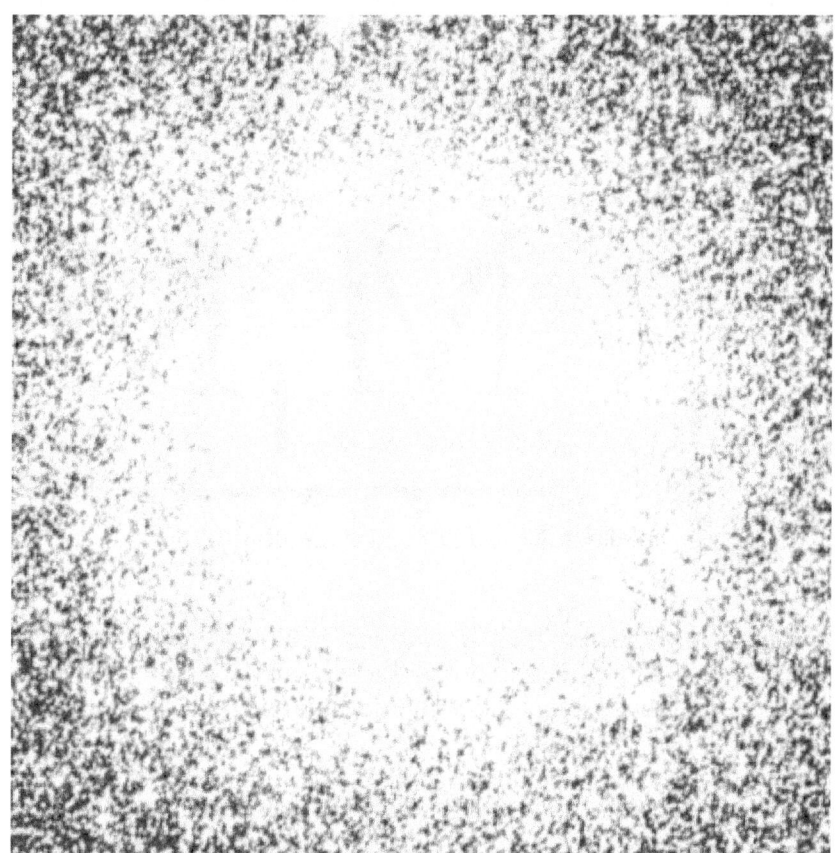

What motivated the moving of dust from the side to form a line? Yes, eventually when the total form was assembled, there might have been one superior structure that accumulated the last bit of dust as the superior structure moved past but for the first ninety nine percent of the time, there was no true incentive to get the particles to move from the side to a line that formed in the centre. Then what made the particles decide the one lot was going to become Venus and just the next group planned on forming the Earth or Mercury for those more closely to the Sun? Why would the smallest of the planets all be the densest while the biggest became the gas structures? They say I am unable to comprehend Newton and they are so correct.

There are so many nagging questions that Newtonian brilliance fail to address. Why did the Moon gather from dust so close to the gathering area of the Earth? Why did the moons of all the planets become the dense structures while achieving this incredible feat with such small mass evoking such poor gravity? Why did the Sun not collect the lot since in the normal formula that applies, the dust had to pass by the Sun on its way to where the location is where the dust then formed planets on the other side of the circle?

What prompted Pluto to become one of almost a double star and not reserve all the accumulation into one unit? Why have the unit, so far from the centre and yet have the dust to form such a firmly compacted unit in two separate structures being so closely together? Why did a giant such as Jupiter allow fragments that we now call moons, go flying about the giant planet in an area so small where the dust assembled as spheres of insignificance, while Jupiter collected the overwhelming dust in the area to eventually assemble the planet Jupiter? What made the planets choose the bands that they circle in at the specific points they chose to assemble as planets? There are a million unanswered questions that I do not find answers to because I have too a poor mentality to understand Newton. It seems the Newtonians never got around to realise these questions, which go unanswered.

Let us investigate what it is in the Newtonian high priest custodian mythology that I am not intellectually suitable to understand.

$$G \frac{M_1 M_2}{}$$

We understand from Newton that gravity is on the increase because the radius is according to his well-informed opinion on the demise. The demise of the radius will set a natural increase of mass into action where the gravity will be on the rise because the status of the mass is growing in prominence.

The radius is on the decline. This idea founded physics for centuries in the past and no one ever dared to question the feasibility of the statement.

$$F = \frac{}{r^2}$$

This meant that the Universe was coming to a conclusion. All that has parted somehow has drawn on the mass that developed and the "gravity" was pulling the cosmos into a "grave". That is what I can conclude from the name but it might be presumptuous of me to conclude Newton's thinking in such a manner. This brought serenity to the cosmos. The cosmos had a purpose, a motive, and a function and even moreover it has a conclusion. At a point the book will close and the curtain will draw. It is where the cosmos will end in a grave and all will conclude into a finality that will end that, which started.

The Sun was pulling the Earth that was pulling the Moon and one day we can rest assured that the Sun will absorb the Earth that would have absorbed the Moon by then. First the inner planets have to give and finally the outer planets will have to give to this force of magic being inspired by the mass that Newton saw everything must have because everything had to have mass. Newton said everything had to have mass because if everything did not have mass everything did not have gravity and with everything not having gravity everything could not be pulling everything else to the final conclusion that Newton predicted. The most important rule in physics is that although God can be wrong, being wrong is just not possible if you are Newton. This statement I put to all those academics that have on their answering machines the Lord's Prayer recited and verses from the Bible but they would rather face criticism coming to God than allow any criticism about Newton.

The Universe contract by the mass it has with the gravity it produced to find the measurable force that keeps the lot spinning and contracting.

Then came an absolute disaster by the name of **Edwin Powell Hubble, the man that will never deserve the Nobel prize, because he screwed everything up that Newtonians regard as sacramental and holier than the Bible.**

Hubble, Edwin Powell

Hubble, Edwin Powell, 1889–1953, American astronomer, b. Marshfield, Mo. He did research (1914–17) at Yerkes Observatory, and joined (1919) the staff of Mt. Wilson Observatory, Pasadena, California, of which he became director. Building on V. M. Slipher's discovery that galaxies had strong shifts to the red end of their spectra, Hubble used the Cepheids in nearby galaxies to demonstrate that they lie far beyond the Milky Way. Because of an incorrect understanding of the Cepheids, this distance was vastly increased years later.

Hubble's law is the statement in astronomy that galaxies move away from each other, and that the velocity with which they recede is proportional to their distance. It leads to the picture of an expanding universe and, by extrapolating back in time, to the Big Bang theory.

In 1929 Edwin Hubble first formulated the law that was named after him after he saw an expanding Universe in his telescope's eyepiece. Hubble compared the distances to nearby galaxies to their redshift, found a linear relationship, and interpreted the redshift as caused by the receding velocity. His estimate of the proportionality constant, now known as **Hubble's constant**, was however off by a factor of about 10. Furthermore, if one takes Hubble's original observations and then uses the most accurate distances and velocities currently known, one ends up with a random scatter plot with no discernable relationship between redshift and velocity. Nevertheless, the relationship was confirmed by observations after Hubble.

The law can be stated as follows:

An Open Letter On Gravity Part 2

$v = H_0 D$ where v is the receding velocity of a galaxy due to the expansion of the universe (typically measured in km/sec), H_0 is Hubble's constant, and D is the current distance to the galaxy (measured in mega parsec Mpc).

Newton produced the idea that mass is pulling mass by producing gravity. The comet tells otherwise.

$$F = \frac{M_1 M}{r^2} G$$

On this accepted formula the entirety of physics rests. I have been on a conquest to fight this lie by contacting academics near and far. At first I thought them to be nice and honest men in search of the truth and while having nothing more to go by, they use this hoax as an intermediate faze until the truth arrives.

Then I wrote another letter in the form of a book and tried to get it published. Not one letter from one publisher came back indicating any interest. When a man writes to a publisher informing the publisher that that person can prove the Bible correct as far as the origin of the Universe goes by using of pure mathematical physics, it has to be a seller. Then some publisher somewhere would show some interest because the only thing I insisted on is to sign a declaration of confidentiality about my work. That is not too much to ask...and not one response from eighty publishers. I did it and the book has the Title of <u>*The Seven Days Of Creation*</u> where I show how the Universe formed in seven cosmic days or seven cosmic periods. All I had to do was to get rid of the misconception that gravity is connected to mass. All I had to do was to show what gravity is and how gravity comes into play. I wrote a book where I show the very point where the Universe started because I found the spot where the centre of the Universe is. At the point where the centre of the Universe is must be the point where the Universe started since the Big Bang was one long process of taking what there is in the Universe away from such a centre point. I include one of the responses I received and the overwhelming majority of publishers did not even reply.

On this accepted formula the entirety of physics rests. I have been on a conquest to fight this lie by contacting academics near and far. At first I thought them to be nice and honest men in search of the truth and while having nothing more to go by, they use this hoax as an intermediate faze until the truth arrives.

Mr Peet Schutte
PO Box

☎ (012) 429-3316
Fax: (012) 429-3449

7 September 2005

Dear Mr Schutte

Manuscript: A cosmic birth...

The Senate Publications Committee of UNISA recently considered your application and regrettably decided not to accept it for publication since the topic does not fall within the purview of this committee. However, this should not be seen as any reflection on the quality of your work. We recommend that you consider approaching a publisher with a more general list of publications who would be best suited to the marketing of your book.

I am sorry that I am not the bearer of more positive news.

With best wishes.
Yours sincerely

Prof Abebe Zegeye
Director: Unisa Press
E-mail: boshosm@unisa.ac.za

Then I wrote another letter in the form of a book and tried to get it published. Not one letter from one publisher came back indicating any interest. When a man writes to a publisher informing the publisher that that person can prove the Bible correct as far as the origin of the Universe goes by using of pure mathematical physics, it has to be a seller. Then some publisher somewhere would show some interest because the only thing I insisted on is to sign a declaration of confidentiality about my work. That is not too much to ask...and not one response from eighty publishers. I did it and the book has the Title of *The Seven Days Of Creation* where I show how the Universe formed in seven cosmic days or seven cosmic periods. All I had to do was to get rid of the misconception that gravity is connected to mass. All I had to do was to show what gravity is and how gravity comes into play. I wrote a book where I show the very point where the Universe started because I found the spot where the centre of the Universe is. At the point where the centre of the Universe is must be the point where the Universe started since the Big Bang was one long process of taking what there is in the Universe away from such a centre point. I include one of the responses I received and the overwhelming majority of publishers did not even reply.

When I explain to them how gravity works, the nice and distinguished honourable gentleman will embrace the first opportunity to try and find a correction for the gross error Newton introduced. I though they were men looking passionately for answers about inconsistencies in the science of physics. In my mind they were all-righteous men that had to harbour the lie in order to look for the truth and at the first opportunity they will at least investigate any suggestion that might pave the way out of the swamp of incorrectness they had to endure all this time. It is obvious that mass has nothing to do with gravity and that there is no tug of war going on between planets. It is obvious that mass is not drawing planets closer because then why will planets escape from certain capture with vengeance? Was I wrong about everything? For six years I took the blame for their lack of interest. I thought I did not come across correctly and wrote a book. Then I thought I did not explain what I tried to convey correctly. Then I wrote another book. Then I had an idea about proving what was never proven before and I wrote another book. Still the silence that came from them was fitting a morgue. I wrote a letter in the form of book. No interest came from them. I got onto the Internet and started contacting academics abroad, in all one thousand five hundred where some was even behind the so-called iron curtain. If anything, their interest was frozen into silence. You may ask why nobody wants to publish any of my work if it is all that revealing? Here comes another falsified myth academics portray as the truth, while it is openly one of the biggest scams there has been, up to the present. Using Newton's formula that inspired all forms of physics we are in a Universe that is about to implode. All mass are pulling all mass onto all mass by the force that mass forms gravity. This was the consensus since the days of Newton and all Newtonian science still drives on this concept. I can prove what causes gravity. I prove what gravity is. I take the reader to the centre of the Universe. I explain time and what time is. I show how gravity functions and I explain the purpose there is in having mass. This I do even later on in this letter you are reading. But as long as those academics that have criminal intensions to depress my work by killing me as to conspire and underwrite such senseless nonsense as telling mass pulls with gravity while the comet escapes into the outer space as if Newton was lying, then they are the gangsters conspiring to defraud the world by falsifying the truth.

Then along came a man that changed the world of science as no one ever did and this remark includes our genius Newton. He had the world of science wheel out of control and brought about confidence tricksters to the world of science as members of the world of science ran to cover and protect the biggest scam the world ever saw. **Edwin Powell Hubble** (November 20, 1889 – September 28, 1953) was an astronomer, noted for his discovery of galaxies beyond the Milky Way and the redshift. Hubble was one of the first to argue that the red shift of distant galaxies is due to the Doppler effect induced by the expansion of the universe. He was one of the leading astronomers of modern times and laid down the foundation upon which physical cosmology now rests. Hubble was born to an insurance executive in Marshfield, Missouri and moved to Wheaton, Illinois in 1889. In his younger days, he was noted more for his athletic abilities rather than his intellectual genius, although he did earn good grades in every subject, except for spelling. He won seven first places[1] and a third placing in a single high school meet in 1906. That year he also set a state record for high jump in Illinois. His studies at the University of Chicago concentrated on mathematics and astronomy, which led to a B.Sc. degree in 1910. He spent the next three years as one of Oxford's first Rhodes Scholars, where he studied in the field of law and received the M.A. degree, after which he returned to the United States as a high school teacher and a basketball coach in New Albany, Indiana. He returned from war to astronomy at the University of Chicago, where he earned a Ph.D. in 1917. In 1919 Hubble was offered a staff position as the director of Carnegie Institution's Mount Wilson Observatory, near Pasadena, California, where he remained until his death. Hubble spent much of the later part of his career attempting to have astronomy considered an area of physics, instead of being its own science. He did this largely so that the Nobel Prize Committee could recognize astronomers for their valuable contributions to astrophysics. This campaign was long unsuccessful and unfortunately Hubble's great achievements would remain not rewarded. He died of a heart attack on September 28, 1953, in San Marino, California. Even though the Nobel Prize Committee decided that astronomy should fall under the description of physics, unfortunately this occurred in 1953 - but Hubble died Sept. 28, 1953, before he could ever receive this prize, or even informed that he should receive it (his wife was informed after his death), to this time the Nobel Prize is never awarded posthumously.

Why all the praise for a man that never received the Nobel Prize one may ask? Hubble gazed through his telescope and saw to his surprise that the region forming outer space was not in decline. It was not contracting it was not reducing. This brought about the realization that Newton's Universe was not collapsing and with that Newton's ideas should have been collapsing but it did not. What this did to human kind was to unleash the biggest and unprecedented scam ever devised to cover the biggest scam any criminal ever devised. Why would I refer to academics as criminals? In the face of defeat Newtonians stuck to their opinions be it to protect previous corruption or present corruption, they protected corruption. Newton said that the Universe was going the other way than what Hubble found was true and because Newton can't be wrong in saying the Universe is going the wrong way, so it then just has to be that the incorrectness must be on the side of the Universe and God directing the Universe in the wrong way. Rather admit that it is the Universe that is wrong than admitting to the impossible such as Newton being wrong…that is unthinkable…to even think of Newton being wrong is wrong …no rather think of it as the more probable possibility, which is that it is the Universe that is going the wrong way and which will implicate that God is wrong rather than making the disgraceful, unthinkable human error to think of Newton and incorrectness in the same thought!

When one goes about and teaches something that person knows is untrue and is never proved, such a person is a liar. When a person later gets confronted by proof that the ideas the person is teaching is disproved by merely looking through the eye piece of a telescope a honest and trustworthy person would come clean and admit to a mistake and go in search of the truth. However, when a person of disrepute is in a situation such as the one I portray, then the person with no morals and no conscience would bring forward more despicable covering up of the truth and more diverting from the truth in order to cover the criminality already in place. This is what happened. This is what happened when Hubble found the Universe expanding at a rate that was measurable. The academics conspired in committing a fraudulent cover-up of a previously fraudulent hoax called the critical density theory. I am getting to that part in a moment or two. First let us finish with Hubble and his ultimate prime work.

Everyone in science was sworn to believe that the Universe was coming together as Newton said. The Universe was collapsing under its own mass and the mass is dictating the force that does the pulling, which acts as the gravity. The main thing is that everything was coming together at the pace gravity determined and this was under the orders the mass commanded. Then Hubble changed everything and confused everyone. He proved Newton wrong!

Hubble saw the Universe growing outwards while Newtonians taught all that were taught about a contracting Universe. Mass is pulling mass and the force that the pulling mass unleashes goes by the name of gravity which the founder of mass and of gravity created. Let me be very clear about one thing and that is that Newton did not invent the notion of gravity. Newton devised a name. There is another person that is much more valuable to the world of science that has that honour. There is another person that established gravity but this person did establish gravity as it truly is, while Newton raped his ideas, plundered his work and falsified his facts. Anyone that sees me as being crude towards Newton should disprove my statement. Prove how Newton corrected the work of Kepler and then prove me wrong when I point out the corruption Newton committed in order to vandalize the work of Kepler as to gain validity to his (Newton's) personal work. If that which Newton did does not come down to despicable fraud then there is no fraud. If academics are not maliciously participating in fraud with criminal intent and do not deliberately continue to falsify facts in order to uphold Newton's fraud, then they must prove Newton's fraud instead of just promoting Newton.

After Hubble, science reached a point where they upheld a contracting Universe expanding like never seen before.

I have another unanswered question concerning elements. Newtonian wisdom tell of hydrogen coming under pressure to form helium coming under pressure to form more compact elements and this process called fusion brought about that gravity pushed elements into denser and more massive atoms. There are in the centre iron $_{56}$ being at the best and the densest elements in a star. If it is true that the gravity is so strong in the star that it can compress say even two silicon atoms into one iron atom, it still does not explain how gold and lead and all the super metals came into being. Where did those come from and where were they in order to fill the centre of something as weak as the Earth? How did those elements form and how did Earth manage to gather the elements?

Everyone was sharing the Newtonian vision of a contracting Universe where the lot would one day again come together and Creation will end where Creation started some time ago. The Universe has mass that is pulling mass towards one another and we are in the centre of an ever shrinking Universe. That is what the lot of us can see… we are forming the centre of the ever contracting cosmos where every Newtonian can vividly see with his or her eyes through any telescope that all Newtonian-minded scientists are sharing the centre stage of the ever collapsing Universe. The Universe is about to end where all mass contracts into one huge lump of material.

Then along came a man that had a good look at the Universe. He looked at the night sky and came to the conclusion the Universe he was looking at was not shrinking but was expanding. Anyone that would look through his eyepiece had to draw the same conclusion because it was as clear as daylight that the Universe was not shrinking even in the least. In fact, the lot was growing apart. In some cases he saw the lot was racing apart. The Universe was growing by miles a second and not shrinking into nothing. The main discoverer had a name, a position of seniority, a title with authority, which prevented others from pushing his findings and his opinion aside. The man was E.P. Hubble, the Master of the telescope and through his mastery with the telescope anyone could come and view that the Universe was expanding at a measurable rate. He saw the expansion was even most rapid and very deliberate.

This unleashed a problem the world had no name for. Everything known to science was at that point devastatingly unknown to science. The world was expanding and not contracting, which made the Universe quite wrong. It is impossible to have any vision about Newton being wrong. Newton could never be wrong because Newton was never proven wrong yet…so if the Universe is out of step with science, then science will correct such an abnormality by finding a way to defraud science and postpone the correcting that the Universe had to comply with, since the Universe owed the Master Newton some apology.

Did the Universe not know that he whom never can be wrong is in name Isaac Newton! Decisive action was needed. At this point I cannot believe that the most brilliant minds were so naïve and therefore I must suspect deliberate deception. Hubble was far too prominent to be swept under the carpet and Newton was found wanting. At that point they put the onus of proof not on Newton but turned the focus away from Newton to what the Academics in physics now present as the factor to carry the blame fulfilling the role as being the guilty party. When will the Universe confirm its incorrectness by affirming Newton's obvious correctness? If they had to admit that Newton was wrong, the most intellectual scientists then had to admit they had nothing to show for all their brilliant work.

Science that was defying the likeliness of a living God stood bare and naked for all to see. They put the onus of proof and converting onto the cosmos. When will the cosmos come clean and prove Newton correct? When will the cosmos admit to a mistake and set its crooked ways straight? When will it meet its diverting from Newton and reach a point where the Universe will finally come to comply with what Newton demands? It is the cosmos that is wrong therefore it is time to find out when the cosmos will correct its manner. To deal with such a task they needed a man with a bigger ego than he had an IQ. They needed a person that thought more of his abilities than his ability to grasp any complex situation. They needed a man that was presented as a genius without ever proving his genius. They had a man that filled the centre of the Universe, which then placed the man in a location from where the man could see the entire Universe. They had just such a man. He went by the name of Albert Einstein. For all the genius Einstein had, Einstein failed to see the most simplistic and tiniest mathematical rule. Einstein failed to realise that if there were insufficient mass at the beginning of the expanding Universe, the growth of the Universe would reduce the influence of such mass as a factor because as the radius grows, such growth will restrict the gravity by rendering the mass progressively more incompetent.

If the Universe is expanding as Hubble indicated, the growth of the radius will reduce the influence value of the mass as every second passes. The mass will become more and more wanting for such a task. Yet, with this obvious shortsightedness of the genius Einstein, the genius saw himself fit enough to calculate and measure something as overwhelming as the Universe. As in the case of Newton, Einstein as an ego driven maniac, saw his abilities fit to measure and master the Universe while his mind was too simple to recognise the most basic principle of mathematics, the principle of relevancies or ratios. What a mathematical genius that turns out to be! While the radius enlarges, at the same proportion, does the influence of the mass factor reduce and the mere fact that the radius increment shows that at no stage further into the future can the mass stem the growth of the radius because the radius overpowered the mass factor already. Unless there is new material entering the Universe at a point, which is impossible, the entire concept is fraud.

The idea was never to admit wrongdoing on the part of Newton and Newtonian science but to postpone, delay and divert attention away from the truth. If there was not enough mass to start with, no dark matter can kick in later on and start a secondary mass frenzy that at that stage will then be enough to bring about the required mass potential that will turn around the Universe from expanding to contracting. To establish a scenario that would hide all deception, they got the man that has a bigger ego than an IQ, they tell the world this man is a genius while the fool does not know the least of mathematical principles because his Master Newton did not know the least of mathematical principles and they got him to measure the Universe. While they did not even have any device (and will never have such a device) through which anyone would be able to see the entire Universe, they set off a scandalous misconception that this Einstein could calculate all the mass in the Universe.

Off course as can be expected, there was not enough mass and there will never be enough mass, because there is no such a thing as mass in the entire Universe. When the deceit played out to the full, they, the fraudsters being the paternity of physics, elaborated on the delusion by trying to find dark matter that is hidden. If the dark matter did not develop enough contraction at this time, there is no chance in the future to develop enough gravity because the factor of what mass supposedly should have is tarnishing and tarnishing as the Universe expands. The bigger the radius becomes the less would the mass effect be.

The community of astrophysics are trying to frame a picture where they set the stage in the way that if the Universe were stretched to a point, the mass would not tolerate any more expanding. The mass will get frustrated in some way and show resistance to the increasingly elastic expanding. The gravity constant (I suppose) must prevent any further expanding. How they ever got to such an argument I never could tell. They surmise that outer space is consistently overall filled with nothing and when this nothing is stretched to the limit, the nothing would resist in growing more nothing or become further nothing and the nothing would stop other nothing to enter outer space in the community represented by nothing. If ever there is a faculty ruled by absolute inconsistency and rubbish as the motto of logic, it has to be astrophysics.

The following table lists statistical information for the Sun and planets:								
	Distance (AU)	Radius (Earth's)	Mass (Earth's)	Rotation (Earth's)	# Moons	Orbital Inclination	Orbital Eccentricity	Obliquity	Density (g/mc^3)
Sun	0	109	332,800	25-36*	9	-	-	-	1.410
Mercury	0.39	0.38	0.05	58.8	0	7	0.2056	0.1°	5.43
Venus	0.72	0.95	0.89	244	0	3.394	0.0068	177.4°	5.25
Earth	1.0	1.00	1.00	1.00	1	0.000	0.0167	23.45°	5.52
Mars	1.5	9.53	9.11	1.029	2	1.850	0.0934	25.19°	3.95
Jupiter	5.2	11	318	0.4111	16	1.308	0.483	3.12°	1.33
Saturn	9.5	9	95	0.428	18	2.488	0.0560	26.73°	0.69
Uranus	19.2	4	17	0.748	15	0.774	0.0461	97.86°	1.29
Neptune	30.1	4	17	0.802	8	1.774	0.0097	29.56°	1.64
Pluto	39.5	0.18	0.002	0.267	1	17.15	0.2482	119.6°	2.03

Every measured kilometre represents nothing. Every mm is one of nothing. We on Earth are 149 X 10^6 kilometres holding nothing away from the Sun. Only they can argue that outer space is nothing with material here and there. If that is the case, then which has more nothing between them, the Sun and Pluto or the Sun and Mercury? The distance between the Sun and Pluto is more, therefore that which outer space is made of is more than in the case of Mercury and the Sun. Therefore Pluto has more nothing between the Sun and the planet than Mercury has between the planet and the Sun. Only astrophysics and all the geniuses guarding the principle of astrophysics can put a calculated value by measure on nothing.

In fact Mercury has a hundred times less nothing between the planet and the Sun than is the case with Pluto. Since my days at school I was always under the impression that a hundred times the value of outer space being nothing is numerically expressed as (zero = 0 x 100 = 0), but where the genius that is such a prevailing part of astrophysics take the stage we find that Pluto can have 100 times more nothing than the amount or distance measuring nothing that Mercury has. The figure containing nothing that puts Pluto at the edge of the solar system is one hundred times more nothing than what Mercury has where Mercury becomes the first planet in the solar system. That is astrophysics. The brilliant minds of the mathematicians hold no rules apart from what they can calculate. Astrophysics is the only department throughout the Universe where normal rules don't apply because with mathematics they can bend all laws as they wish…in fact Newton started the trend with his deceit. Only the guardians of the policy of astrophysics can know why the undetected dark matter will start producing gravity to change the expanding to contraction. Would the fact that it is detected, change the influence it established? Or is it merely to extend the cover up and allow the deceit to linger until the following generation. There is no mass and anyone that says there is mass, let such a fraudster then explain why all the planets irrespective of size or density, spin around the Sun at the same speed as all the others. Let them prove that the Universe acknowledge big and small and let them show how Jupiter can move at the same pace as does Mercury and Pluto while Jupiter is so many times more massive than the other two mentioned. More condemning evidence is yet to come because the astrophysics tricksters did not leave the corrupting of evidence just at that.

The fatherhood of physics never once diverted from acknowledging that Newton's contraction is the prevailing thesis on which the cosmos is built because they accepted that Newton used unlawful arguments and to cover up Newton's fraud which they still use to this day, they then proceeded with further criminality when producing the bluff they established with Einstein just to fool everyone in the normal public. Without ever recalling Newton's contraction theory that is obviously not working or admitting doubt about Newton's testimony to the effect, physics accepted the Big Bang Theory. The Big Bang theory opposes whatever Newton might have implied. The physics paternity however finds it wise to still advocate Newton while admitting to the Big Bang event. Newton said the lot is contracting. Go on and marry that with the Big Bang that says everything is expanding. You can't promote both except if you can define why we would see the two merge.

The Universe comes from a point the size of a Neutron. That makes the radius parting the Universe, infinitely small. It just about removes the radius as a factor. At the very same implication it takes the pulling of the mass (if there are pulling forces converted by mass) to a level it will never again have. As soon as the distance between the objects holding mass started to grow, the power and influence of the mass factor started to diminish in the same ratio. If the mass that applied in the past, as it does at present and were incapable of contracting the Universe at any time throughout the past and up to the present, then that mass factor will eternally and forever remain unable to unify the cosmos by contracting the Universe. Then you may ask what is the story? Read on and you will learn how far Mainstream Physics strays from the truth and how big a cover up the paternity is protecting.

According to science the Universe started with singularity. Quoted directly from the Oxford dictionary of Astronomy the following:
The definition of singularity is as follows:

Singularity: a mathematical point at which certain physical quantities reach infinite values for example, according to the general relativity the curvature of space-time becomes infinite in a black hole. In the Big Bang theory the Universe was born from singularity in which the density and temperature of matter were infinite. The average daily temperature was "$10^{\alpha\beta}$ to 10^{34} K".

Then the second "day" the daily average temperature came down to 10^{34}K and 10^4K. That is fine, but if the temperature was in Kelvin, then what was 0^0K. In order to make sense of the scale used, there must be a minimum to secure a maximum otherwise the maximum can just as well be the minimum and is only advocated to impress humans applying Earthly standards. By using a scale as $10^{\alpha\beta}$ to 10^{34} K, it places the lower temperature at a modern 0^0K to make sense of standards. If that was the temperature the standards were lowered, compromising something to gain something, because something had to grow larger for heat to reduce. We know space grew larger bringing heat down to reduce heat levels at that day.

Being the onlooker, the viewer has to maintain one position. From that position some particles would be circling a centre point, as the particles would be coming towards the onlooker. The other matter would be circling the centre point while rushing away from the onlooker.

At the very end the single dimension may come into the dynamics but where the single dimension comes in, the factor of zero is removed. If there is space, there is a flow of light and a flow of light has to produce lines in relation to angles forming space between them. Something must be present to confirm space because there is an absolute difference between being in space and no space to be found. If there was a line that formed nothing, that one line that forms nothing would completely destroy the other lines' chances of ever forming a triangle, let alone having all lines and they then have a total being zero. As shown in the example, no line can form zero and therefore no mathematical equation as far as it extends to cosmology can ever bring about zero as a number. While there is space present there has to be three dimensions relating to each other by time and in three dimensions there has to be three lines in relevancy to each other by angles formed holding space in (at least) six opposing sides. Removing one line must bring about a flat Universe and that then will constitute nothing.

Cosmology is about light flowing by means of lines indicating space obeying the rules enforced by time in motion and light flowing dictates crossing space and across space light is using lines. The book: *An open letter Announcing Gravity's Recipe* is dealing with the subject finding singularity by removing the concept of nothing from outer space. By diminishing nothing, one uncovers singularity and the effort brings in a new perspective not yet introduced.

For your benefit I will shortly give a summary by which I hope to interest you in reading the manuscript. **Compressing space produces heat**. Releasing heat will bring expansion, bringing about space. We call such a release of heat an explosion. In other words heat translates to space and space concentrates back to heat. The one is a product of the other where space forms expanded heat, as heat forms contracted space. Space forms when heat expands and heat forms when space cools.

They are quick to show the time that was applying at the time being some thousandth of a second or the heat that was present being numbers we have no name for. The other side of the story they ignore completely. They ignore the other side of the story because to acknowledge it would mean that their promotion of Newton comes down to madness. If you reduce the radius applying at the present back to what it was at the time of the initiating of the Big Bang, you must also increase the influence gravity and mass had at that moment by the same number you are decreasing the radius. That is pure mathematics and the most basic physics of all concepts.

The shrinking radius will increase the effectiveness of the influence of the gravity that the mass can produce by the margin of the shrinking of the radius. If the Radius was infinite at that point, then that means the gravity was eternal. With the entire Universe being as big as a Neutron, the Universe was the size of an atom. If the Universe were the size of an atom and the mass within that Universal atom could not prevent the Universe exploding into immeasurable atoms, then it would not be able to retract all the atoms into one unit again. If there was not enough mass to start the contraction, there can be no contraction of mass that is producing the gravity at this stage. If the gravity is of such a nature that it allows a continuous growth of the radius, then the radius firstly cannot be zero as Newton suggested and the extending of the radius proves there is no contraction in the way Newton led everyone to believe. If Newton's mass contracting mass is true, then on the other hand it must have resulted in an implosion as that which can never repeat again. With Newton's formula of

$F = \dfrac{M_1 M}{r^2} G$ forming gravity, then the Big Bang is just not possible because

from that formula the Big Crunch must respond.

One can derive Hubble's law mathematically if one assumes that the Universe expands (or shrinks) and that the Universe is homogeneous, meaning that all points within it are equal.

For most of the second half of the 20th century the value of H_0 was estimated to be between 50 and 90 km/sec/Mpc. The value of the Hubble constant was the topic of a long and rather bitter controversy between Gérard de Vaucouleurs who claimed the value was 100 and Alan Sandage who claimed the value was 50.

Hubble also suggested that the clusters of galaxies are distributed almost uniformly in all directions, although more recent studies show that clusters are combined into huge super clusters of galaxies: at this new level, however, the distribution appears to be even. He was the first to offer observational evidence to support the theory of the expanding Universe, presenting his findings in what is now known as Hubble's law. With Milton Humason, Hubble classified the different types of galaxies including irregular galaxies, three types of spirals and barred spirals, and elliptical galaxies. Included in his writings are A General Study of Diffuse Galactic Nebulas (1926), Extra-Galactic nebulas (1927), Spiral Nebula as a Stellar System (1929), The Realm of the Nebulas (1936), and The Observational Approach to Cosmology (1937).

This puts a lion amongst the pigeons and the lion will devour any cat that could enter the pigeon den. The Universe was upside down and this was latterly. What everyone thought was coming, was in fact going. What was reducing was expanding and what was to become less became more.

What would you say were the reaction launched by Newtonians? Newtonians corrupted corruption by enlisting the biggest fraud ever concluded. This outmatched

Newton's by a billion to one and Newton's effort in corrupting Kepler's work already surpasses nothing ever cooked up by any gangster ever.

The Academics in charge of official policy never held Newton in contempt. They never reinvestigated Newton. They never flinched an eyelid at Newton's claims. Not once did one raise an eyebrow against Newton's work. The reaction was that if the Universe did not prove Newton correct, then when would the Universe start proving Newton correct? If the Universe was expanding, when is the contracting coming in place? They did not say wait a minute that makes Newton a fool. No, they said the foolishness of the Universe must immediately end and start adhering to what Newton predicted. The Universe must come right and it must do so fast. The Universe must attend to Newton's statement and get a grip on its act. The Universe must show its critical density immediately so that Newton should not be ashamed of the actions of the Universe. They never said once that with the Universe expanding that means that Newton actually stands to be corrected. No…they said the Universe is wrong. God and his Universe must be wrong because Newton is correct. Never once was the idea of incorrectness associated with Newton. Never did any person thought of Newton in terms of being disproved. The reaction was condemnation to the cosmos. The mistake the cosmos made has to be corrected. There has to be a density mismanagement committed by the cosmos. The cosmos is inadequate, faulty, deficient and incompetent. The Universe lacks the mass required to substantiate Newton. There was a critical density to be investigated in which Newton would be vindicated. With the Universe at fault at this stage to prove Newton's absolute trustworthiness, then it is up to the Universe to prove when the Universe will become trustworthy. When can one trust that the Universe will correct its mismanagement of mass and start contributing to Newton's correctness? The issue never went the way of questioning Newton's legality in the statements about contraction that Newton made…the disputing of the concept legality went the way of the Universe. If the Universe can dare to show expanding in the face of Newtonian contraction, then when will we call the Universe's bluff? When will we show that the Universe is up to no good and when will the Universe face its flaw and correct its incorrect ways? When will the Universe stop the farce it is busy with and begin to act responsible? The name they came up with to enlist the rehabilitation of the cosmic disgrace was the critical density.

If the cosmos was wrong then when will the cosmos become right. When will the cosmos surrender to what is correct and start to abandon its incorrect approach to the physics of the Universe?

They could not dispute Hubble, but they could dishonour his dishonouring of Newton by not honouring Hubble with the Nobel Prize. They could disgrace the achievements of the person that showed the tenacity to disgrace the cosmos by indicating the cosmos dared to not enlist Newton's rules. By never honouring Hubble, they could disgrace his disgraceful discovery and at the same time they could reinstate the honour of the cosmos by finding the critical density of the material that will supposedly start to rectify the unbelievable behaviour the cosmos dared to show. They will find out when the cosmos will agree to adhere to Newton's gravity law.

An Open Letter On Gravity Part 2

The Academics did what Newton did best. They reinvented the truth to reinstate the Newtonian lies. They conceived more corruption to cover Newton's fraud.

They employed the services of the man that filled the centre of the Universe to calculate the mass of all the material in the Universe as to account for any shortfall there might be to bring a final conclusion in the contraction Newton instated.

The person that did the investigation had to be in a place where such a person could see where all the material in the cosmos was. That person had to view the cosmos from a point that gave the person the advantage to see as far south as possible while also being in a perfect position to see as North as the Universe can go as to find the view of all the mass that the Universe holds between point south and north. At the very same point that person had to have the ability to see east and west as far as the directions can go so that the person can find where every morsel of material is located. The person had to be just as far from the top of the Universe as the person was from the bottom of the Universe to find the view where every crumb of material is hiding. After all, the person was given the task to find and calculate all the mass in the Universe and that was first and foremost then to locate every square inch of material that held mass or could have the ability to provide the gravity that will bring about the contraction there has to be to validate Newton and vindicate the cosmic disobedience about Newton.

The person must be in the right position from where the person could glance at the entire Universe. That person must be in the centre of the Universe because only from the centre of the Universe can the person view all the mass there is and that might be hiding in the Universe. The person had to think himself as filling the centre of the Universe which from there he had to be in the academic centre of the Universe which will allow him the vantage that all the light he might use to see the material in the Universe will be heading straight to him. If one light photon passes him, he would not be able to bring that particle that the light represents into his calculation and we know that some galactica seems so far distant that they are represented by one photon. Therefore, only a person that fills the centre of the Universe will be able to see all the light that is coming by gravity to the centre of the Universe and being in the centre of the Universe, will place the person in the centre of all light moving. That is the most critical position the person has to have and that is the most ardent precondition that the person has to achieve to be able to calculate the mass in the entire Universe.

Fortunately, at the time, there was one person available that met all the requirements and criteria for the task. He was what was perceived to be the centre of intelligence in the Universe and he was in the country that has the idea amongst its inhabitants of forming the centre place in the Universe and he was part of the people that was vain enough to think of themselves to form the centre of the Universe because he was representing the rights of the Man that formulated the centre of all physics in the Universe, which was Newton. And because of such importance, Newton was being the man that brought the entire Universe into disrepute because the entire Universe was at fault in its showing of being in disagreement with the laws Newton (where Newton is the man representing what is correct and what is incorrect) instated. So we have the most brilliant mind, being Einstein the intellectual at the centre of the Universe forming the physics centre of the Universe, which is American physics in a country that thinks it is in the centre of the Universe, which is the United States of America, with the people that hold their position as the centre of the Universe which is the American's view about their role in the entire Universe, representing the man in the centre of all physics being Newton, where he placed in physics mass to hold all gravity to form that which preserves the centre of the Universe. What a hell of a lot they were...and now everyone knows where the centre of the Universe is...it is where Einstein found his critical density theory in the heart of America in the centre of American physics.

They have no other person for this masquerade than Albert Einstein in person. Einstein fitted the profile to a T because Einstein had the image and the flair and the reputation that was required. Remember they were not looking for a result. They were trying to establish a time delay. They did not try and find why the Universe was not doing what Newton said the Universe was doing. They were not placing the incorrectness at the door of Newton. They were putting the Universe at fault. They were in search of when Newton will become correct and not why Newton is incorrect. They tried to establish when the renegade Universe will address its mistake and come into line by correcting the mistake God made in sending the universe in the wrong direction. Placing the universe at fault and not Newton, gave the opportunity not to address Newton and his miscalculation but find the opportunity to present a Universe that is correcting the mistake God made. In that they placed all the apparent blame on the Universe by disclaiming Newton from any resolve.

The critical density is the biggest and most elaborate case of fraud ever perpetuated by any group of people. They did not say Newton was incorrect. They said they are on a mission to see when the Universe will correct itself and prove Newton correct. Newton remained correct while they gave the Universe the chance to mend its ways. If there was not enough mass to start with, why will there be enough mass in the future? If the Universe is growing, that then means that the mass in the Universe is not pulling the Universe into contraction. If it is expanding, it cannot be contracting. Only an over bearing egomaniac with an ego outweighing his common sense by a margin of many times to one, will take the opportunity to calculate the mass in the Universe. Step outside tonight and see what there is to see outside in terms of mass. They put the exercise to a formula in order to calculate what no man can even presume. If you wish, you can read the following example I offer to prove what elaborate criminal scheme the entire venture called the critical density is and to which fraud it amounts to. Here follows a part taken from a web site that tries to prove fraud by elaborating on the fraud by implying more corrupt fraud. See how they place the motion over to contraction by putting the motion ($\frac{1}{2}mv^2$) equal to the supposed contraction (**GMm/r**). Hubble found no trace of motion towards contraction when he found all movement flowing in the direction of expanding.

A simple, non-relativistic, derivation of the critical density can be performed as follows

Recall that kinetic energy of a body of mass m moving with velocity v is $\frac{1}{2}mv^2$, and that gravitational potential energy of a body of mass, m, at a distance, r, from a second mass, M, in a radial field is GMm/r.

In order to find the escape velocity of a spacecraft from a planet, we can work out that the spacecraft will escape if it has enough kinetic energy that gravity will be unable to slow it to a standstill. Using the two expressions above we can write

$$\frac{1}{2}mv^2 \geq GMm/r$$

The kinetic energy of the craft is converted to gravitational potential energy while it escapes. If the two are equal, it would stop at a very great distance (infinity), if the kinetic energy were bigger, then it still has velocity when at infinity.

We can do the same calculation to see if a galaxy is able to escape from the attraction of all other galaxies or whether it will be stopped, turned around, and caused to fall back. If this latter situation is the case, then the future of the Universe is finite. The galaxies, which are at present travelling away from each other, will stop, turn around, and head back towards some common centre where everything will be compressed, perhaps out of existence, 'The Big Crunch'. If, however, there is not enough material in the Universe to stop the galaxies, then perhaps the Universe will go on forever.

Hubble's law shows that the velocity of a galaxy is proportional to distance,

$$V = H_0 r$$

Where H_0 is about 5×10^4 m s^{-1} Mpc^{-1}, 80 km s^{-1} Mpc^{-1} or between 1.587×10^{-18} and 1.62×10^{-18} s^{-1} depending upon whose figure you accept!

So the kinetic energy of a galaxy can be written as

$$\frac{1}{2}m(H_0 r)^2$$

The mass of all the material inside a sphere of radius r is given by

$$M = \frac{4}{3} Pi\, r^3\, Rho$$

Where Pi is the average density of the Universe.
Substituting these two expressions for the first equation above gives

$$\frac{1}{2}m(H_0 r)^2 = Gm(\frac{4}{3} Pi\, r^3\, Rho)/r$$

This can be simplified to

$$m(H_0 r)^2 = Gm(^8/_3 \text{ Pi } r^2 \text{ Rho})$$

and rearranged to

$$\text{Rho} = 3H_0^2/8 \text{ Pi } G$$

One cannot seriously place the expanding Universe into any situation where the contracting already applies. The spacecraft does not represent the expanding Universe because that is placing Newton accurate above all facts. It is deliberately mismanaging all concepts in favour of a predicted and a decided outcome without finding a need to allow the facts to cast a result. The spacecraft is part of the Earth and by trying to escape from the Earth it divorces the hold the Earth has on the craft making the craft the captured. In order to become part of the Universe, the spacecraft must renounce the capture to be on equal terms with the capturer as the capturer releases its hold on what was part of the Earth. The Earth was in control of the craft and in this case the Earth was the capturer and the spacecraft being the captured. It is a divorce that is going on and not a contradiction.

The above my friends are hogwash. It is what represents the utter most thoughtless and mindless arrogant mismanagement of all facts. If the mass was not up to the grade to pull the Universe into contracting, it is incorrect to presume the formula will apply. By finding that the Universe is growing in size, one should question the authenticity of the formula $F = \frac{M_1 M}{r^2} G$. This formula can only apply to a shrinking or contracting Universe. To try and integrate the formula as pronounceable in the case of the expanding Universe, is committing fraud. What the Newtonian brainpower in their overwhelming geniality totally misses as usual is that the cosmos is expanding inwards and that is not the same as contraction but more about that in another book. That which is growing, is the validating of the lesser space-time.

$F = \frac{M_1 M}{r^2} G$ In the using of the formula such as Newton recommends the radius determine the influence of the mass.

$F = F = \frac{M_1 M}{r^2} G$ In mathematics to put any number in relation to the value of another number is to agree that the top part of the formula will find a suitable and applying measure by the measure and the size of the bottom part.

$$F = \frac{M_1 M G}{r^2}$$

If the bottom part is bigger than the lower part, the bottom part will be a fraction of the top part. If the top part is smaller than the top part, the bottom part will be a fraction of the top part. The bottom part and the top part will always and without reservation have one of the two smaller or then a fraction of the other, except in cases where the two are equal. In such cases there will be a unifying number coming about as one. Putting two factors into a ratio or a relevancy, places the one value in charge of the other value while the other value is dependent on the first value. Not one of the two then can be dismissed as nothing or having no value because the other value holds the measure of the first value.

$F = F = \frac{M_1 M}{r^2} G$ Appreciating that the top is sizably more than the bottom value will put the top part in a numerical superiority and the answer will not be in any fraction, which includes a mathematical fraction of less than one.

$$r^2$$

Having the radius growing, means the top part of the formula that represents the mass, is shrinking in influence.

Edwin Hubble proved that $F = \frac{M_1 M}{r^2} G$ is not applying because the cosmos is not shrinking. To further use an already falsified formula is cooking the books to a point where the books being cooked, start to burn.

By enlisting $F = \frac{M_1 M}{r^2} G$ you are placing the mass and the mass as a multiplied unit with the gravitational constant in relation to the radius value that applies. That means if the radius does not shrink but it grows, the mass does not reduce the radius but it promotes the radius into growing. By incorporating the incorrectness into what one then tries to establish, is committing fraud by falsifying the facts derived from fraud even further. There simply is no formula as $F = \frac{M_1 M}{r^2} G$ so then there can be no integrating the formula to whatever form $\frac{1}{2}mv^2 >= GMm/r$ is required. The cosmos proved that $F = \frac{M_1 M}{r^2} G$ is a farce and Newton should be investigated. The Newtonians elaborated on Newton's fraud by investigating the Universe of fraud. There just is no mass found in the entirety of the Universe.

Read my book and you will find that the Universe is not space but the Universe is time. There is no space in the Universe, for through expanding the Universe placed time in space. As time develops, it changes density into space. The Big bang is about substituting the development of time into forming space. Time is space because space came about as time progressed. You cannot calculate the Universe in terms of ($\frac{4}{3} \Pi r^3$).

Guess how surprising was the result that Einstein produced. Einstein shockingly and most surprisingly confirmed a constant defiant Universe. Einstein produced results confirming not only Hubble but also that Newton will have to wait much longer to be proven correct. A new plan was devised to vindicate Newton and his correctness. If the cosmos did not play ball, then the Brainy Bunch will cook up a real witches brew that not even the cosmos can dare to defy.

A search went out to find dark matter. Dark matter will result in correcting Newton if nothing else can. If one can't see dark matter, then no one can prove dark matter and while no one can prove dark matter, then it also becomes true that disproving dark matter is impossible. It worked with the instigating of the original fraud that Newton invented. Not ever proving Newton and his mass that creates gravity by attraction between the mass factors also brought about that not one person ever thought of disproving the mass deception. With the mass being impossible to prove, it also makes the mass of dark matter impossible to disprove. Then on what grounds do I lay my charges of the biggest hoax ever created to defraud tax payers out of their livelihood by stealing tax money and diverting the proceeding of the tax collector to establish a criminal inspired corruption never yet before experienced by man?

Kepler introduced space a^3 growing by the time T^2k that allows the space a^3 to move. That is time T^2k allowing space a^3 to be. For space a^3 to move about, it will use the motion that time provides in the rotation of a^3 as well as the displacement of a^3. In space there is no mass because in space there is no proof of mass.

$$F = \frac{M_1 M}{r^2} G$$ If there was not enough mass to redirect the expanding Universe into contraction, then there can never be enough mass to redirect the flow of the cosmos. What will produce more mass if there is not sufficient mass at this point? Why would the dark matter be gravity dormant and what will enlighten the dark matter into activating gravity? If it is mass and the mass is supposed to establish gravity, the mass then has to establish gravity whether the mass is dark or light. What will make matter that is visible more active in creating gravity than matter not seen by man? Why will dark matter come into gravity later on, as it at presently holds the gravity dormant? Why will the dark matter at present not form sufficient gravity by mass but will later on become energised and jumpstart the cosmos into contracting? If the mass is there, the mass should charge gravity. If mass does charge gravity, why would dark matter play hide and seek and hide their potential gravitational abilities for a later date?

$$F = \frac{M_1 M}{r^2} G$$

The suggested formula indicates that it is mass that produces the gravity by which the force pulls structures and thereby it reduces the distance between the objects. The mass is proportionally charging gravity. The mass is responsible for the force of gravity by the measure that the mass has. A lot of mass will charge a lot of gravity and a little mass will charge less gravity.

The mass is there in relation to the establishing of the force and forcefulness of the force proportionate to the amount of gravity applying by the production in relation to the mass. If the mass is there, the gravity is there. If the gravity is absent, the mass is not present. If the mass cannot reduce the gravity, what then is the point of trying to establish when the gravity will come about, since the mass is obvious lacking as a quantifiable amount to charge enough gravity to contract the expanding Universe? If the gravity is not available, the mass is not sufficient and therefore, if the mass is not sufficient in the first place, how the hell will the mass become sufficient later on? The mass is not an elastic band that allows expanding up to a point where after it will reduce as the elastic energises.

$$F = \frac{M_1 M}{r^2} G$$

The formula does not compensate for any such suggestion. The notion alone is one huge farce inspired by criminal minds to cover up flaws in the top banner bearers of science's world of physics. The whole issue goes against the grain of physics and mostly against all mathematical principles. If they were incompetent novices trying to address a school project that has a mathematical inscribed theme that goes way past their abilities and they are in far too deep water to tread, then yes I can find some degree of honest miscalculating and incorrect judgement of mathematical founded facts. They hold the top notch of all mathematical insight. They father the laws of mathematical principles. Therefore, in that light they woefully and deliberately avoided the truth and rendered their honesty to distrust. The fact that they skipped inspecting Newton and created a farce to mislead any thought the public may have about the mistake, shows their deliberate action to avoid the truth about the matter and carry on in criminal intent. I challenge anyone bearing the title of professor in physics to prove otherwise. Let any academic in physics show where I overstep my boundaries when I charge them of deliberate acting with the motive to corrupt the truth and avoid the true impact of Hubble's findings.

$$F = F = \frac{M_1 M}{r^2} G$$

If the mass was prominent from the start in relation to the distance separating the two objects, then the contracting of the object will reduce the radius parting the objects to become a factor of one that has no influence on the formula.

An Open Letter On Gravity Part 2

$$F = \frac{M_1 M G}{r^2}$$

If the mass grows and the mass keeps growing, the mass will reduce the input of the mass on the gravity. The further the mass goes apart from the mass, it shares gravity with, the less the influence will be of the mass exerting gravity. If the gravity was insufficient from the start, then it will remain insufficient as the expanding progresses. The gravity does not grow as the radius increases. It is the other way around and as mathematicians of notoriety they knew it. That makes their criminal intent even more appalling than what it would have been if the actions of criminality came from the midst of others, being persons in a less trustworthy disposition. However, Kepler gives all the answers they tried to criminally cover up.

What we dealt with here is only the tip of the iceberg. When one starts to dissect the inner workings of the Critical Density Theory and find their way of thinking about how the operation will carry out its purpose, then it is clear how much the theory proves how the madness can grip the idiot's mind and unbalanced thoughts go out of control. If the mass stops the molecules in the process leading to turn about of direction of Universal flow, it will stop the less massive first and this will have the more massive molecules plough into the slowing as well as stopping smaller molecules. The more massive molecules would look like trains colliding with cars and bicycles and removing the bicycles from the face of the Universe. At that point a second Big Bang will come about that totally destroys everything that was not destroyed by the first Big Bang.

A question no one ever asks the Brainy Bunch and therefore no one ever answers in any polite conversation amongst the brilliant intellectual Academics, is if the black matter is hiding in the dark and if there are stars formed as black matter, then what has the light that shines got to do with the mass it has, that does not present as gravity. Why is it that because the matter is black or dark, such black matter has not activated the gravity the mass should produce? It has mass, so where is the gravity? What has the darkness got to do with the absentness of the gravity it is not producing? Why does it not at this instant, use that mass it already has just being there, to apply gravity? What is holding the gravity back that the hidden mass should and can produce?

Surely the fact of having mass in the first place with the mass already there, shining or not, that mass should generate gravity? Why will the mass not produce gravity just because the mass is dark? How does mass and gravity interlink with the darkness where the darkness disallows the mass to establish the gravity that the mass should establish and why is it absent just because it is dark? It is the purpose of mass to establish gravity, whether the star shines or doesn't shine. When there is a lie, such crookedness becomes hard to sustain with logic by fooling, irrespective of what the fraud incorporates. The mass should at this moment deliver the measured quantity of gravity to either pull the Universe into contraction, or not give sufficient support in establishing the requires gravity to get the Universe contracted. Why would the mass, being dark, wait for the gravity to generate in order to set free the necessary gravity that would later reduce the Universe, as the gravity is still not responding on the request of the mass just because the mass is hidden from our view?

Can those fraudsters really not think about the fraud they contemplate and commit? Are they really that stupid that they do not realise what menace they create by diverting from the truth in such a manner? Or do they truly think that the rest of the human race is so mindless and senseless that no one but these gangsters and crooks can think? Do they really overestimate their abilities by underestimating the abilities of the rest of the human race?

This is what happens when a lot of criminals can go about, as they are allowed to work unchecked. As they are blindly trusted by all those unsuspecting, honest persons in the general public holding no criminal intentions, they never think that these Academics are never being controlled, which means the Academics can steal billions from the coffers of the tax paying public. That they then also do when finding funding for the most bizarre research that is so fruitless it falls in the category of the insane. One such a venture that comes to mind is the monumental fraud of research into alien life coming from somewhere in the outer space. This is only one of so many that they use purposely on many levels. In other instances the fraudsters apply the same modus operandi by inventing some bizarre mathematical formula that must supposedly serve some purpose and then they unleash this fraud onto the public at large albeit to try and cover up Newton's fraud or to defraud the public in other ways.

The latest book that is supposedly written by Steven Hawking is a perfect example of the degree of mad corrupt and idiotic methods they employ to commit such fraud. They invent a formula. The formula in theory is meaningless because the practicality of the formula can never be tested and must therefore be accepted. The reasoning behind such a formula is as unrealistic as the critical density crime venture, which I just explained and has as little practical function as does the critical density theory. As I show, they go about to bend and cheat mathematical principles and logic at will to corrupt the truth as to find proof for their meaningless venture. By applying even the least of logic they prove callow. But since they deal with honourable people they take the public to the cleaners.

The people forming the public are too scared to ask questions about the sanity behind the reasoning because the people feel incompetent in reading the mathematics. That uncertainty, these evil gangsters use most shrewdly to their financial benefit. Behind this fear of feeling incompetent that the general public has, is a rational of not asking questions because they do not wish to feel stupid and this the academics in astrophysics exploit for their criminality while the gangsters continue with their evil exploits.

The criminals calling themselves astrophysics academics then realising this flaw that the public have and then go on to commit gang rape on the unsuspecting public, which are also those they see as witless beings they can manipulate and control at will. I challenge those fraudsters to sell me that lame idea that the latest Hawking book explores and see how I plough their fields to uncover pure bullshit in front of a TV audience.

I call them criminals because notwithstanding the position any person has, when such a person pretends to be what the person is not, or to have what the person does not have, or to launch a project that the person intending to benefit from the results and the research of such a project but in launching the project the person intentionally has to spread untruths and corrupt lies wilfully, then this makes that that person intends criminality. By telling not the truth, one tells a lie. By spreading lies, one commits fraud and deception. There is no small medium and large way of deceiving a person because untruths are lies. In doing so, they intentionally divert to truth to sustain a lie and such actions are criminal. Even not admitting to the fraud is a way of committing intentional diverting of the truth, which is the sustaining of criminal behaviour. Any behaviour not being in line with the truth is committing criminal activity and that is the narrow and the broad of the lot.

That which Newtonians confuse as gravitational pulling is the legacy of the Roche limit applying between all cosmic objects. Gravity doesn't pull but the Roche limit places heat in the form of electricity that, when being on the limit of what singularity would permit, does liquefy the smaller structure.

If what I have said and the proof I brought when I said what I have said, is not sufficient to convince anyone that there is a serious mistake in the Newtonian approach to science, then whatever anyone will ever say will not be enough to convince the doubtfully amongst us.

My arguments and debating has taken me a full circle from the start where I too posed this remark. I have discussed the mistake about science in length. I pointed out how and why the mistake is present in science. I have shown that the mistake can be named up in one word: "Newton". He raped and plundered science from the only truth, as Kepler advocated it. Kepler showed why the Universe has an infinity connecting to the eternity there is. Kepler showed that infinity being k^0 connecting to a^3 goes by means of k. Eternity or time in motion or then gravity T^2 connects to space a^3 or forms space a^3 by placing space in relation to or connecting space to infinity $T^2 = a^3 / k$, **where singularity then can be located at $k^0 = T^2 \times k / a^3$**. Using this mathematical expression in this manner solves all the riddles one might find in the Universe. This is so much more sensible and truthful than Newton's madness.

We can even judge what a Black Hole $k^0 = T^2 \times k / a^3$ is and not put it down to gravity "gone mad", which is as bizarre as giving a star death or think of a star dying. That is gravity $T^2 = a^3 / k$. That is also the formula representing space-time $T^2 \times k = a^3$. That is time $T^2 = a^3 / k$. That is what Newton never new because Newton was not up to the task to understand what Kepler said what the Universe was. Again I put it in as simple terms as I possibly could: the mistake can be named up in one word: "Newton". But that is as far as science is concerned.

An Open Letter On Gravity Part 2

There is another mistake we use in mathematics. There is no zero in the cosmos because Kepler said $k^0 = T^2 \times k / a^3$, **and this expression renders the factor we put in as zero useless.**

Kepler helped me to resolve other unresolved matters but it was only possible by using Kepler's work.

I too am well aware that at first glance you will immediately arrive at the opinion that the theme of the letter has to be considerably below the standard of an intellectual Master such as you must be, due to the position you hold, and because of that, the normal research work you do. Nevertheless, I hope that this writing may spark interest even at such a low academic level and grade in scientific sophistication and development because I am about to prove that I discovered:

1. The location, the position and the value of **singularity** as a factor forming space-time;
2. Finding **space-time** by dissecting Kepler's formula in relation to valuing singularity;
3. Finding space-time, **proving space-time** and **aligning space-time** with gravity;
4. The **working principals** behind and manifesting **of gravity** as a cosmic occurrence;
5. The **Roche limit** and explaining the resulting of a law coming about from singularity;
6. The **Lagrangian system**, how and why that becomes the building form of the Universe;
7. The **Titius Bode law** and I show mathematically how gravity comes about from that;
8. Proving the **Coanda effect** and the producing of gravity through reproducing space-time;
9. The **sound barrier** by proving it **is gravity** generated **by motion** in space becoming independent where motion creates independence. Breaking the sound barrier is the motion in space duplicating space by crossing over gravity borders. It is $a^3 = kT^2$ where $(k > T^2)$ or $(k > T^2)$.

Newton said that it is the reducing of the distance between the objects that would bring about the irreversible reducing that will end in a total demolishing of the radius that is between the cosmos structures, but instead we find the gravity applying in outer space is one of the instances where gravity provides an orbit circle that gravity seems never to complete as the orbiting objects follow from closing any circle that is leading into a following circle up to where the circle is completed in cyclic precision. That is not the gravity that Newton identified, although Newton admitted that there is a presence of a centre forming a point in the middle between the two objects. He was unable to know what caused or even explain the presence of the Coanda principle, which forms such a critical part of my theory. The formula concerning cosmic balanced gravity however leaves no room for the admitting of such a point and by not leaving a possible inclusion of such a point in his formula Newton did by such gesture in principle repeal his admission of such a centre. This had me cast doubt on what is taught at institutions of learning. It motivated me to venture back to an era before Newton came to influence science. I came to acknowledge Kepler as I came to understand Kepler. The acceptance of what I understand of Kepler involves much more reading into what Kepler said by finding what Kepler did not say in the way that he did say what he said, than reading about what Kepler said as it is written in precise detail and to the letter used in his statements. He never directly stated what he said. Again I must stress this point: When I refer to what Kepler said it most likely means reading into the part that he did not say when he was saying what he said but I accept that he meant to say what I am reading and translating from Kepler as part of what he did not say but meant to say. I have to read more with my mind than with my eyes. This comes as a result of interpreting Mathematics to the verbally expressed. I had to learn to read with my mind and not my eyes and I found that that is the manner in which one has to approach cosmology. From the first time I discovered what manner one should use if one wished to read into Kepler's findings, I saw Kepler was all about uncovering the unknown. Realising that, the conclusions I drew by reading in such a way cemented my better understanding of Kepler's work, which then helped me improve my insight into Kepler's work as it increased my understanding about cosmology several fold. This helped me to realise what implications were to be found underneath Kepler's discoveries. From my realising what approach I should use, it helped me to improve my cosmic realising by using the method of reading Kepler and from that I could come to appreciate what Kepler introduced.

Only then did it bring insight and proof to me as a student of Kepler and this proof I found by dissecting what Kepler did not say instead of what he did say, which I now present to you with this letter, you being a superior intellectual person. Kepler said $a^3 = T^2 k$ and that correctly translates to a mathematical expression $k^0 = a^3 / T^2 k$ which in the verbal statement in English translates that Kepler said that there is a **space a^3** which is **equal =** to the motion in **the time duration T^2** thereof between two specific points which holds a relation onto a centre k^0 where from there forms **a straight line k** that is centred on the spot where space begins from k^0 **that produces k** as well as producing the circle, therefore that spot $k^0 = a^3 / T^2 k$ has hold k^0 at a value of having the least space. The line **k** is centred onto a spot where space begins specifically at k^0. This point not only produces the line k^0 but represents also the space that forms the eventual circle T^2. Therefore from the centre holding k^0, k^0 leads to **k** that forms the roving space a^3, which is rotating at a distance **k** where T^2 forms the outer limit of k^0. Mathematically $a^3 = T^2 k$ will be $k^0 = a^3 / (T^2 k)$ because $k^0 = 1$. But $k^0 = 1$ also presents the single dimension where all factors are a product of one. If one can locate k^0 one will find singularity. That is where gravity is because gravity is strongest where space is least. Then that suggests that gravity is strongest at k^0 because space is least. That is gravity because that is what keeps the orbit ting objects in orbit but also that is what Newton completely missed when he changed Kepler's work. Newton failed to recognise gravity as the only ingredient in Kepler's formula. He admitted he missed this because he admitted he did not know what gravity is while Kepler explicitly showed what gravity is. Gravity is what keeps the orbiting objects orbiting. $k = a^3 / T^2$ is **distance^1 = space ^{3}/ time^2** forming from a pivoting centre k^0. That is a cycle and moreover it is a cycle formed **by space / time**. What Kepler said is that space is a^3 in **motion T^2 k.**

That says **space³ (a³/)** relates directly to **time²** that uses the symbol **T²**. This is also what I refer to when I say one has to read what Kepler did *not* say when one wishes to see what he *meant* to say. Kepler introduced space³ –time² long before Einstein's date of birth appeared on any calendar although Einstein is credited with the formulating of the concept of space-time and giving it a name. Going even further Kepler stated that the space **a³** is on the move **T²** around in a circle at a distance **k**. That is what that comet we are discussing is doing. The space³ (Comet) is circling the sun using a radius **k** to establish the cyclic time² as a period of continuous motion and continuous motion is gravity. That reads much more correctly and closer to the truth than what Newton predicted what according to him (Newton) was happening in space. **Remember in this statement I am separating cosmic principles applying from the way that gravitational principles apply on Earth.** I distinguish that which is the rule in the cosmos from what we find ourselves trapped in on Earth. The two just don't mix. I am removing cosmic physics from normally accepted physics because the gravity concerned is not the same.

The proof I bring is real, however simple it may seem. It has none of the mind-blowing complexities normally associated in the presenting of investigative analyses of Astronomy. I realise the information in this book carries the arguments in a childlike manner which are very simple to follow, and for that in the past I have been blamed over and over again as being unprofessional. In my answer to that I can only reply by using another question: Are only professionals adequately equipped with minds that make them (the professionals) the only ones able to think? We being part of the human race are all thinkers. Everyone as a human being can think. Every person on Earth is a thinking thinker that uses his brainpower by exploring thoughts mainly and normally to his or her personal benefit. It is what we think about that produces the results of our efforts by which we accomplish whatever we are thinking about. I have met professional Academics that I found foolish as much as there are other cases where the so-called amateurs can credit themselves with much wisdom and insight. Albert Einstein as a patent clerk was that much but to name one. Please understand that I do not compare my achievements or myself in any way, shape or form with the likes of a Master such as Einstein, although I speak my mind when not being totally in agreement with some of his or other views. My unsophisticated retracing of Mainstream physics concerning the Big Bang in detail helps to reinvestigate established principles and moreover investigate proof in the light of modern evidence. In principle I distinguish between Kepler and Newton in that Newton is one hundred percent correct concerning gravity on Earth but as far as outer space forms gravity, the conclusions of Kepler and Newton do not match and they had totally different ideas about what they saw in gravity. I am in disagreement with some basic principles that science acknowledges and I divert strongly from all accepted roads Mainstream physics follow. By my doing that, those who are considered and accepted as self-proclaimed members of Mainstream Physics have categorised my views in the past as incoherent. That I do not accept. I admit that my line of thought is extraordinary and controversial but only to Mainstream science and not to the standards laid down by nature, since the concepts I follow start at the beginning, and I take Kepler at the point where modern cosmology began and in that mindset I re-evaluate Kepler's work. I start by tracing a new approach as to what I see Kepler found. The main condition of my investigation is to establish a divorce between what Kepler said and what Newton thought to add to what Kepler said. It is this divorce I create that Mainstream science finds repugnant or even in some person's opinion repulsive. I believe the repugnancy does not come from or is not manifested in any part of my work to the letter as such, but rather what my work suggests and who is doing the suggesting. To my view in cosmology such adding to Kepler by Newton was unnecessary and it diverts Kepler's work away from cosmology. But as the generations moved on, Newton became religiosity in the mind of science wherever science was taught. To students there is little or no choice in the matter since the only choice left to them is one of understanding by forcefully accepting or die an academic death since Newton is academically accepted without asking questions or raising an opinion. For the second choice, the less accepting students are greeted with a Dear John good-bye letter sending them off into the unknown sunset that such a future outside physics will bring them. That is brain washing.

From studying Kepler, I saw that we have to gauge what we find in the Universe. What we find is not what we realise with our eyes but what we observe by using our minds to translate from visions coming from our eyes to our minds. We have to test the part that we are seeing much more than merely accepting what there is to see on face value. We have to not only see what other life beings blessed with much less insight most probably also should see. We must stop using our eyes in the same manner as animals do and start seeing with our minds, as humans should do. Being the superior evolved species that we are gives us the ability to read into that which only we can see and that we only can see by using our intellectual mindset.

By seeing with an intellectual understanding what there is to see when we see what we can observe, we should therefore have the ability to be in understanding by looking at what we can see but moreover understand that which we cannot see. It is the same as playing chess. See what should be moved instead of noticing objects not having the ability to move of their own account. This I first found to be true about Kepler's work and when I started projecting this method of observing what the Universe is, as it scattered most previous perceptions I found that using the new method brought along answers so fast I could sometimes hardly keep up with the interpreting thereof. But as is the case with Kepler, so is the case with the entire study of cosmology: One should see what there is about the cosmos which is unseen to us and then we may find so much more in the cosmos unseen to us representing that which we cannot see and that which we cannot read because we have to learn to read what is not written in light. Armed with this realisation, I then proceeded from that point by further arguing and debating the full implication of Kepler's contribution. Kepler placed cosmic structures in relevance to one another and so does the Big Bang Theory. The backbone of the Big Bang is that relevancies apply in dynamics and such dynamics are placing all structures without any reservations independent from each other. As the Big Bang progressed, everything contained and locked inside the Universe is in the same Universe that was at the beginning and therefore the Universe will remain the same and everything inside will always be the same, however, the relations that the elements comply to, bring across new relevancies with new positions to fill. Everything that was, still is but it is the relevancies that change as time progresses and become space. The father of the Big Bang concept is a person by the name of Father LE MAÎTRE, GEORGE ÉDOUARD (1894 -1966) who was a Belgian priest and cosmologist. He was the first person to embrace the fact that the universe expanded from an infant stage. His model of an expanding Universe (1927) was superior to that of W. de Sitter in that it took into account mass, gravitation and the curvature of space. Similar models were proposed in the early 1920s by the Russian mathematician Alexander Alexandrovich Friedmann (1888-1925), but Friedman compiled various such possibilities. Lemaître argued further (1931) that the quantum theory supported an origin in the explosion of a 'primeval atom' or 'cosmic egg' into which was originally concentrated all mass and energy. As modified by A.S. Eddington, Lemaître's model provided the springboard for G. Gamow's Big Bang theory. In the wider picture of science in general a lot changed to just allow such turnabout in thought since the day of Isaac Newton. From Newton's attraction and contraction many things came into place that allowed change in the most hardened minds. Accepting facts about the Big Bang concept is quite radical. By promoting expansion, the Big Bang theory contradicts gravity and our accepting of the Big Bang has to change all other concepts. By accepting the Big Bang other changes are also involved.

Instead of studying the true value and contribution of Kepler's laws, an Englishman going by the name of I. Newton placed his own interpretation to Kepler's laws, and in doing this, he wilfully destroyed the principle working of the Creation. Saying this I hear the alarming hooters announce Newtonian dismay. In the past my experience was that all the revered Academics lost their appetite for any further investigation of my work. That is sad as much as it is regrettable. Through Newton's tunnel vision, he applied his own misinterpretations to the correct presumptions of Kepler and through the Newtonian tunnel vision Academics did not move an inch away from repeating the same procedure. In the past it was this that had Academics shying away from me because at the point where I raise criticism of the Newtonian viewpoint, I am rejected. The point where I declare my suspicions concerning their accuracy and the correctness of their theorising, which is where I should then be raising doubts about their way of thinking, is the point where instead I raise their suspicions about my way of thinking. This is what caused the rejection of my criticism of Academic Newtonian science and evoked their criticism of my views in the past, instead of them following the logic by investigating what I said. Their rejection of self-investigating had me and my work rejected to a point where the applecart lost its wheels on every occasion. It is where Academics read my remarks and what brings (seemingly in an instant) wrath to Academics. I say this because I realise that reading my remarks or hearing me remarking about this notion brought much resentment on their part and if the reader at the present moment is a Newtonian, boiling his/her blood. It is blood boiling because I believe they see my remarks as belittling what they feel they have accomplished. This is not the case but still my remarks have the same effect on the Academic as pouring icy cold water down the back of his shirt. I mention this because I know it has happened many times before and if possible I wish to avoid this response. Therefore I ask you kindly to please be warned about the negativity you must feel towards me where you are the Newtonian and I am not.

An Open Letter On Gravity Part 2

Before you lose interest in reading this letter any further please allow me to finish. In the past Academics thought me to be presumptuous and that normally became the point where all the Academics find their interest vanishes. That should not be because if Newton's work is as utterly accurate as those with faith in his work believe it is, then every aspect about Newton should stand above any and all reprimanding or any form of doubt causing a notion to reprimand. The testing of Newton's work should withstand all testing notwithstanding the person or the prominence of such a person's social or academic standing in the Academic society or even the prominence that such testing will deliver. From what I see about Kepler's work, it is a flow of circumstances that leads to Academics neglecting Kepler's work and the realising of the theory I suggest is not forthcoming due to my personal brilliance. I do not consider myself to be the brilliant in any way as to be the one that can remove the verbal splinter from the eye of the Academic. Yet…if there is a splinter what else should I then do…Newton reduced the implication that Kepler's findings hold by introducing the law of gravitation. He then went about and changed it to three laws of motion. It is clear that while he formulated the laws on motion he missed the way Kepler introduced gravity as space a^3 coming about through motion T^2 and that gravity is space a^3 within space **k** within motion T^2. Newton also missed the fact that gravity is at its strongest where motion and space cease to be. This is most important to recognise about gravity in one of the two forms it has. I. Newton generalized Kepler's first law, verified the second law, and showed that the third law should be amended to the form; $4\pi^2 a^3 / T^2 = G(m + m_p)$. In this, the value of "**T**" and "**a**" are the period of revolution and semi major axis of the orbit of a planet of mass m_p about the Sun of mass m, and G is the gravitational constant.

It should be clear to any person investigating Johannes Kepler and his work that Isaac Newton hijacked Kepler's work and any time there is the slightest referring to Kepler about the research Tycho Brahe and Johannes Kepler did, such referring to Kepler always lead to and always include the mentioning of Isaac Newton changing the work of Johannes Kepler. It is as if the World never could acknowledge Johannes Kepler because the work of Johannes Kepler would be completely wrong and misleading if it were not for the intervention of Isaac Newton saving the skin of the less admirable Johannes Kepler. This comes in the midst of everyone realising that Kepler used the information he received directly from the cosmos. I do stress this on many occasions throughout the letter because the embarrassing part is that Newton changed the work of The Universe and not of the man called Kepler.

Should you, reading the letter entertain the opinion of Newton and feel any urge to defend Newton you should ask the question "who is standing corrected?" Is it Kepler or is it the cosmos that gave Kepler the information he concluded? The cosmos supplied all the information by using mathematics, which Kepler then had to translate. But Newton destroyed the accuracy by altering what the cosmos said and directly by adding to that which he (Newton by name) thought that the cosmos left out. This set a precedent by Newton in cosmology and also set a trend, which was retained in all future cosmological development and it lasted in cosmology for three hundred and fifty years. In this book you are reading, I am about to show that such practise should no longer be accepted in cosmology. In the process the world of Mathematics developed and the world of cosmology stood still for almost four hundred years. Faculties contributing to cosmology and feeding off cosmology improved as much as they developed, but when cosmologists see the Roche limit in action in the lens of the Hubble telescope and refer to the event as "stars blowing bubbles" being the ultimate response coming from those persons who are supposedly the Masters of cosmology affairs, then the truth of what I just said comes down on you like a ton of bricks. Everyone having any remote interest in cosmology will find they are being very disillusioned by such "official" testimony about the evidence the Ultra Wise report on. This book is about showing how great Johannes Kepler was and how enormous his work was. It will show he preceded all ideas of everyone that came later and officially introduced the novelty of such ideas. Back during the time Kepler was introducing his work, the stature and the magnitude of his work was beyond any person's understanding (including Isaac Newton) and this prevailed for most of half a millennium. I do not say I am the brilliant one to uncover Kepler in the face of everyone failing that came before me, but as I am not a Newtonian, such bias was not part of my repertoire and denying me the fortune of being a Newtonian added to my fortune of realising Kepler.

Yet as you will notice, the work I contribute is much below the sophisticated norm of modern investigative research and the levels that modern research accomplishment demands to better the effort of the understanding ability in the splendour that investigative research work should deliver in view of our modern times. It is only pure neglect in science circles that moved science past Kepler. Not seeing and therefore not investigating through almost half a millennium has paved a road past the inferior levels that the researching of Kepler's work holds because it was rocket science four centuries ago but the brilliance of it has faded since then. My contribution holds no astonishing flair that may add to science in general. Only failure to notice what I see on the part of those truly brilliant can explain my being able to present my contribution in investigating Kepler. Only by their passing such degrading levels of the Academic establishment in the past and the present can bring the blame for such an obvious discrepancy because any involvement in the work at such an inferior level as that which I bring cannot interest and excite a salted Academic and when thinking about it, the idea is totally unthinkable.

This letter, although it is on this inferior level, is about correcting this tendency and has in mind the effort to put in writing what would place Kepler in the greatness and glory he deserves. As I already said, if Kepler was wrong, then the cosmos was wrong about facts and applying relevancies and tendencies in the cosmos. I yet again wish to reiterate, we should never for one moment forget that Kepler received his information directly from studying the cosmos so how could the cosmos stand corrected? In spite of all the brilliance attributed to Newton nonetheless if Newton had the mind to change Kepler's work and my saying this includes all persons agreeing with such changing by Newton of the work of Kepler, those persons admit that he or she or Newton never took any time to really and truly investigate what the cosmos told Kepler. From my reading into the work of Kepler I prove gravity, the Titius Bode law, singularity, space-time, space-time relevancy, the Lagrangian system, the Coanda effect and the Roche principle, the sound barrier, the principle behind the Black Hole. The precondition for my ability in doing so is that I have to remove Newton's opinion about Kepler's work from Kepler's work. Whenever cosmology comes into question and all the phenomena, which I mentioned just now remains unexplained and by that token alone it shows to what degree cosmology remained undeveloped. Whenever there is any mention of Newton, Kepler is never mentioned. But the reverse always applies. Mainstream physics holds the opinion that Kepler may only have an opinion if Newton can change the opinion. Kepler gave space-time, gave gravity, gave singularity, gave the Plank theory, and gave the theory on relativity but no one ever found Kepler's work deserving enough to launch any investigation such as I did. I be-laboured this because of what revulsion my rejection of Newton unleashed. This is one barrier much unnecessary but it has been an insurmountable barrier thus far.

I researched the work of a man that is most exceptional and therefore should be placed much more prominent in the allocated position his work has in the history of mankind. His contribution in the gathering of information that furthered the entire human species in their accumulation of knowledge as well as the human understanding in cosmic affairs, stands second to none in comparison to most others whilst most people are not even aware of the full implication of his work. Whilst recognising the work of Johannes Kepler, Mainstream science bluntly ignores the impact of his work, and in that they miss the full vastness of the wide influence of his work. Newton shrouded Kepler under a blanket and every one since kept Kepler there. It is therefore almost absolutely realistic to say that what you are about to read in this open letter sent to you for your attention was never yet printed in the near or the far past although the work has been with us for about four hundred years during which time it went unnoticed. It seems to me that any research predating Newton never came into use or into practise. My investigation of Kepler's work brought about a conclusion that no one yet arrived at concerning the findings of Kepler because no one scrutinised Kepler's formula. Kepler found planets rotating around a centre but Newton saw a circle and added what is mathematically required to indicate such a circle. Newton added a mathematical $4\Pi^2$ to the formula of Kepler and removed the distance symbolising measure that Kepler introduced using **k**. On the other side Newton changed the symbol of **k** by using the symbols G (m + m_p). This is just a longer and probably a more detailed manner of indicating **k** and better defining of **k** but it symbolises precisely to the point what **k** stands for nonetheless. I wish to draw your attention to the matter of Johannes Kepler's findings that Mainstream science considers as resolved and closed for many a century while it is not.

While Kepler and his work were considered a closed case at the same time not one of the following principles was yet successfully proven but I believe I have accomplished that goal. I first started my studies in the field of Cosmology as a spontaneous development of my natural curiosity spawned from childhood interests in the field of cosmology, which I developed even before I went to school. The studies were a reaction (I would imagine) that was part of my personal childhood development in how I was forming a personal concept of a lifelong interest that followed me into my future. At first I conducted all my earlier studying mostly on the basis that inspired me to find out more about what made the Universe tick, with no intention ever on my part to reach a point where I would be writing books on the subject. At first I was investigating cosmology on a part time basis. This went on for the best part of twenty odd years (*as* time and *when* time would permit). Then in later life with my health deteriorating I committed myself to more intense investigation and my effort developed onto involving a study using time that is only permitted by a person when that person is involved in such a quest on a full time basis. That quest has now been going on for the last seven years in full devotion and if one includes all the years invested on my part including the twenty odd years before, part time, then the time I have spent in completing my theory when adding all, it comes down to almost twenty eight years. This is to say that I did not come to realise what I am about to introduce on a light-hearted conclusion. I mention this because I wish to ensure the reader that he/she should have no doubt about my most sincere commitment in producing a cosmic theory on matters concerning the start and the working of the Universe during and before the Planck era. At first I began by arguing that there is something that is blocking our progress. There is some barrier preventing humans passing a threshold whereby our understanding will pass such an obstacle. If there were any way that anyone may break through that barrier which is preventing normal research to go pre-Big Bang, it would be accomplished by finding the barrier whereby the vision we use to focus would pass such a limit. If we wished on progress in our pursuit of the very first cosmic moment, then we have to find and cross the barrier that blocks our view. We have to look deeper and in another direction should the desire driving us be strong enough to commit us to reach into the very birth of the cosmos. We have to rethink the strategy we use. Max Planck was one of the most brilliant men of all times and even he, notwithstanding all his personal brilliance, accomplished little. There are parts missing in what we have and that which we have at our disposal to use because if there was no such an obvious barrier then the Wise-Men involved in science would by now have found the way to break through the seal that is locking us out of the critical past which will uncover the origin of the Universe's infancy stage. I went about trying to find what everyone since Adam, (meaning all of the rest of mankind and myself) were missing throughout the ages of speculating and interpreting while philosophising about whatever we find inspirational. The obvious we saw; that was clear. Therefore I had to find a route that would lead into the not so obvious which all of us were missing, notwithstanding the best efforts of the best qualified to accomplish such a breakthrough. My efforts involved trying to accommodate that which was in the cosmos available to use by the cosmos in all phases of development. If I had any hope of finding the answer, such an answer had to be simple because I am not very inclined to unravel what is deemed as complicated. The simplicity had to be locked in what was not yet understood about that which was in the cosmos as it formed part of the process used in forming the cosmos. My realising this brought me to focus not on that which we understand. There is not a lot we actually understand because even gravity is very poorly understood. In fact gravity is so poorly understood that there is not one person alive that can claim the prestige of understanding gravity and among the dead there is even less that can make such a claim. There are several phenomena that are presented in nature and acknowledged by science but also discounted by science and therefore not presented as accepted science. By admitting that which we have available to us to use concerning our research of cosmology in an attempt to better our understanding of cosmology, is useless to use, then one realises that not having what there might be makes what we already have useless. It then is useless to use what there is as part of the big picture we are trying to paint because what we use is not really part of the picture. This leads one to believe that the picture of the cosmos Mainstream science is painting, is being painted without painting a full picture.

In my first attempt to understand the full picture of what science was painting, I found so many colours missing there was no picture painted that anyone could appreciate. This is what made me decide to go on researching the 'unknown' in the hope it might clarify the 'known' and as the book unfolds, you as the reader may agree that I was correct in pursuing the misunderstood and rejected phenomena. Finding the missing phenomena helped me to place the phenomena mentioned above in a theory where the principles also mentioned above form a part of the overall gravity used in binding the Universe. I believe what is in the Universe is not able to be coincidental because of too many influences contributing to what there is - notwithstanding the fact that this is the manner science uses when they refer to the Bode law. What is in the Universe has a role as it had a role, which is the same role that phenomena *has* had and in future *will* have. This is establishing a very new idea about the working relationship between particles and in explaining it by using Kepler's studies. Redefining the work of Kepler's views brings a new Universe to light involving new concepts that are based on old principles but principles in updating man's view about cosmology are very new in that capacity. Through that new vision I was able to come to realise what the reasons might be why Kepler never saw it fitting to include the measure of Π in his formula. I do not suggest his neglect thereof was intentional. Nevertheless the formula he devised without using Π proved that there was no need for the inclusion of Π since his figures brought about a correct answer in the final end result leaving a well-concluded fitting answer. The numbers he produced brought about a specific space $\mathbf{a^3}$ contained in a circle $\mathbf{T^2}$ at the distance of $\mathbf{k}$ from a defining centre. Thus the calculations did not require the use of Π to find a meaning. In that Kepler did not see a need to include Π. I would not go as far as declaring with absolute certainty on his behalf that he did it deliberately. However, there never arrived such a necessity. It is prudent to agree on whether or not such a need is necessary, because if one is agreeing about such changing not being required, a new Universe emerges. The circle that Kepler discovered came about without ever forcing Π into the frame because it is clear that the circle formation came about as a natural consequence and came spontaneously delivering an equation while he was working. In this book I prove that the reason for adding Π to the rest of Kepler's formula is unnecessary. This unnecessary addition is because when going one step further in the investigation one will find that $\mathbf{k}$ and $\mathbf{a}$ and $\mathbf{T}$ are symbolising the same value with the only difference being that each one represents a different dimension to our six dimensional or six sided Universe we enjoy. In fact I shall show that Π replaces "$\mathbf{a}$" and "$\mathbf{k}$" and "$\mathbf{T}$" and that Π is the true value that should be replacing each factor as to indicate the correct value to the sides nominating Π. We humans work on a numerical base, using ten as a basis where we count to nine and re-establish a new decimal numbering line by adding a nought behind the number in value. This is using the numerical basis of ten, which I suspect we took from ancient knowledge about cosmology and not from using our fingers and toes as the earliest calculating processors. In this letter there is unfortunately no room to explain my suspicion but another fact I do prove is that the cosmos uses Π in the cosmic numerical basis as a means to measure and quantify. Therefore in fact the

Kepler formula should read instead of $a^3 = T^2 k$ as it does, it must be $\Pi^3 = \Pi^2\, \Pi$ where I shall show that Π represents singularity wherefrom the entire Universe sprang from Π and by forming as $\Pi^3 = \Pi^2\, \Pi$ it is confirming that space is equal to the motion thereof. Kepler's greatest achievement was showing that the cosmos is space–time $a^3 = T^2 k$ while time is the motion of space in space. The value of Π is the primeval and most basic of measures applying as an accepted cosmic legal value that the cosmos used exclusively at the very beginning as well as it does still today. The measure of Π in the Universe, values particle development that brought about all development ever conducted in the Universe. Only after this stage did the rest come including mathematics and went on to freeze spilled singularity into frozen material. Reading this statement may sound suspiciously senseless but as the book unfolds the sensibility will become apparent. The full implication of such a statement will become clear when one dissects different facts coming from studying Kepler. My discovery of this fundamental basis of legal valuing ensured me again that there was no need for someone of the likes of Newton to add Π in any form to the work of Kepler because Kepler discovered the ultimate Π in the Universe, the Π giving the Universe form and gravity. The concept of Π that is the only single form of all other forms available that can by duplication of Πs assemble the value of gravity. When replacing the symbols with Π the facts of the Universe become self-explanatory because the most basic form that forms the cosmos has a definitive and uncompromising value.

But getting this far took me down roads overgrown by ignorance and which I had to uncover myself as if hacking away miles of overgrowth with a machete chopper. All of the disbelief science showed to my work in the past and their refusal to see past Newton made any and all attempts on my part as bad as they could be, strangling and smothering my attempts to announce the uncovering of the newly found insight on my part.

For decades I tried to come to terms with the inability there is in science to explain the cosmos in real terms, when using the science of official reputation. That which there is makes a mockery of science because the undisputable clues left in the cosmos makes what little correct explaining there is available, seem like a comedy of errors, when it is mixed in with all the other near Dark Age errors we still use after so many centuries that provided countless opportunities to revise the old muck. By applying current accepted Astronomy as such, the phenomenon found all over the cosmos is still beyond the explaining ability of Mainstream science. This is true and it is a shame because it also is an undeniable fact in spite of the vast knowledge and progress in other forms of science taken in the manner science uses when it approaches cosmology. Cosmology truly lagged behind while the understanding and advancing of physics, mathematics and chemistry as subjects were flourishing. By comparison I saw how little there was available in explaining cosmic phenomena and how much improvement in understanding the other departments such as chemistry, electronics, medicine etc. could offer as results were coming about from research. Even where there is a little explaining available in cosmology it turns out that such explaining is confusing to say the least and at best it highlights the manner in which science is applying double standards. For decades photographs were the only progress forthcoming as an addition to improve the meagre field in cosmology and that improvement was artificially stimulating cosmology. By providing a false impression of advancement, everyone missed what and how much was missing…To the connoisseur desperately looking for more than the obvious stirred in with some out-dated misinformation dating back to the Middle Ages, it all seemed as if it was a picture portraying the ridiculous to make the sublime look good. The pictures only proved the opposite of what progress in cosmology will represent. In truth and as such in cosmology the cover up that was hiding the lack of progress about the science of true cosmology was only forthcoming in the improving of electronic optical telescopic advances and spectroscopic progress. There were only photographs carrying beautiful pictures which pleased the less informed except the photographs did not bring progress to cosmology at any intellectual level by promoting insight. The explaining that the photos demanded about the subject had the opposite effect of installing hope because what it did do was underline what lack in any notable progress there truly is in our understanding of cosmology and laws in the cosmos.

While such Hubble telescopic images might seem to be as clear as daylight, it was more than clear that there was little academic value to them. To the person in need of more stimulation than being impressed with pictures of God's marvellous Creation and the sightseeing that always accompanies such pictures, such persons always felt very disappointed. The pictures did give satisfaction to those more easily impressed, but the rest of us seeking knowledge accompanied by understanding the images left us despondent. Although they leave the vast majority in total amazement, there are those less impressed about not knowing the 'why' and the 'how' in such amazing pictures. I am aware that the group I fall into may be the greater minority and the majority may only demand the portraying of the images, which is what that "easily satisfied" group demand. The rest of us rouse with anguish at the lack of information about what is known, and what lies behind what those pretty pictures are conveying. Nevertheless, there can be no real progress in scientific understanding about the images portrayed by the Hubble telescope and others, if no one is able to show the slightest clue of a deeper understanding of what is going on in the Universe. Everyone is almost breathless waiting for commentary by the most informed, which accompanies the magnificent cosmic portraying of God's Creation. When we are portraying the new images, we should also be investigating that which we see the cosmos is at that moment portraying. The lack of actual believable explanation coming from investigating by means of telescopic imaging should impress one and all, but the impressing must not be based on the colours in the images but the sensible information attached to the image investigated. It is **that** which we wish to see. What we wish to see must at least be accompanied by scientifically backed information, which provides the proven understanding coming from science. When science is employing new explanations with such photos it should also be discarding senseless baggage carried over from the past. Most images contradicted Newton and for saying that, every Academic I ever came across in the past ostracized me. That bothers me little! I know I cannot possibly be the only person absolutely discontented with what Mainstream science accepts as science. Here I refer to the out of date theorising Mainstream science still accepts amongst many others as how they suggest stars and planets are forming. One cannot promote cosmology in honesty and advocate scientific fact whilst dishing up such fairy-tale nonsense to students. Moreover I hold the opinion that amongst Academics in particular there must be many if not most that share my personal serious doubts or have an inclination to share some of them. This I say when considering the overall doubtful picture painted about what there is and what one believes there should be. I just cannot believe those forming the most intellectual group of mankind are unaware of the mismatching facts seen over the broader picture because the contradiction and lack of a plan, makes what there is so very doubt provoking. Newton dismissed the formula Kepler presented as all factors forming motion. That is where the apple cart derailed.

In honesty we have to realise that we cannot dismiss the whole formula that Kepler produced as being motion. It is so much more than just motion. It is $a^3 = k / T^2$: That is what Kepler brought into civilization for all time to come. He saw space a^3 being in isolation due to the time it uses to move T^2 claiming such space forming independence according to the lines **k** indicates. Let us look at the factors in more detail before we proceed with the rest of the book.

a^3 symbolises a mathematical interpretation of implicating the three-dimensional space.

T^2 is representing the period or time that Kepler suggested we should use to calculate time that holds the orbiting planet in direct contact with the space in relation to a very specific centre.

k is the space taken from the centre to the end of the line from which the planets must have grown if one accepts the Big Bang growth of particles and the affect of the Hubble constant on all cosmos material. The specific value about the centre is most important because from the specific centre, gravity always applies the strongest influence.

One cannot justify Newton's dismissing of Kepler's formula as that all factors only contribute to the motion indicated because that is misleading. We all accept that the true cosmic form *would be* and most probably *is* a sphere. Everyone accepts the universe as a whole as a sphere…but why would the sphere form? What would be the reason why the original form that we devote to the Universe would take on a sphere as a natural form? Apparently our imagination grabs the sphere as form. In all natural events the gravity in that space which stands apart and independent from all other space takes on by cosmic pre-casting the sphere as form of shape … **it is because gravity chooses the smallest space to hold the strongest force**.

I am of the opinion that gravity is about dismissing space to the advance of heat increasing in such a specific and concentrated space using the concentration as measure for the heat as well as the space holding the heat in space. According to Kepler that is what he found to be true. Space a^3 will always be circling space around as T^2 in any position from the centre **k**. That is what Kepler said when he said $a^3 = T^2$ **k.** Kepler indicated space a^3 will forever fight for independence and show separate individuality in remaining apart as identifiable cosmic components by means of motion.

An Open Letter On Gravity Part 2   Page 555

Every space will cling to independence indicated by **k** through fighting off the integrating of another coverall unifying unit by applying the motion of T^2! The problem we have to solve in this letter is what will the cosmos use to secure such independence between all particles? What sets space apart from the rest of space? First we have to admit that Kepler was the one that introduced the following.

Kepler gave us the answer to the following but no one ever took notice!
Kepler was the one that discovered **space / time** as $k = a^3/T^2$
Kepler was the one that discovered **singularity** as $k^0 = a^3/T^2 k$
Kepler was the one that discovered **gravity** is holding **space-time** relative by the measure of distancing **k** as $k = a^3/T^2$ and $k^{-1} = T^2/a^3$

Everyone able to read mathematics has to realise that Newton suggested collisions between cosmic structures must eventually come about as gravity erodes the distance separating the cosmic structures multiplied by the product of the mass of both structures from both ends. Newton said the multiplying mass of both structures destroys the distance between the structures by using the eroding force of gravity in the square. The cosmos then must end in a Big Crunch with all material joining together, but that joining is not forthcoming at all…and that only indicates how much insufficient understanding there is on offer in cosmology by the educated–to-be-wise-about-these-matters. There is precious little available to explain about their field of cosmology amongst the ranks of Astronomers. So…let us return to the beginning of cosmology before everyone became oh so wise and see what there is to see.

Finding an academic with those qualifications was my biggest quest thus far. At first one may think of such a mistake as trivial. It seems minuscule small and should not bother any person anyway. From the onset, the mistake seems as insignificant as it is small. The mistake came about with the culture of education and the mistake in itself seems harmless. When admitting that, one must also admit that any pioneer that got lost and suffered privation by ultimately succumbing from starvation through an incorrect travelling direction, made the very first part of his ultimate mistake by a mere glance in the wrong direction? How harmful does looking in a specific direction seem, and yet such a mistake leads to his ultimate capitulation. The traveller's mistake might start by taking his first directional flaw with that first incorrect step only when possibly putting his foot skew. Or he could have turned his face. It is the outcome of mistakes granting the mistake notoriety and not the start of the mistake. Thus, please keep that in mind when you may find my next declaration somewhat silly and most likely overrated, but my mentioning is about the start and it is the start I introduce and not the end of the mistake. **Lines mathematically cannot start at zero because there is no evidence of zero as a factor in mathematics.** The question in need of answering is this: **What will the length be of the shortest hypothetical line imaginable and moreover, should you disagree with my statement, what would the total overall length be?**

The value of Kepler's space, which he indicated as a third dimension a^3 does not depend on indicating a structure a^3 that is in rotation T^2 but only needs one position having a constant of some sorts. Any point where **k** may indicate a position, one will find a value matching a^3 and the matching location will fit T^2 at that point. That is the relation there is in the solar system between all planets and the sun. The sun always indicates the centre and the planets always indicate the rotation. But $a^3 = T^2 k$ is only producing a relevancy of three dimensions that is equal to two plus one dimension.

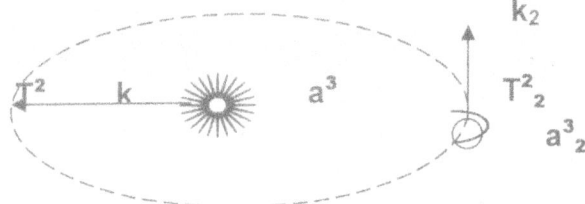

From the sun there are three points moving between two points from one point to two other points giving the six dimensions we find in space. It is space in time or space converting space through the movement of time. It is a location of a point in the third dimension a^3 that will move according to the second dimension T^2 that will implicate k as a reference in the first dimension. It is about dimensions in reference to one another.

Let us take it from a point where the sun provides a centre as one starting edge of k then that centre k will provide a line from the centre and the line k will provide three spots in a formation that produce a structure by the square T^2 of the dimension. Not once did Kepler indicate size as a contributing factor to a^3. That means every single point that k indicates there are three positions a^3 implicating sides of a double dimension. In the same manner is k not limited to distance or is T^2 lesser by size. $k = a^3 / T^2$ That is what Kepler said. There are three dimensions a^3 between any two points T^2 flowing as time from the centre of the sun, which is indicated by the line k.

The implication of the relevancy produced by the use of the formula $k = a^3 / T^2$ brings about that when dividing T^2 into a^3 there is k left. The fact is that a^3 is a three dimension (^{3}) of single k (^{1}) showing one or T^2 is two dimensions of k being the one dimension, it means that k is a part of space a^3 or T^2 which is time. It is the same thing in a double dimension or space being a triple of k then k is one factor and k cannot show a position of zero. If $k = 0$ then there is no possibility of $k = a^3 / T^2$ because $k = 0$ then $0^3 / 0^2 = 0$. That does not make sense. Mathematically space cannot be zero because those being of the opinion of space being zero or nothing must first prove mathematically that space is zero. Moreover they then must prove mathematically how does zero grow through the Hubble constant. By translating Newton's vision of the circle in completing a cycle would become zero through rotation…well that does not count the use of the formula a^3. If k cannot be zero, then k could not start from zero. With $k = a^3 / T^2$, no point can be zero because k shows space $a^3 = k T^2$ is no reference to the volumetric mathematical formula used to calculate $a^3 = 4/3 \Pi r^3$. Nor does it show the use of the circle in the second dimension being $a^2 = \Pi r^2$. In the case of the Newton formula, the circle factor becomes the square as indicated by the duration of the time T^2. The factor standing in for the line which normally would be r and then be the square value, is in the case of Kepler not the value indicating the square. That means Kepler never indicated a circle of mathematical procedure but said mathematically the distance of the planet from the sun k holds space a^3 in relation to time T^2.

Lines mathematically cannot start at zero because there is no evidence of zero as a factor in mathematics. Should you disagree with my statement the question in need of answering is this: **What will the length of the shortest hypothetical line imaginable be and moreover, what would the total overall length be in that case?**

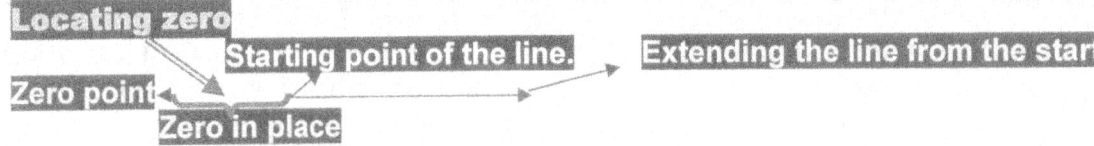

The fact of form proves that the sphere captured all sides that can possibly influence the sphere. The sphere therefore holds **k = a³ / T²k** within the boundaries designated to the sphere. When a body is placed in a location on the outside of such spherical borders, then the lines forming points of any object seem to float in any direction. There is no control one can establish which will secure movement in any specific direction of preference except by releasing heat to counteract the required motion in a specific direction of choice. We all have seen what happens to any object that comes into the border area of a sphere. The object suddenly is motivated by motion to follow a specific designated direction and the motion leads the object to move towards the centre of the sphere. It is as if the support of the six opposing sides has lost one side where the sphere took over the control and movement starts in the direction of the Earth centre. The support of one side is literally removed by the centre of the earth where Einstein claimed the strongest gravity is and the motion of the object starts in that direction. There is no pulling on the object but there is a removing of space by the centre of that specific point leading the object and the space it is in as well as the space it carries to move to the centre spot. In the sphere the borders the sphere holds are deliberate and very distinctly placed edges forming a specific distance from the centre. The centre is also proven beyond any debating. The centre of any sphere has to be at the very point where space completely falls away. That will put that space at that point in the single dimension and centre is the single dimension.

The shortest possible line (hypothetically) must be so short it must have **an initial and ultimate point sharing the same spot.** Any theoretical line being the shortest possible line cannot have the line holding the initial starting point at point zero and advance from there. If it used zero as a start, the zero part would not count, because the line will only start at a point past zero where the line then will start. Zero ultimately means not existing and then that point, as a start does not exist. When the line **has a beginning and an end at the very same spot** and it wishes to extend the position as to further the possibility it has, which direction should it favour? Extending the line in any one direction will favour one direction without any clear reason not extending in other directions. The only mathematically sensible option about extending will be in all directions equally in order to give a meaningful non-bias flow of mathematical equilibrium.

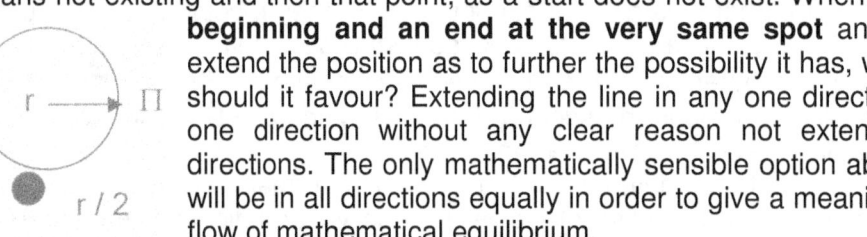

When **the circle reduces**, the **value** located to **r** will become implicated because **r determines specific size. Not so in the case of Π, because Π** in the true sense only **indicates that the circle is a square without**

corners and therefore Π **dictates form and not size**. By **reducing size** only **r comes into contest** and will point to such reduction. By **reducing** the circle **radius r by half continuously** will lead to an **infinite small circle** but Π **will remain because the circle as a form remains** even being infinitely small.

The shortest line in the realm of possibilities must have a start and finish holding one spot and such a line will also be a dot or a circle. Not favouring one direction puts all directions at equilibrium meaning that any form of what ever might develop from such a spot with the end and the start being in the same position also has to be a sphere because the flow outward will be equal in all directions. This reasoning prompted me to look for singularity in such a spot because if the prime spot from which all came was a spot holding all, then the spot must hold the shortest line but more prominent it will hold the smallest form including the smallest circle or for that matter the smallest sphere. One possibility that the shortest line or smallest spot can never have is having a starting point on the zero mark. If the mark of zero holds the start it must also hold the end because the end and the beginning have the same position. If the position of zero then is the beginning, the end will also be zero leaving the line or spot without an end as well as without a beginning. Such a spot will constitute all of nothing.

The size only depends on the fluctuation of r in the square as a component to the circle or sphere but that does not affect the form by indication Π in any way there may be. The conclusion from this is that no line can start at zero because that will be a mathematical impossibility. A line or spot starting at zero would therefore be shorter than the shortest line possible. For obvious reasons can no line, or any line grow or extend from zero because such a line must then quit zero and become something, thus abandon its original value. That would mean the start of the line has a different value to the end and a line holds conformity throughout. When any line is starting from

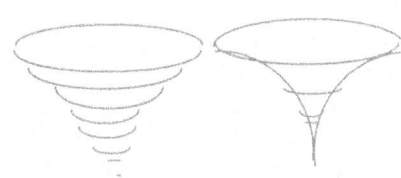

point zero, it can never leave zero because of the influence of being zero disqualifies any possibility of growth. If the line then had to grow in all directions at the same pace, the line must therefore be a circle or being three-dimensional, a sphere. Flowing from this fact is that in the universe there can be no zero point or unfilled space. In the case of the growing sphere, the value of the circle is Π, and that is where creation started. That gave me the clue where to start looking for singularity. One would find singularity in the value Π and the value Π will be in all things rotating in a circle. You might wonder how does that apply to the cosmos and moreover to gravity?

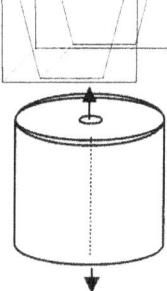

By reducing r indefinitely to the tune of half each time, r would become infinitely small, beyond human calculating means, however as mentioned in the case of the smallest dot holding one spot, r would become insignificant beyond human comprehension even, but never reaching zero and still Π would remain intact and dictating form.

An observation coming instinctively to mind one may recognise is that the form reminds rather explicitly of a natural phenomenon such as a hurricane, a water whirl and even the shape most commonly favoured to express the cosmic object referred to as a Black Hole. The similarity may be more than coincidental. Let us consider the statement in the reverse.

Anything occupying space in the cube will apply r, notwithstanding the name used confirming the shape or r named as length width or height, it is all just a straight line bringing about the cube with all its other names that may find attachment to a specific form but nevertheless still remains only a six-sided cube with connecting lines applying different angles changing in some cases.

The normal perception is that any circle growing spontaneous would grow by the radius, which is r. That cannot be the case because r is an indication of a straight line. By growing with the aid of a straight line, the influence that would have on the circle would result in many circles following one another and not a continuous growth. Gravity is the dimensional changing of space holding r as reference to the sphere holding Π as the reference. In order to generate spin that is producing time in all matter that is occupying space, therefore it is creating dimensional change, where Π has to be a factor indicating the possibility of spin. The answer must be in finding Π, and thereby locating singularity.

$\Pi = r$

in constant directional change as time flows through rotation

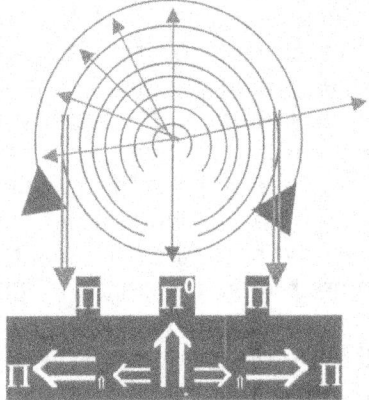

Due to the spinning nature of such a point with all surrounding the point will be alternating direction favouring change every second and in that the value to such a point can only be Π because of its constant changing. Using r would specifically oppose another r from every angle because the use of r will bring about a static relation to the previous and following instant and therefore it will cancel the constant spin flow.

There is a clear method to pinpoint the positioning of singularity going from Π^0 to Π where such positioning requires space on either side of singularity forming the border set by singularity.

The new direction pointing to a new location in relation to the previous point will oppose the previous point it had in relation to direction considering the centre point.

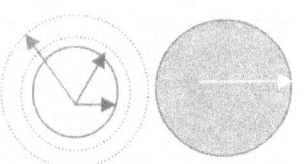

In order to form a concept of that which I am referring to, then think of pneumatics. In Pneumatics (air compressing) we can associate such compressing with the use of r as a pressure indicator and in the case of hydraulics one would rather think of using Π because the firm state that Π offers will form something more compact than even solids can offer. Therefore in air the substance can compress and become more compact but in the case of liquids the density is such that liquids cannot compress into a more dense state that it is in liquid form. Liquids act even tougher than solids specifically because it uses such relevance in the application of Π and not r bringing conformity evenly.

The circle to the left would come about from a straight line r growing and influencing the appreciation of Π, but to influence Π would lead to a breakdown in r as Π and r are different entities. The circles to the right shows a continuous growth by extending Π every time and since Π is the same part as the previous Π, only extending that billionth of a millimetre each time, the circle will be truly continuous without any signs of a break.

In the circle using $r^2Π$, the r has to have distinctive qualities placing it as a factor apart from Π. Where the growth shows no separate distinction but a continuous flow from the precise centre to the precise edge, the flow would become in relation with Π depicting the circle and Π replacing r as reference to any point on the circle. By using r, distinction in the circle is possible but by using Π there is no distinction possible.

When working with concrete and heavy metalled solid objects r would show as a crack distinctly parting solid structures, while Π indicates a continuous flow of solidness giving the material an overall and continuous structural strength, yet engineering never recognised this difference. By confirming Π the circle employs singularity in all components and therefore proves to be a much stronger support as building choice than other shapes.

The triangle, the half circle and the straight – line has two things in common, they share 180^0 as a mutual value and they are part of singularity.

Looking at the effect of gravity, it shows the precise quality of no distinctive point, as gravity never seems to end at a point but flows all over affecting all that holds a position in its sphere of influence. The gravity coming from China meets the gravity coming from America at no particular spot but intermingles without distinction.

Using the concept that gravity applies Π as the circle factor, Π as well as $Π^2$ replacing r^2, the replacing by Π brings two values as Π and $Π^2$. That I found is the case with gravity and will be apparent when explaining the sound barrier as well as the Four Cosmic Pillars. In order to create a distinction I remained using r as the indicator of the cube or non-circle that has vacant space and by vacant space I refer to non-solid structures. In the solid structure I use Π as a value for reasons that will become apparent in due time.

The value of singularity stems directly from the law of Pythagoras or **Pythagoras** is the result of **the average of singularity. With the shortest line being a dot, all lines must start from a position implicating** Π. A circle is a square without corners implementing Π and a half circle is therefore a triangle without corners. The corners are the factor that confused every one in the past. When replacing the value we normally attach to a circle being r with Π, the law of Pythagoras becomes quite meaningful and mathematical.

By placing a connecting circle on the sides of the triangle, half a circle forms. By implicating Π as a relevancy and not the straight-line r, two values of Π applies to each circle and the straight line is no longer r, but is $Π^2$. This will bring about that each circle holds half the square value implicated to the allocated conditions applying to Π in that specific instance. By adding the two half squares forming the two half circles and then calculating the square root of the total that then forms the average diameter, an average of Π in the connecting line will come about. As both lines are the straight line forming singularity coming from one line being Π, the connecting line then must be the average of the two lines as $Π^2$. That is what **the law of Pythagoras says.**

No object can be in two spherical quarters at the same time, but has to alternate in alliance to the space in accordance to time rotation.

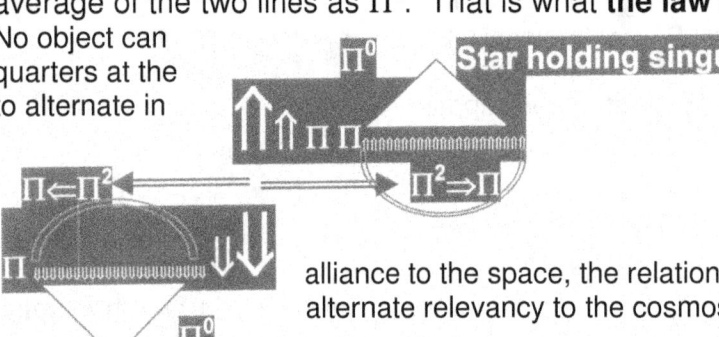

To alternate in alliance to the space, the relation of time in space has to alternate relevancy to the cosmos.

A STRAIGHT LINE, TRIANGLE AND HALF A CIRCLE WILL ALWAYS HAVE EQUALITY IN DIMENSIONAL CAPACITY PROVIDING EQUILBRIUM BEING 180^0 BECAUSE EACH ONE SHARES A COMMON DENOMINATOR IN SINGULARITY TO THE VALUE OF Π. As the straight line averts a zero, it holds another straight line in place to set about such an averting where the two lines will always carry a relevancy in elation to progress (the triangle) and a common denominator in the start from singularity. This concept we apply as the graph or the vector.

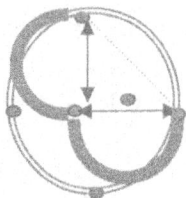

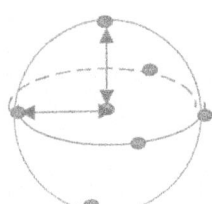
In the sphere there are never only one direction implicated in movement. Movement are always in relation to the centre position because as a line goes up it also goes in or out. When a line goes north or south, it also comes towards the centre or going away from the centre.

There is always relevancy present in movement. As this moving indicates direction, it also applies Π^2 for indicating value forming the time factor.

Because every moving line represents one quarter of the sphere in relation to the rest of the sphere and the line also indicates the relevant position between the point indicated and the point in the centre, it is a relevancy of singularity in progress. By connecting the line, as Pythagoras will suggest, the singularity within the sphere becomes a specific indicated value representing one half circle.

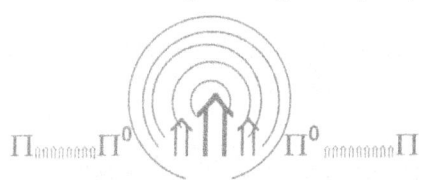

As the meeting of r points to a very distinct different r in direction, such a point of meeting opposes the other points in meeting and will lead to destruction of the form Π in any event of any value changes by Π changing to Π^2 and r. Space-time is a four dimensional position of the universe where the position of an object is specified by three coordinates in space and one position in time.

In every sector, the directional flow will provide a distinct meeting of Π linking r to Π^2 and this allows the time component in the rotation.

There is so much speculation about singularity, but in the end singularity is so easy to explain. Singularity is one. Singularity is in the space where all space is one.

Singularity means one. Singularity is the entire range of forming one in as much as $1^0 = 1$, $1^1 = 1$, $1^2 = 1$, $1^3 = 1$, $1^4 = 1$, $1^5 = 1$, $1^6 = 1$, $1^7 = 1$, $1^8 = 1$, $1^9 = 1$ and so on.

That means all the philosophy and the debating and all the speculation is down to one figure being $a^0 = 1$, $b^0 = 1$, $c^0 = 1$, $k^0 = 1$, etc. and therefore we have to look where any factor becomes a single unit contained as the exponent of zero forming the factor equal to one. It is where space ends in infinity and where space as we know space is, has no space in the Universe. It is where space is = 1.

Locating and finding Singularity

In the **precise middle** of all **objects in rotation** is a precise centre dividing the object in sectors that will **start the spinning initiation** from that centre point. But the spinning object **will have a middle point**, a very specific **centre point that does not spin** and only holds Π as a specific value. One value such a line **cannot have is zero** because **zero does not start any** line and therefore the **value of the line must be infinite**, just as described in **accordance** and by **the definition of singularity**.

An Open Letter On Gravity Part 2

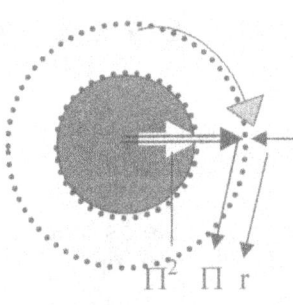

That point albeit hypothetical, is also as much a reality none the less and is where that point **must be standing still** because every line **running from that point** in **opposing directions** are also **in opposing directional spin to each other.**

In instant of characteristics characteristics it which they are rotation. The fact opposing considering the spinning motion in the fraction of time in the detailed instant, every aspect of rotation will turn in every change in time. Although the points had the same only seconds before, they oppose the had just before and just after the very second in and to which they relate by similar points also in of the graph proves my point in quarterly dimensions and values.

With singularity placed in infinity within the centre of every rotating object every atom and its relation to its surroundings including other atoms form space-time diverting from the point holding singularity as far as rotation goes because every object holds three relative positions in as far as where it was, where it is and where it will be in relation to singularity providing time. I elaborate on this else where.

Any point will be opposing itself within the **rotation of 180°** where it **then changes every aspect** of its **previous flowing** characteristics it had or **will once again have** in **360°** from there. While in rotation from the viewpoint of a bystander it all may seem static and never changing but to the object in spin every next instant in time will be diverting from every aspect it had every second passing, and the direction it held in relation to the direction it held the previous millisecond will totally be incompatible with the direction it holds the very next millisecond of rotation.

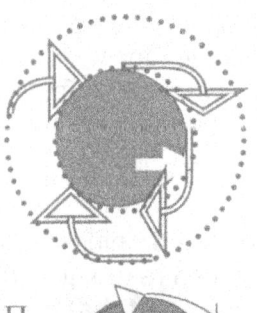

This is why we can use degrees measuring the circle by (6^2) (forming the square relating to matter through singularity) X 10 (square if space) = $360°$, however it is always in motion. That proves no point can be static or constant, though it may seem that way to outsiders. Although matter is matter, matter can also be anti-matter and moreover form its own anti-matter at the same time. This degeneration of structure is very likely to occur with overheating. Revaluing Π to Π^2 will bring about a new contact point where Π meets **r** forming another relation in Π^2.

Time is the **changes in relation** where Π **contacts a different r** notwithstanding the many r points there may form because **every r constitutes a different value** to the universe through other ratios and relevancies brought about **by heat and light. Time is the duration it takes Π to rotate between any two given points of r** and therefore must always amount to **a square (T^2)** moving from point to point through the **cube of space (a^3)** in that **duration of time (k)**. With that it proves **Kepler's a^3 (space) $=T^2$ k (time in the instant of motion)** but motion must continue through a specific value in space where the space-time is maintaining relevant equilibriums throughout singularity connecting.

In the circle $Π^2Π$ the space surrounding the rotating object will also extend by Π as the concentration of the spinning motion draws or drags on past $Π^2$ extending the influence of $Π^2$ by the value of Π. This extending of $Π^2$ to accommodate Π we refer to as the atmosphere, but physics apply to this extending in the normal fashion. From the spinning motion Π does not stop at the end of the solid structure but the influence of Π extends and this then becomes the atmosphere. The influence of $Π^2$ stops at the end of the solid structure but the influence of Π extending plays a most dominant role in the cosmos, although not yet recognised and that factor is most crucial to a better understanding of the implications of laws governing the cosmos.

With the circle being $Π^2Π$ the $Π^2$ will reflect the circle in the square with Π forming the extending of $Π^2$. The extending of Π will not end immediately but will carry to the surrounding space the circle influence through rotation. The influence immediately above the circle will have the biggest influence and reduce gradually as the value of Π reduces in the leverage that the space has on Π and a gradual but definite change from Π to r will affect the extending of Π progressively more. The decline of Π will follow the same contour of the circle at 7^0.

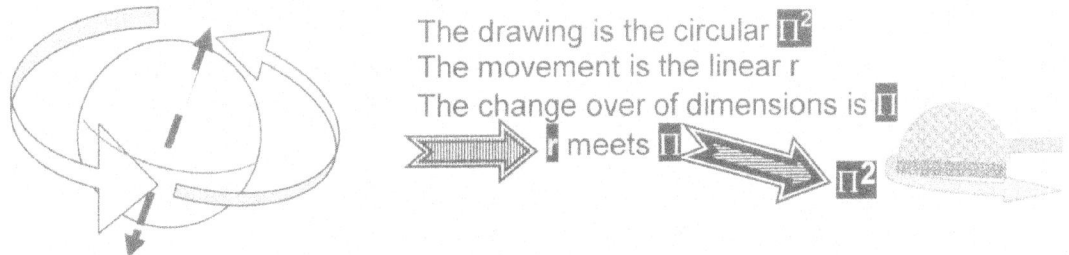

The drawing is the circular $Π^2$
The movement is the linear r
The change over of dimensions is Π
r meets Π ⟹ $Π^2$

From this line of reasoning I dismissed the theory of the presence of a force being gravity but rather consider it as a dimensional changing contributed by the spin of the Earth and the spin comes from singularity located in the centre of the Earth.

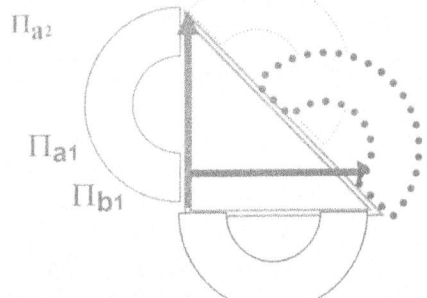

From there it influences singularity in the triangle flowing through to the half circle. It is an interaction between circular and linear motion as the value of Π continuous past $Π^2$ (at the end of the solid) and every cosmic structure holds an individual and specific singularity. The field where Π extends we call the atmosphere having a value of 21.991 / 7, which is Π.

$(Π_{a2} \times Π_{a1}) + (Π_{b1} \times Π_{b2}) = (Π^2_a + Π^2_b) / 2 = Π^2 =$ **gravity** **and that is proven by Pythagoras.**

An Open Letter On Gravity Part 2

Gravity is the average movement of matter through space in time determined from the position where matter in the sphere meets space in the cube from a point of Π to a point of Π^2 In this the figures of 2(5) = 10 (space) stands related to 7 from singularity as (matter).

Singularity: is a mathematical point at which certain physical quantities reach infinite values for example, according to the general relativity the curvature of space-time becomes infinite in a black hole.
Space-time is a four dimensional position of the universe where the position of an object is specified by three coordinates in space and one position in time

Being at... Π Going too...

Singularity =Π

Coming from =Π

With no line starting from zero because there is no zero as a mathematical fact, then all particles hold the point of infinity and not merely the Black Hole.
Where singularity holds a position in the centre of any and all rotating objects as a value of Π merely applying movement (in the form of atoms) qualifies space-time. It does not only fit the space within Black Holes but it fits all singularity becomes part of all the stars the largest cluster of matter.

all matter to be description of stars where from the minute to

From that argument one may conclude that all stars will become Black holes depending on the gravity increase they may generate.

Through rotation encircling the point of singularity and matter is (1) coming from, (2) being at, (3) as it is going to in one movement in relation to the specifics of the centre point being singularity all matter then qualifies to form space-time.

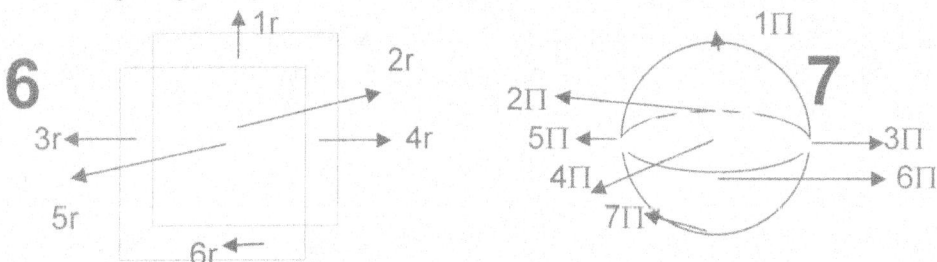

In the sphere there is no radius but only the extending of Π from the centre Π in six opposing directions relating to one another by the square but remaining Π because of the unity the matter holds in relation to space. It is not possible to draw a precise line that would form a precise ring and not cut some atoms in parts. Because there will always be an atom disallowing the precise positioning of the circle, the circle continues on a solid basis holding Π as a positional reference and not r. In every sphere there then are the seven Π relating in precise dimensional and positional equality forming equilibrium to the centre Π as well as to one another by 90^0 and 180^0 implicating the dimensional positioning. Therefore the sphere holds $_7\Pi$ and the cube holds $6 \times r^2$

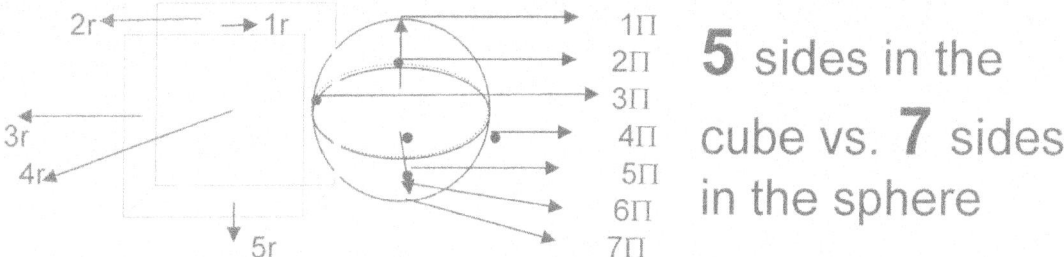

Where space comes into contact with the sphere, the cube loses one of the six dimensions it has to the more dominating seven contact points of the sphere whereby the seven contact points working in equilibrium will dominate the six dimension-cube that is loosely connected by r bringing about that the cube then has 5 sides to the seven of the cube. This means that in the cube the "bottom falls out" and without a "bottom" to support objects they fall to earth. Remember that a body "floats" in space, but at one specific point it starts to "fall" to the earth. That is gravity and it is a dimension change much more than any force. I shall explain this last remark later on. That too is the Lagrangian system with five cosmic structures holding relevancy to the centre structure where the centre structure stands in for seven positions diverting from singularity and the orbiting structures standing in for five positions in space.

With the dimensional change from space in the cube to space in the sphere a relation of 5 to 7 comes about depicting gravity. The principle of 5 sides in space relating to 7 in the sphere holding matter forms the basis of the Titius Bode and the Lagrangian principles.

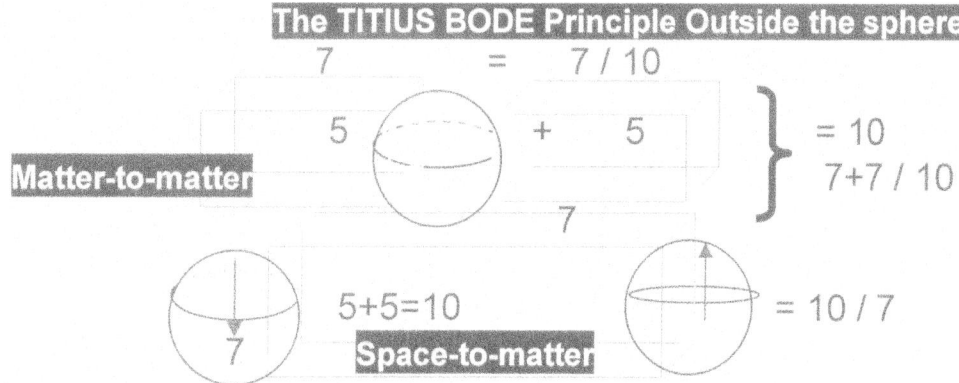

The Titius Bode law is an extending dynamic deriving from the law of the gravity dimensional factor where the space factor in a square of ten relates to a matter factor in the square by half (half since nothing can be in two places in the universe simultaneously) of the matter factor of Π^{7+7} or the square of space (10) relate to the matter factor of 7. From such a point every other point will be opposing any other point not pointing in the direction to which the first point is pointing, whereby it extends the direction it holds. No matter what the point is or where the point leads, such a point holding a specific direction will be unique in the direction it is rotating because at that or any other specific point wherever, it will be directing not in the direction it spins but in the direction flowing from the centre point outwards.

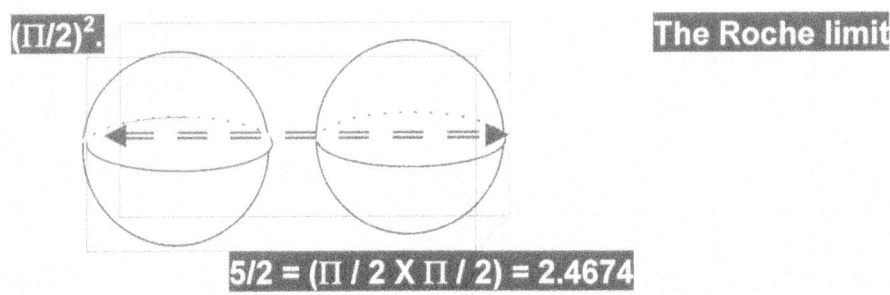

In the Roche limit the space factor provides space to a solid structure and therefore the value of r is replaced by the value of Π bringing about a square in half of Π. The cube holding 5 to either side removes the shared connecting point that is between the sphere and the cube, allowing the extending of Π to indicate position to space. Where Π extends to lock onto the next sphere's extending indicator, Π has to connect to Π forming the square of space and translating that to the half of Π being $(\Pi/2)^2$.

The space between the spheres divides in half, but because of the extending of Π and not applying r as ordinary mathematics will suggest where Π replaces r the singularity extending from Π^0 will be half of Π in the square of $\Pi = (\Pi/2)^2 =$ **2.4674.** In this lies the dynamics why planets have a positional (be it rather a dimensional) relation of 7/10.

The TITIUS BODE Principle inside the sphere

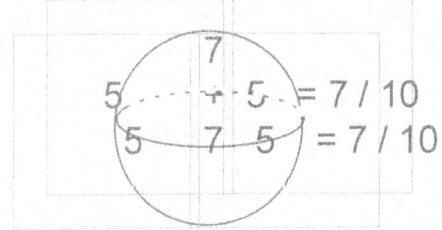

In the action of the inseparable drawing closer and moving closer, gravity finds the dual value of linear and circular gravity. There is no separation of the two factors acting as one but both have different applications and values in the unit. This is the result of singularity having three parts acting as one but giving three distinctions in application. From the star holding a dominant point or most valued point in singularity, it affirms all three other structures, each holding singularity individually and in a compliment of 5.

The network of individual singularity not only provides spinning through governing singularity in the sphere but also provides spinning in the geodesic throughout the cosmos linking all matter to matter in a network no one will ever come to understand in full. In the sphere the four squares forming the triangles linking the lines to the half circles hold space in time maintaining singularity of different assortments. In view of the matter-to-matter Roche factor where the factor consists forming relation between particles occupying densified space-time of where ($\Pi / 2$ X $\Pi / 2$) relating to the foursquare triangle, the value of gravity Π^2 comes in position as $\Pi^2 / 4$ X $4 = \Pi^2$

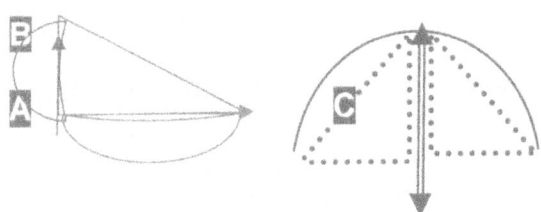

BC EITHER RELATE TO AB OR AC AT ANY GIVEN TIME OCCUPYING SPACE AS MOVEMENT DICTATES DIRECTIONAL CHANGE THROUGH DIRECTIONAL FLOW

With the normal extending of singularity it will always form the triangle in a half circle whereby Π relates to the cube by 5 points to either side of the line singularity forms. Thus there is 10 standing related to seven and visa versa. By calculating the 4 squares in the circle with the dimensional changing of space (5) becomes the twenty.

The normal flow will allow singularity extending to 10Π but when singularity blocks another sphere in singularity the two will form a joint value and by this joining the larger will dominate the space as well as the time of the lesser taking control of the surface and the atmosphere. Through this the Roche lobe comes about with all its other dynamics I describe farther on in the thesis. The principle is the same, which we know as the conducting of lightning and Jupiter uses it extensively to implement this action.

In the Roche all three components that singularity consist of, apply

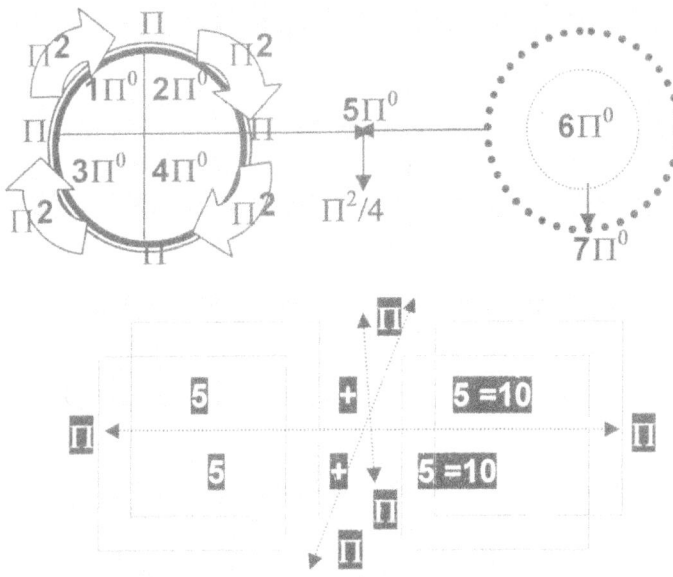

In the Roche limit the straight line forms part (1) and the half circle is part (2) and the triangle forms part (3) to singularity (4) Holding 5 points outside singularity.

The influence of singularity as the extending of Π into space links Π^2 to r and forms $2(5)+2(5) = 10+10=20$

From the position of singularity there are different values in Π where each indicates a position.

Singularity holds five dimensions inside and five outside singularity as matter and space forming space-time. The ten dimensions I named the atomic relevancy is also showing the double value of singularity as singularity extends into as well as beyond space. The atomic relevancy is $(\Pi^2+\Pi^2)(\Pi^2 \times \Pi \times 3) = 1836$ that is the mass relation between the electron (3) and the proton. Proton = $(\Pi^2+\Pi^2)$ Neutron = $\Pi^2 \Pi$. The atomic relevancy holds the dynamics of singularity control.

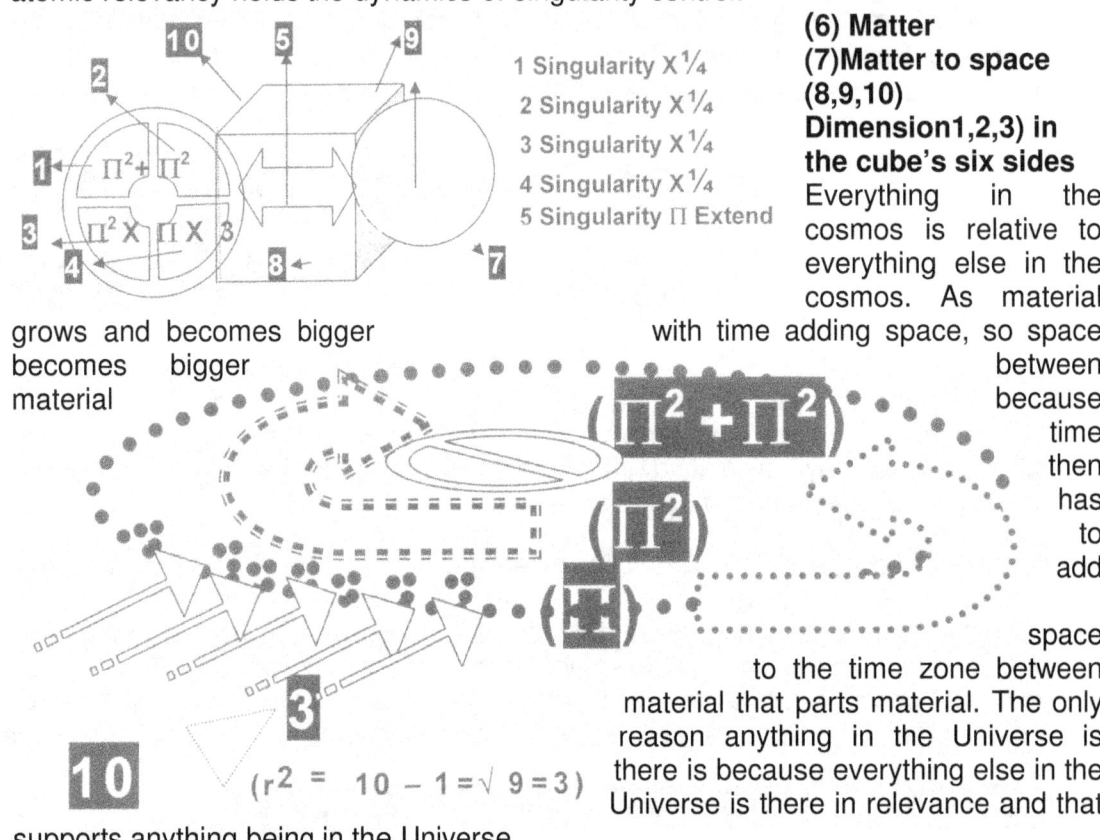

(6) Matter
(7) Matter to space
(8,9,10)
Dimension1,2,3) in the cube's six sides
Everything in the cosmos is relative to everything else in the cosmos. As material grows and becomes bigger becomes bigger material with time adding space, so space between because time then has to add space to the time zone between material that parts material. The only reason anything in the Universe is there is because everything else in the Universe is there in relevance and that supports anything being in the Universe.

The diameter of the cosmic structure holds the value of r and singularity holds the dimensional value of Π meaning that the radius or diameter (r) extends to become the diameter multiplying the value of singularity. But since r already consists of the square of space holding a definite positional relation with the value of singularity being Π the diameter comes into effect.

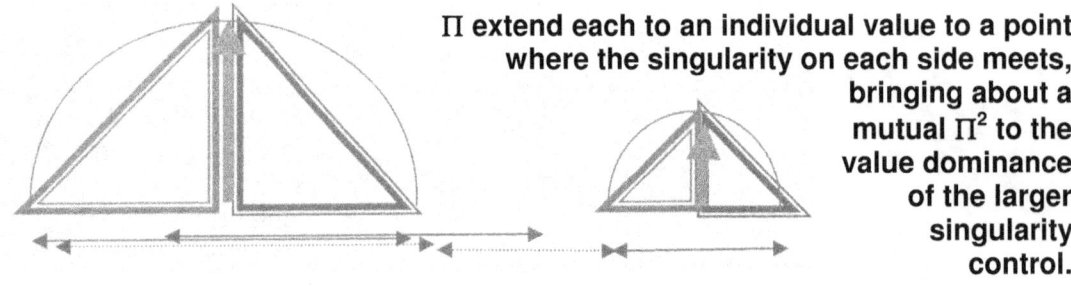

Π extend each to an individual value to a point where the singularity on each side meets, bringing about a mutual Π^2 to the value dominance of the larger singularity control.

At this point the equality of the straight-line dimension to the triangle and the half circle holds prominence, as a straight line, a half circle and a triangle are dimensionally equal. The common denominator will bolster all factors to an equivalent ratio.

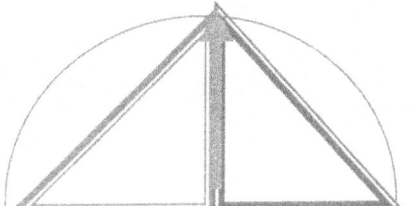 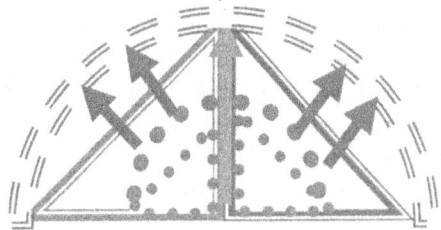

When singularity by the straight line increases the singularity by the triangle, it will also bolster giving equal potency in singularity by the half circle. As the singularity of the major component revives the lesser singularity to equality, the triangle in singularity will match the performance and so would the half circle respond in precise ratio setting equilibrium in order. The major partner's singularity in the straight line excites the minor partner's singularity in the straight line affecting all other aspects holding singularity in both objects to match equilibriums in all aspects of singularity. That is the Roche lobe.

From this the lesser partner will fill by the extent of the larger partner and as soon as equilibrium sets in, the growth will duplex to matching in both accounts, normally to the fatality of the lesser partner, as the lesser partner will be capitulating under the strain of the dual. In that way the inner planets came in place as I explain in part 7 of the thesis.

The Titius Bode configuration in accordance to orbiting formation holds a slightly different explanation to the explanation that applies to cosmic structures surrounded by space. It is moreover the individual singularity in maintaining the major singularity, which sustains the governing singularity providing equilibrium in space-time.

Not only does atomic individual singularity maintain self-preservation, but in doing that it also sustains a governing singularity holding structural composition and forms within a cluster of matter for example a star. As there is between stars so there are in the same manner a mutual or bonding singularity between atoms in stars, which we see as fusion.

Every quarter provides a distinct value that indicates the progress of the flow of time from the one point Π to the next point Π.

An Open Letter On Gravity Part 2

Any changes occurring in Π will lead to an unequal triangle providing two different values to r and will alternate the link between r and $Π^2$ bringing about a different form (Π) and time ($Π^2$). When singularity forming the lines of the triangle is not in equilibrium, the triangle will destroy the matching of half circle.

The sectors provide individual singularity as a means in sustaining governing singularity by which provision comes through maintaining governing singularity of the required spin in maintaining cooling. If this process did not apply, there would be no connecting individual singularity to major singularity.

I shall show what exploding double stars have in common with the Hubble Constant and how that fits in with the life story of Creation from when Creation started to Creation ending.

At the heart of bringing about the solution to one of the greatest Astronomic riddles one will find a child's toy… the riddle of Einstein's singularity pointing to the position where the cosmos started so many billions of years ago.

In the spinning top, matter would always relate two three positions as does Einstein's space-time declaration require.

All atom particles forming the matter composing the top would have to relate to the centre and two other positions.

Being the onlooker the viewer has to maintain one position. From that position some particles would be circling a centre point, as the particles would be coming towards the onlooker. The other matter would be circling the centre point while rushing away from the onlooker.

All atomic particles would have to refer to coming or going as much as circling around. This comes about as no atom can be ignoring any of the three positions it is aligning with in direction. Some would be on route from North to South and others would be on route from South to North. It is a response to an ever changing directional re-aligning with the centre, holding the centre to a specific location in relation to the position it previously owned and would own the next minute, changing constantly and in according with time location. All matter will have to adhere to any of the two directions, which in fact is actually four, but it also changes dramatically in moment. In the centre a line MUST form separating the comings from the goings and again the goings from the comings, where no matter can be located.

That line is too small to hold any atom, sub- atom particle or matter of any kind. All matter is either on the one side, or on the other side, but never can be neutral. It even gets more complicated because another line forms locating matter in groups. The one group is relating a position from the "centre point" holding "back" and "front" running through "the centre" where the other line is relating from "side" to "side" running through the "centre point". The fact of the lines is that "they are there", but we cannot see them. Try as you may, no one will be able to calculate the very position that forms the lines, but as they change all particle characteristics, the lines are a reality as the spin of the matter is real. Being too small to hold atoms, they then therefore must become part of singularity, where singularity is a spot in the centre with two lines crossing the spot at an angle of $90°$. That is the basis of singularity, and since all the positions still relate to a centre of a circle, forming a part of a spinning circle, Π must form the basic value. The matter will hold a time related position and cannot be in an eternal position.

The lines and the centre spot hold an eternal position as the position never changes. Therefore the circling matter is either in time coming, or time going, but can never stand still. That is the reason why time will forever relate to a square of coming from as much as going to, but never becoming one. Being one, time then must be standing still relating to singularity in the single dimension of $t = \sqrt{1 - (C^2 - V^2)}$. Time in a relevancy of t will be eternal one, and all forms of whatever matter can or will take on where that then is holding space forming time can never enter that line or dot.

In that there is one specific group in relation to coming towards the centre while coming towards the front.

The following group is rushing away from the centre while coming to the front.
The third is heading for the back while heading for the centre and…

The fourth is rushing away from the centre while heading towards the rear.

This reconstruction of the points and lines forming singularity is the very principle on which the graph finds legitimacy. The graph has four quarters in relation to one centre point.

The Brainy Bunch holds the view that in a graph the line crossing amounts to breaking the zero mark. That cannot be the case.

An Open Letter On Gravity Part 2   Page 573

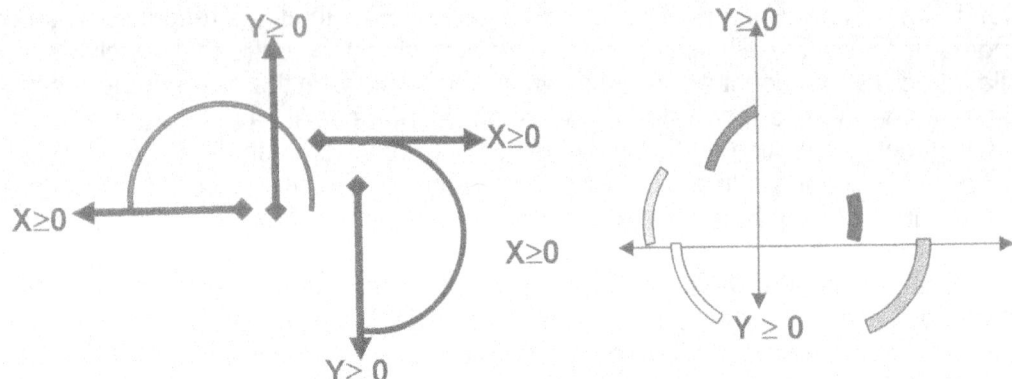

If the graph broke a zero mark, then a new and totally unrelated line will form on the opposing side bearing no relation to corresponding with the line in the previous quarter. The new line may start from any point and lead to another point holding no resemblance to the opposing side. What will prevent the line from finishing at a point marked say three and starts afresh at seven because there is NOTHING between the lines. The nothing will provide releasing and detaching all corresponding relativity there may be through disconnecting. Through nothing there can be no resemblance or corresponding because the point of start need not even to continue at all. Nothing will release any connection and if such a connection may come about, such a connection may just as well be very co-incidental and may never be used in calculating accurately.

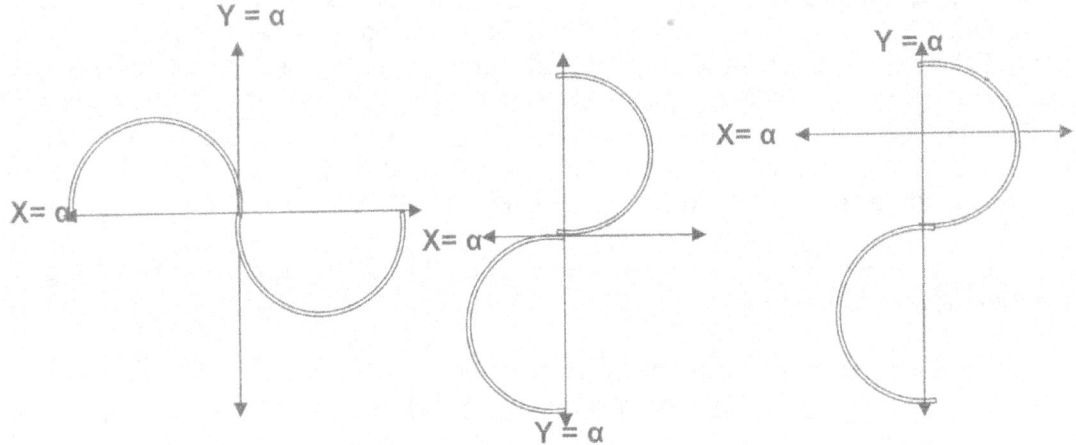

Experience taught us that there is a definite precise and secure corresponding that can only result from a direct connection in as much as the line being the same line. In that case the line must then come down to infinity and release from infinity as the same line still connecting to a point of re-bouncing to either side. The graphic cross is the results of singularity applying opposing sides but still maintaining connection through the application of Pythagoras that will connect and always bring about a direct relevancy. The graph does not hold zero because information derived as result of a relation prove a contact remaining when the line crosses singularity without applying detachment.

Academics, whom I have contacted in the past, treated my work with great scepticism. That forced me to find more proof and every time I went back to the basics again and brought about further proof IN THE HOPE OF ACCEPTANCE to prove my case. At first, I did not wish to criticize Newton or science, but eventually through the persistence of some academics, I was forced to do just that.

It is not that I wish to replace Newton in the least, but when Newton formed his formulae, the background of science was very patchy and dark in comparison to what science knows at present. I have tried to publish the book through the commercial route by finding a publisher. The response I received was that of publishers not willing to be involved because it falls outside the boundaries of accepted science. The battle continued as I then took a more direct approached by going to the Internet. This was no solution at all because those that seek information on the Internet did not make head from tail to what I was saying, and those persons I wished to draw their attention do not look for information on the Internet.

I have spent too much time, money and effort in this work to dismiss it just because people do not understand what I am saying. I am too sure about the correctness of my case to let go.

Aanplasing, verplasing, versnelling and inperking

As this book is a translation from Afrikaans originally, some terminology and expressions I had to revise to accommodate my Ideas. Where I could, I used modified English words to express a thought or an idea. One such a term is gravity. I had so much criticism about the word, which I feel I do not deserve. There is a certain notion clinging to the idea represented by gravity. Gravity links to a force that is all compelling, but I do not agree with SUCH A <u>COMPELLING FORCE</u>, such as the word gravity implies. The gravity I introduce, works on two principles, but gravity to Newtonian standards is a single force. When I refer to gravity the normal reaction is that I am referring to the force, which I deny. By declaring that gravity <u>THE FORCE CONTROLLING ALL OF THE UNIVERSE AS A STANDARD CONSTANT IS NON EXSITANT</u>, I bring the wrath of the scientific world upon me.

When I make the statement that there is no gravity, every person considers me mentally unstable. Of course there is a movement of energy keeping all objects attached to the earth, but gravity implies work, and with that work principle I disagree, because it is NOT WORK, THAT IS A COSMIC BALLANCE THAT STARTED AT A POINT OF ETERNITY AND WILL END AT A POINT OF ETERNITY. ONLY LIFE, STANDING ALONE AND DETACHED FROM THE COSMOS AS THE ENERGY THAT MANIPULATES SPACE-TIME CAN COMMIT WORK, THE REST IS A BALANCE RUNNING CONCURRENT FROM TIME'S BEGINNING TO TIME'S END. I HAVE UNBELIEVABLE DIFFICULTY IN RELAYING THE DIFFERENCE BETWEEN LIFE WITHIN THE COSMOS, AND THE COSMOS WITHOUT LIFE. In the entire universe there is no work, it is a balance running concurrent through time and space. In some cases the balance shifts to favour space as where in other cases to favour time more. But in all of that shifting, a continuous balance strikes every aspect of space-time.

This applies new ideas never brought to light before and the new concepts clash with the conventional names that science applies to current ideas. I had to divorce the science ideas from those I introduce and the only way was with new terminology. I have to start implementing the newly created terminology, which will apply to the rest of this book. This stems from my lack in ability to find suitable words in the English language that would define the concepts as they are, in order to establish the difference in meaning from the current words, which convey the existing misinterpretations (or if you wish, to my view incorrect applications).

Firstly, we start with the word *densified*, which is not a normal English word, but a word I had to produce in order to make a comprehensible statement. The correct word that applies is concentrated, that much I do know about the English language. But concentrated has not the correct meaning or the expression that I would like to bring over. Concentrated can apply to any substance, be it gas, liquid or solids where one of the ingredients becomes more than the rest of the ingredients. In that way, matter as a solid substance produced from the eternal substance which is heat, cannot be concentrated. Nothing in the entire universe can compare with the density of pure heat that spins at a rate in which that very heat can produce a value and which has a density far beyond anything else. Therefore, I chose to use the concept of concentration in a position where it makes a lot more sense.

A star is concentrated space-time, but there is a huge difference between a star's concentrated space-time and the value of pure matter. When a star does therefore become densified space-time, it can only be at the end of the Big Crunch eternity, witch I prefer to call moment Omega; that is when space becomes eternal and time becomes Zero. In this light I chose to call matter densified space-time. Densified space-time should therefore be in a definition where matter or substance has reached a point in density that will last one eternity, but has no limit. Concentrated space-time, on the other hand does have a limit, which is at the point where it becomes densified space-time.

The second word I created is *Aanplasing,* which is the ongoing redirection of heat as in matter to heat as in time and that connects to a circular deepening of the separation that matter undergoes transforming to time as it discards heat for the cold of fusion. Later (I hope) it will be clear enough for every reader to comprehend and to distinguish between the various factors that bring about **aanplasing** as should the reasons be clear why I prefer to have created this new word.

In this case however, there was no English word that could merely be altered and then be re-applied. With a choice to my disposal I chose to alter an English word as was possible with densified. A more suitable word that relates to a better meaning in the case where I brought in **"densified"** would have been the Afrikaans word **"verdigting"**, where "verdigting" stands in relation to "konsentrasie" (concentrations). The fact of the matter is that I am not wilfully forcing Afrikaans down the throat of the

Anglo-American and in the case of density I was able to adapt and modify a known English word that could adopt a new concept. Unfortunately in the case of aanplasing using an English word would mean that there is no liberation from the "misleading" focus that depends on gravity, nor can it liberate the feature of this "misconception".

As for the Afrikaans words: **aanplasing, verplasing, versnelling** and **inperking**: there are no such words or concepts in existence that the precise meaning can derive from the English written or spoken language. Should any such words exist, the misconceptions that remains connected to the original English words, would not bring justice to the concept which I wish to apply to convey the meaning that lies behind the correct value of the thought. If I stuck to the word "gravity", the concept I wanted to introduce would forever remain confused with the Newtonian application. To that end the new realization would then never come across in the way I intend it to be.

The R that I use in the formula has nothing to do with Radius as a term, except when used, to calculate the value of a circle or a sphere. The R is derived from the Afrikaans word **Ruimte,** which means space. In the Afrikaans word: **"Ruimte" the "u" and the" i "** is used in conjunction, which is pronounced the same way as " ai" is used in English words such as in **p**ai**n, dr**ai**n, tr**ai**n, v**ai**n, r**ai**n, etc**. So spelled incorrectly it should be pronounced as **"Raimte" and the "te" is pronounced "huh". The T stands for the word tyd, which incidentally is time.**

This is what I named negative space-time displacement, which results in **verplasing. The word verplasing is pronounce FHERPLHASHING (FHER – PL –HA – S – H-ING),** which means to "relocate" without destroying or changing the composition in any way, as the object is moving away from a certain position.

The word **aanplasing is pronounced AHNPLHASHING (AHN –PL – HAS – HING)** and literally means to relocate without damaging or destroying the composure or structure of the objects, as the object is moving toward a certain position. This very same value was previously mistakenly confused as being gravity. It is the effect on matter where space-time is in motion and matter is motionless or "standing still".

Both **aanplasing** as well as **verplasing** cause time differentiation and matter's structural re-valuation. That means the duration of time is re-valued and the space compromised. The excelling of the time factor is versnelling and the reduction of space is inperking.

The **word versnelling is pronouncing FHERSNELHING (FHER –SNEL – H-ING)** and means to speed up. If one is placed in a star it would seem as if time inside the star is accelerated while time on the outside of the star would come to a standstill. This concept is explained in far more detail, in a later stage in the book. The **word inperking means reduction or containment of the structure or scaling it down to a different size without penalizing or altering the shape in any way.**

Both inperking and versnelling is how matter relates to change **in space (inperking)** and **time (versnelling)**. The generation of the heat is within the structure and relates to time in space.

Inperking: This stands apart from the idea of curtailment because in curtailing. The *In* part is pronounced as in English where the *per* one pronounced in the same fashion as the sound an English sheep makes BHE placing a *H* sound before the *E* with *king* already explained. Sorry but that is as far as the Afrikaans lesson goes for the day.

Inperking: This stands apart from the idea of curtailment because in curtailing something or someone, means that object's or person's movement or moveable motion is deprived. This then brings over the misconception in the accepted notion of an expanding universe. In due course I shall explain the concept, but inperking involves the same value that was there at first and will be there in the end, only the location in the balance shifts to favour one or the other part of the same coin. Because of the fact that no such thing applies when space-time "accelerates" (versnel), and where this brings about inperking, it does not apply. Instead, all functions and factors still apply when inperking becomes valid, therefore the meaning of inperking becomes more applicable and this word describes the process much better. Inperking relates to time, where the duration of time extends, but not the value of time as such, as time applies in the cosmic sense.

One should realize that the entire atom as well as its surroundings, including all other surrounding atoms, are reduced in space-time volume, so the atom is not actually curtailed, nor is its surrounding which means the word curtailment does not really apply. All aspects of occupied and unoccupied space-time are in reality, re-focused down in the true sense and above all, remains to the precise relative relation value it had for one entire eternity where the relation between such times, only refocus. However, scaled down neither would apply, because that would not refer to the time involvement, which lies at the hart of this revaluation. Where less time applies, inperking would be more severe and where more time applies, less inperking will apply.

In this reference to time, one second would remain one second to matter inside the star but the duration of that second, compared to geodesic time validation would appear to stretch enormously. All words in the English language by implying its dictionary meaning, will inevitably lead to more language confusion, seeing that the explanatory meaning does not cover time enhancement and space reduction, heating and slowing of time lapse. By introducing a new word to the reader, I hope to screen out any misconceptions. Hand in hand with inperking, goes versnelling. When the reader encounters the concept of inperking, it should accompany the idea of versnelling.

Versnelling: It carries exactly the same meaning as acceleration, but the meaning or concept connected to acceleration implies to matter as the matter increases its own positional change in space and time. That is not the impression I wish to relay when referring to versnelling, because it is exactly the opposite of that meaning. In this, the actual meaning is more applicable to the true connection. These I must explain carefully, not to convey confusion. When a person stands outside an explosion of some sort, the time lapse seems instantaneous, quicker than the senses can relate to. However, inside the explosion, time is almost standing still.

Any person, who is inside such an explosion, would relate to time on the outside as being instantaneous. Whether this statement is accepted or not, the truth is that a person in an explosion cannot die, although his body is shattered in a million pieces. Such a person is sealed in a period separated from the period he and we live in. This I explain at a later stage. The time duration slows down immensely, but from the outside, it accelerates immensely. Therefore, time versnel to the outside of where everyone relates.

I wish to bring over the fact, as just been said, that the concept we have, is quite the opposite. Versnelling implies that the motional increase lies with the transfer of space-time, regardless whether matter occupies it or not. As aanplasing (not gravity) and versnelling bring about inperking (not curtailment as the body remains free to do as it wishes) the space-time that the body occupies and the surrounding sphere are in constant state of versnelling. The increase in motion has an effect on the matter, but the matter stands weightless as its specific density applies the time in that particular space.

Verplasing: This word is preferred to that of displacement, because although the matter in motion is displaced, time and space are implicated in the process. Verplasing is in fact the transferring of newly created magnetic space-time by matter, as a body composed of atoms has to replenish the space-time it occupies in order to maintain its position, place and structure in space-time in time in space, according to its geodesic positional allocation within the star's space and in time. Verplasing comes in effect as matter progresses in position, but the time-affect of verplasing that it has on matter, comes into real effect when an object reaches Mach $_3$ depending on its shape and altitude. In short: **_Aanplasing_** is relatively where matter is in a geodesic motionless position as space-time carries the motion component of the two values. This means that aanplasing is relative to positive space-time displacement.

Verplasing on the other hand has to do with the motion being with the newly created space-time in relation with the matter and the geodesic space-time remains relatively motionless. In both cases inperking and versnelling is a consequential result of the process. The difference is in the application of the time component itself.

A practical example of the difference between aanplasing and versnelling is as such: a body in **_aanplasing_** is in example a skydiver falling towards the earth and **_verplasing_** is where a body, such as a rocket is on a trajectory path as it fires into space. Both bodies will comply with the linear and circular displacement, but the circular displacement will relate oppositely in each event.

An Open Letter On Gravity Part 2

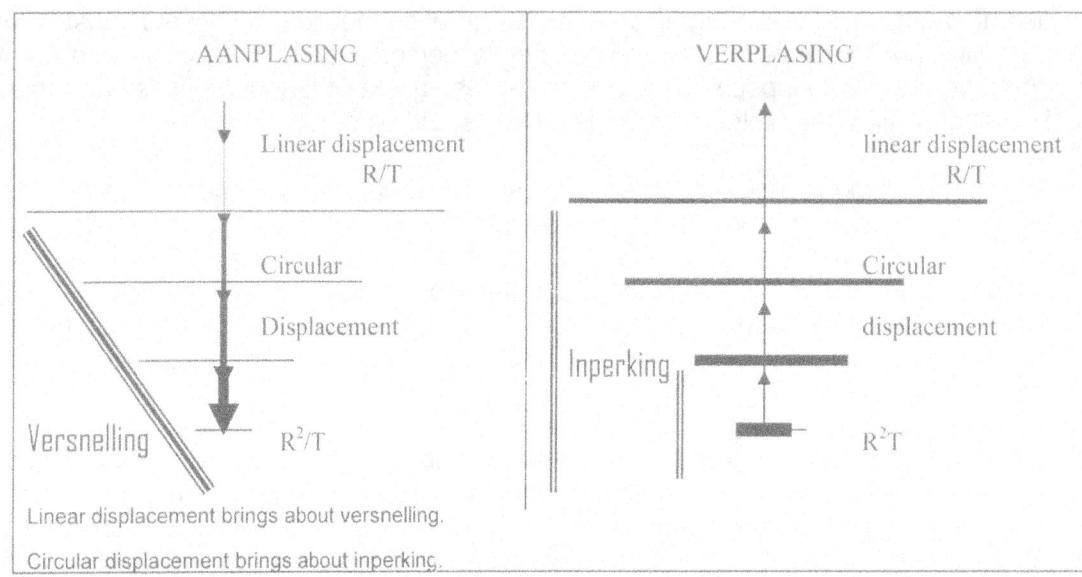

This aspect, Newtonian science disregard in as much as they disregard that there is any connection between the atom and the cosmos. I deny the fact that a star can have winds, although winds are as close to that concept which the earth can provide. As you will later see, winds are the transferring of heat, but so is electricity and lightning, and one cannot call lightning wind. Neither can one call lightning electricity. It is altogether a different product of the same transformation of heat, but the applied principle separating the products of heat transmission stand in total different areas.

It is as much Newtonian to believe in mass as it is Newtonian to believe in life (maybe once upon a time) on Mars. It is also just as Newtonian to believe there is life throughout the Universe and weekend visiting between them and us, no matter where they are, is very possible and even more probable. If you want to be Newtonian you have to dish out mass to wherever mass does not belong and you have to put life wherever your fantasy takes you. If you are good at these two qualities, then you are a good Newtonian. Then it is your Newtonian rite to be able to fiddle with time and to mess with the idea about time as much as you are able to do as long as you have no inclination what time is and how time applies. You can replace time as much as it may please you as long as such ideas of you controlling and man-handling time will bring you standing next to some bizarre form of extra terrestrial life that does not exist except within the Newtonian imagination and then with that over exposed Newtonian imagination holding such non proven life forms living in the Newtonian imagination, you may then proceed to devise methods by stupefying time to the point where you can contact whatever life that can only exist within the stupidity of Newtonian thought that Newtonian stupidity placed anywhere in this limited Universe. Amongst the upper Newtonian intellectuals there is a mad hype about reversing time. Newtonians claim on such abilities are only possible when the Brainy Bunch have their mathematical wisdom on cosmic matters run riot and these multi- intellectuals have psychotic concepts of rearranging time in space as they alter space by chucking away the time they do not require at that moment. How they will do that in practise is not there to be understood by simple minds such as us forming the labourers and being in the embodiment of the lower forms of life. We forming the other types of humans that is blessed with mental incompetence such as I am, will according to their superior wisdom, not grasp their superior thoughts when they explain these time travelling details as the concepts of understanding falls in the minds of only by the supremely intellectually gifted Newtonians. Only the supremely intellectually gifted Newtonians can grasp such high thoughts where the rest of us slow witted thoughtless creatures are drowning in our stupidity and all of us holding the positions being another excuse to be called humans, we are supposed to be accepting and the matters are not to be argued by us low witted life forms, we forming those that are incapable of understanding intellectual mathematical facts. Us forming the mindless masses and the mentally retarded many and all other intellectually restrained amongst us considered by them to be the uneducated mindless, must accept that they, the more able and intellectually advantaged are the more humanly inclined to possess the godly gift of thought because they form the better Educated Brainy Bunch, the members in a form of human kind and notwithstanding being bodied as a human, although also they have the God given rite to think on our behalf. It is remarkable that every one of those supremely gifted Super-Educated- Wise- Amongst- The -Wise wishes to reverse time by going back into history to do what they apparently neglected to do but never develop an urge to speed time up going onto the future and haste their oncoming date to die. They are willing to travel back in time and redo what was done before and visit was is known to have happened but never is there one that wishes to travel faster in time and reshuffle the future. They never wish to go ahead and reach the day the Super-Intellectual-Time- Traveller is going to die. We know all about the spectacular machines the Brainy Bunch mathematically create where some even wish to employ a multitude of Black Holes and reduce space to squash the daylights out of time. I wonder if they ever thought of using their eternal wisdom to clearly observe the full complexity of time. Have they ever brought the big Bang into there time concept and thought how time connects to the Big Bang. Has even one thought that the Hubble law is the flow of time and that by going back in time in order to return to the past, one has to reduce the **k** factor of every atom in every star and every star incubating galactica through out the entire cosmos? Time concerns the entirety of the cosmos and not the point our super Educated holds the centre of the Universe! It involves a far more complicated process than just move down some imaginary double Black Hole that creates an artificial space whirl everywhere they so wish to install such a gravity defying Black Hole crushing Space whirl.

By using Newton, one cannot even begin to explain any one of or the combined effort of the above cosmic phenomenon that are all over the cosmos and forms all the laws in the cosmos. Newtonian definition cannot even recognise any of the principles but only Newtonian science are taught to students. Newton is an institution. Newton is not science or facts but an institutionalised concept overriding reality, overbearing all sane wits and short circuiting mental concepts. Any student that ever wished to go against Newton has to be prepared to go against the entire Physics establishment. I have personally experienced this in all its implications and I am not even a student by a long yard. That student will be fighting a fight that he will lose and his defeat will not come on the grounds of the other side showing implacable merit but mostly on the fact of the lacking of such merit. That student will suffer a blowing defeat merely because he is contradicting the establishment where the establishment takes on the person of Newton. Contradicting the establishment is seen as challenging sanity and who is the establishment, it is of course Newton in full glory? Let's test one definition about spin for instance *Spin is an internal property of elementary particles, related to, but not identical to the everyday concept of spin.* The most incompetent politician could spin a better yarn that says more than such indiscretion. This says they have no idea what his should say because they have no idea what spin is. Spin is the duplicating of material through the motion time brings in relation to singularity and that I can prove through Kepler. Space a^3 duplicates by = spin T^2 bringing a^3 into relation k with singularity k^0. You, the reader may decide which makes more sense. The academy is set up in a manner that no student will ever find any means to challenge the establishment in Newton because those in power will have the student legally removes from the system. If those in absolute power are corrupt then the corruption is absolute. If those in control are not controlled then those in control goes out of control. There are so many misconceptions and when challenged those facing the challenge avoid the challenge by dismissing the challenger. Those dismissing the challenger has all the authority to dos so because vested in them are the absolute power to be or not to be. Then this student coming from a normal school and advances into a higher establishment such as a University has a quarry with a statement Newton made, but that helps little because the institution is not built to tolerate criticism about Newtonian misguided conceptions? What are his chances of finding a sympathetic ear? There is no chance because Newton is the institution. It is not what Newton said that became the institution…no; it is Newton in person that forms the institution. If and when any Newtonian made a statement on behalf of Newton then every one listening heard the voice of God speak from the mountain onto Moses. Then when a student comes and challenges this structure called physics or Newton doing such insanity will get him nowhere. I might be uneducated and I might be light years away from the educational upper class but if someone such as I whom are just a motor mechanic could see and read Kepler and distinguish between Newton and Kepler and sift the rubbish of Newton from the facts of Kepler, then how much more must those in super intellectual position have such ability to see what I see. Everything the reader was introduced to thus far in this book is merely common sense. There is no spectacular brain bashing life-altering implication of concepts going into madness and beyond when a person ventures into the everyday facts when reading what I say. My entire arguments portrait pure common sense and is so simple that every one has the ability to realise. It is so simple in reality that even I can understand space-time. Then those that are presumed to be brilliant and those carrying the touch of the intellect on behalf of all mankind does not even lower their esteem to try to contemplate what I have to say because they have the principle and the attitude that I, being part of the lower class of the human race and being a very typical example of how inferior a product in human evolution there ever can be and here I with no education personifies what man is not able to achieve, such low life such as I come and address them on their institution namely Newton. The in their mind I deserve to be put in place and get the treatment I deserve so richly. If that is the treatment they gave me, I can only presume what the fate will be of any student that does not immediately support and hale the ideas Newtonian concept shower on the world of physics. I might be of the lower form of human development, but I am not mentally retarded to the point where I shall presume that I am the first to have the ideas I Promote, such as those I share with you the reader. There had to be others that came before me that have had the audacity to contradict Newton! What happened to those and to them that came before me contradicting Newton? It seems that they were silenced as I am silenced this far and that it seems they went off to go and die quietly where they will not disturb Newton. This must have happened to those that went before me and had my ideas because Newton, the institution is still standing in strength.

I dare you to fathom the next statement that is as Newtonian as the institution Newton impersonates. This is a definition of a wormhole, which is a bizarre unproven unidentifiable concept supporting insanity and confirms how madness can become a disease. They define a wormhole, which is another name for a space whirl as follows: A Wormhole is a thin tube of space-time connecting distant regions of the Universe. Wormholes might also link to parallel or baby Universes and could provide the possibility of time travel.

If this is not the gurgle of the insanely mad and the mentally rotten vomit of insanity then I am sure glad that I am a simple-minded low class idiotic mechanic. This statement tells me so much...it tells me Newtonians have no idea of reality...Newtonians have no ideas of truth and Newtonians finally admit to going mentally pear shaped.

A Wormhole is a thin tube of space-time connecting distant regions of the Universe. Have they seen this or have they got a picture of this? They have no idea what gravity is... yet; they base everything on a baseless fact they establish that is not present anywhere but in their minds and in the imagination of Newton? What can produce such a tube since it has to be material in an accumulating presence that forms gravity and gravity takes on the shape of the curvature of space-time where the curvature of space-time is the result of singularity establishing a round shape implicating the sphere? What will produce a thin line except their bizarre interpretation of mathematics gone mad? Why would space-time abandon the curvature that forms space-time just to link their stupid ideas with the bizarre?

Wormholes might also link to parallel or baby Universes and could provide the possibility of time travel. I think they have learned so much they forgot the basic definitions or they should start to reinvent their concepts by changing the terms they apply. There is just one Universe because that is what the word Universe implies. They have no idea what space is because they think that space can contribute in any form their bizarre way of using mathematics would allow. They don't even know what space is and this statement is the best proof about that! They have no idea what time is because they think they can skip time. In what will they be during the time that they do the skipping of time? Do they think they will disappear into the magic of mass! They have no idea what space-time is. They have no idea what the Universe is. They have no idea what science is. Thank God He made me as stupid as I am because at least I know all the above. Statements such as their bizarre definitions serves as vivid proof that makes it is so obvious that they so apparently have no idea about anything of the above. All they smartly can confirm is that the Universe is made up of nothing and they extend that nothing to everywhere.

Concerning this Wormhole idea, I wish to introduce you to the chambers of the darkened, the lightless channels running through the aw inspiring mind of the Newtonian by taking a sample from another book I have written called **_"A Cosmic Birth Dismissing Nothing"_**. The wormhole is some supposed **"Black-Hole"** that supposedly lines up to perfect match with another **"Black-Hole"** and the spin in the two forms a vortex that not only links the two **"Black-Holes"** but crosses space and time and matches two Universes. While are reading, please remember that a **"Black-Hole"** is something not even the photon can escape from and a photon is the smallest and therefore fastest observable object we can see in our Universe. This is because the gravity inside this **"Black-Hole"** far out accelerates the speed of light. Not only is it that the gravity supersedes the sped of light but space has gone so small it crushes the photon into a thought of what was. Not even the smallest we have in our four dimensional Universe has any chance of entering a Black Hole unscathed. However the Masters-Of-Mathematics-And-Brainpower-Walking-On-Legs finds that the mathematical abilities they have outweighs their mental sanity they don't have. Now, on the one hand we have this Miracle-Worker-Brains-on-Legs-Newtonian-Calculater-Pumping-Mathematical-Machine-Genius-Walking-amongst-Us-Mere-Mortels that knows how to compute any calculator and they are driven by the force they created and named mass that supplies them having such creative intellect, with the wonderful stuff others know as fantasies and by being the Newtonians such as they are, they see themselves as a finder of truth and fact. This remarkable Mathematical driven walking form of human intellect is on a mission to find not one but two **"Black-Holes"** that is not only lined in formation but is connected and so much so it connects Universes. It seems to me that when he is unable to find any well connected and aligned **"Black-Holes"** his mission will be to find two **"Black-Holes"** and then go about in a matter of fact manner by lining the two himself. When finding one **"Black-Hole"** the incredible intellectual super-scientist will venture on and locate another, where he then seemingly will find a way to drag the one **"Black-Hole"** to the other **"Black-Hole"** and line them himself in accordance to his calculating abilities. I being as simple minded as I am without such admirable mathematical genius to compute with find mane questions I have no answers to and to me being as small as I am, find those questions locking any progress that the genius may have. Where is he going to find a **"Black-Hole"** and how is he going to travels to the **"Black-Hole"**?

The closest **"Black-Hole"** is God knows where, and I am not blasphemous, with that remark. Only God can determine where about the closest **"Black-Hole"** is. Finding it is problems, which I have too little, too see my way past but our wonder-kid-genius-Newtonian find that not to be a very serious problem. After all, I suppose he is tasking with him his faithful calculator. When the-scientist–in-search-of-the **"Black-Hole"** locates this object, the-scientist–in-search-of-the- **""Black-Hole"** then intends reposes the object firstly, then this Man – of-Men is going to relocate and then reposition this object to a new location and position that suits the-scientist–in-search-of-the- **""Black-Hole"**. Remember, this object is so powerful light cannot escape from it, the-scientist–in-search-of-the- **"Black-Hole"** intends to find a way of re-aligning the cosmic position this object holds. Remember also, this object is so powerful, entire galactica may fit into it as to produce the accumulative gravity required to generate the Black Hole to a standstill. Yet with his calculator in hand only with nothing that can stop him and that nothing is just what he can create and counter create as much as he wishes for he is the Newtonian, therefore that Newtonian with his calculator is unstoppable on his mission. After this enormous feat, he intends to locate a second **"Black-Hole"**, wherever that is. One should keep in mind that according to Newtonian misconception even when finding any ability to leave our solar system, then if that impossibility was a possibility then travelling to the next neighbouring galactica would take two million years just to get there beaming across space at the speed of light. I am absolutely sure that one will not find a **"Black-Hole"** lingering around between us and our closest galactica neighbour, so just getting there at the speed of light will take much determination for it will last hundreds of millions of years to pass the closest galactica that has a **"Black-Hole"** as a neighbour. Then it must take him at least one hundred thousand times that time to locate his first of two **"Black-Holes"**. The fact that it will most probably take him the best period lasting a thousand million of trillions of years to reach the first place where he then may find such a **"Black-Hole"** does not trouble our genial Master in the least for he is a Newtonian armed with a computing calculator on a daring mission with nothing to stop him. The fact that he cannot reach the speed of light with earthly manufactured crafts does not bother him in the least. The fact that he cannot see a **"Black-Hole"** and therefore would not know in what direction he should leave while heading in the direction he cannot observe does not bother him at all. The fact that he most probably will only last through the first eighty years of his voyage (if he should be that fortunate to reach such ripe old age) and then will retire to a death, as all of us have to do, all these mentioned problems goes past this cool cat like it is no problem of his concern. After e a Newtonian armed with a computing calculator. The fact that there is not enough fuel on Earth he can take with that will allow him to complete this journey does not worry him in the least, because he has his calculator with him that has the ability to mathematically solve any and all unforeseen problems arriving, such as this tiny fuel issue. Still us with the smaller version of what is known as a human mind do realise that the Master-of-Thought may concern these tiny problems with dismay but notwithstanding and nonetheless our Computing genius will have to treat these all so slight problems with more contempt because in order to find ultimate success he will have to overcome or by pass the stonewalling abilities that these issues may present in order to reach the real serious issues that will land real devastating problems that he cannot by pass or solve by denial.

Finding the first **"Black-Hole"** brings the need of a tow Truck large enough to take your average garden variety **"Black-Hole"** on a tow for billions of light years in his quest of finding a space large enough to fit in two **"Black-Hole"** with some room to spare. This large tow truck will have to be built on the Earth and then have to be lift off the Earth and will have to be transported with the vehicle that is taking our scientist–in–search-of-the- **""Black-Hole"** there. The fact that the tow truck alone must be many millions of times larger than the whole solar system including the Sun goes past and very much unnoticed to our local hero which is the scientist–in–search-of-the- **""Black-Hole"**. But this means little to our brave scientist–in–search-of-the- **""Black-Hole"** for his computing calculator works in numbers and digits and has little scope for such small problems and moreover since our genius is on a quest to solve real serious issues like time travel, how then can a person going on such a venture concern his great mind with issues such as fuel and vehicle sizes? Even the fact that he will never be able to land that much of the required material on a surface as small as the earth goes past his great mind with so much other genial thought to carry. The fact that he needs to build such a vehicle on the Earth without demolishing the Earth totally, that does not seem in the least to concern our brave hero even slightly. All that complications with the extra tow truck adds to his entire fuel problem but that he can solve as easily just by never thinking of it and also not mentioning the problem. That also goes for the manner in which his Newtonian ways find solutions by avoiding questions from which he then does not have to arrive at answers and in that way his Newtonian manner avoid the other problems as well that is why problems too small to mention leaves him with the mind of a giant unconcern. Let us now be generous and not small-minded and grant our scientist–in-search-of-the- "**"Black-Hole"**" leverage by allowing him some marginal success. Let's give him the benefit of locating his first **"Black-Hole"**

After locating this as well as his second **"Black-Hole"** because if it is that easy to locate the first **"Black-Hole"** then locating the second **"Black-Hole"** must be just as easy. With him having found his second **"Black-Hole"**, he plans to shift its position to align with the first **"Black-Hole"** that he also found but just a few billion years earlier. The alignment has to have an incredible degree of accuracy in order to create a tunnel in time between the two objects. Even my simplistic mind tells me that. After all this major engineering of locating and shifting two **"Black-Hole"** through, most probably, millions of light years, aligning them to the most accurate position Mathematics can ever accomplish, the true test will start.

He then attempts to
1. Move him down a funnel OF THIS **"NOTHING"**, **from** which even light cannot escape once it entered this tunnel of darkness,
    Therefore at a velocity exceeding the speed of light he will steer what ever he wishes should take him through the "Black-Hole-holding-NOTHING",

2.   To the centre OF THIS **"NOTHING"**, where the space-time is so limited, the entire Earth will occupy less space than whatever space the atom of a BREAD CRUMB occupies when it is in the centre of the Earths core he has to…

3.   Locate this invented tunnel between the **"NOTHING"** parted by no space at all,

4.   Move down this tunnel of **"NOTHING""**, where light cannot even travel, because light moves too slowly and the photon takes up to much space,

5.   Find a passage out of the second **"NOTHING"**, in all the surrounding darkness where light cannot exist

6. Where he then has to move MANY TIMES FASTER than many times the speed of light in order to escape and reach the outside of the "Black-Hole-holding-NOTHING".

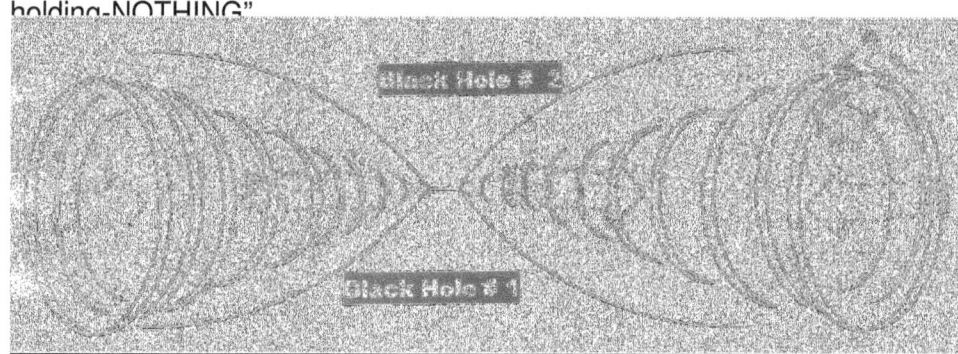

At the one side we have an intake of space established at C^2 as the space accelerates down the funnel. Then on the other side we also have a similar scenario with space entering at C^2. The fact that the space will come together at an opposing speed of C^4 does not detour **Our Super-hero-of-cosmic-proportions** because he intend to travel down the one funnel at C^2 into a tunnel being the size of "**NOTHING**", pass through what ever blocks the escaping light and space so that it can never again return back to the Universe it came from, then when entering the "**Black-Hole**" from the one side entry, go through some corridor, which **Our Super-hero-of-cosmic-proportions** is going to create or establish beforehand and up through the funnel as **Our Super-hero-of-cosmic-proportions** will come out on the another side representing the entering passage of another "**Black-Hole**". He should know being a Newtonian that the second biggest Newtonian of all time in as much as Einstein in person declared that "**NOTHING**" can go faster than the speed of light. (The biggest Newtonian of all times of course was Newton in person!) But **Our Super-hero-of-cosmic-proportions** is going to attempt a cosmic speed record of C^4 because going against the flow of space upwards while the space is coming down, down the second space-whirl with space travelling at C^2 he will need to reach a velocity of no less than C^4 to pop out on the other side. The speed off C^4 is only **8 100 000 000 000 000 000 000 kilometres per second.** Well...I surely wish him all the luck he would dearly need because there is no way he will establish enough momentum to reach that velocity, with or without the aid of his faithful calculator. All this he is going to achieve after towing the two "**Black-Hole**" for many millions of light years across space to land them both at the rite spot at a location where two "**Black-Holes**" can fit back-to-back! What astonishing determination and rock solid will power does **Our Super-hero-of-cosmic-proportions** show? He is a man amongst men! He is a calculator managing computing wizard of Newtonian proportions. He is the-scientist–in-search-of-the- "**Black-Hole**" the latest Superhero Hollywood has not yet noticed and therefore gave everlasting fame.
He is coming in at C^2...<u>and then he plans to pop out on the other side doing round there about in the region of approximately C^4</u>

We all are aware from Einstein's figures that he proved that the speed of light cannot exceed **C** being 3×10^5 km / sec. Only nothing can go faster. Everything else can't,

An Open Letter On Gravity Part 2   Page 587

not even by one percent or one kilometre per second. That is the speed of light and that is what it will stay, unconditionally without arguments or exceptions…except where by some fluke it can jump to C^2 as one will find **in $E=MC^2$**. Not **C** plus some or twice time two times **C**. Surpassing **C** in any form proves to be impossible because going past **C** is going past time and that is not possible. But when calculating atomic bombs exploding abilities we find that suddenly, according to the man that said the speed of light could never exceed **C**, declares that the speed of light can jump from **C** to C^2 by never giving any explanation of why that will be possible. **It is** unexplainable, which makes it very Newtonian and therefore acceptable. But still it cannot double although it can go into some square of the figure it cannot surpass. Therefore one must accept that another jump will become possible, not by going to a number of twice C^2, no that will be impossible, but it must then be possible to jump from **C** to C^2 and then to C^4. Why that will be possible I don't know because I don't even know how it is possible for the speed of light to jump from **C** to the square of **C**, without touching twice or four times **C**.

Not once did any of the **Academics responsible for official opinions** waste their precious time on the minor stupidity of the common class questions of the feeble minded lesser intelligence such as those I did ask. With their most dynamic Intellectual superiority they will never scoop so low as to worry about detail and minor inconveniency about how they will get there or find the suitable Black Holes or moving them across space and securing the accurate positions. For that they have calculators that they know how to use. All these common thoughts by the lesser minded morons such as what I present with my feeble attempt to ask time-wasting questions is reasoning that stoops well below their mental abilities, therefore they much rather prefer to match the space-whirl inventor's brilliance as they all reach for the calculator to prove or disprove his calculations! In the light of such bizarre reasoning, how can any person not find their attitude horrific? By the way, this is the sane and clear mathematical logic of the brilliant atheist, which can find no sensible reasoning for the existence of a Creator.

Whenever a photograph is taken of one of the world's most brilliant mathematicians / scientists/ whatever, the photo is more a poster portraying his or her brilliance in the field he or she represents. Behind this person of the moment will be a black board with the most breathtaking mind boggling mathematical equations, man can produce. In other cases, there is to the background the most impressive factory, which spends four hundred thousand million dollars, a year to catch and weigh a neutron or to "commit" "cold fusion" or "bend light" or attempt to achieve any other most impossible task they can dream up, in order to prove to the world their personal intellect. They have not the slightest inclination of solving any real cosmic problems for hey have no inclination to ask what gravity is or how does mass establish gravity. They are on a crusade to test the Universe in establishing the fact that the Universe does have the ability to create problems that will prove to be magnificent enough to match their intellect…for only something as wide and as big as we have in the Universe can find any ability to have anything that can test their super superior intellect and their ability to use their mind by thinking out problems that only they can solve. Time has arrived where John and Jane Dow should stop allowing these chance-takers, to scare them witless and robbing the tax payer blind by bullshitting the daylight out of everybody with a lot of legal deception and lawful corruption.

A time ago I watched CNN broadcast a program where a United States Senator asked the cream of the US scientist to bring forward an explanation in reconciling the Bible with science. All the scientists looked amused and sheepish, but no one could produce an answer, and no one had the guts to stand up to the senator. This horrified me, as the Senator asked the most intellectual question man can ever raise, while in response to the Senator's supreme question he received the most horrific and un-intelligent responses man can devise, namely: **"WE DON'T KNOW."** The answer **"WE DON'T KNOW"** was not permitted in my classroom during the years I was a Tutor. If you do not know, you **THINK**: and that you continue to do until such a time that you do know! As much as this reply shocked my senses, and filled me with disgust and despair, it served as a reminder why I dreamed of completing such a book since child hood.

It is an honest attempt to show all fellow believers, like the Senator, how truly sheepish science, in their outlook is, and how brilliant the Senator's question was. King how one would reconcile the Bible with science is the equivalent as finding the ultimate truth…and in all modesty; I have found a means to match Geneses one in the Bible and the scientific explanation of the cosmic birth. I so dearly wish I could somehow obtain the name of the Senator, to prove that his question is one of the most brilliant ever asked and by not being able to answer the question, science proved how far they are off the mark in any understanding about cosmic science. I hope when the reader read this book, he or she will be able to recognize the deception and trickery the **Academics responsible for official opinions** play on them. These the-scientist–in-search-of-the-"Black-Hole-holding-<u>NOTHING</u>" go about announcing ways to "travel at the speed of light' or "to conquer the neighbouring galactica"; "colonize other solar systems and planetary life" or "the discovery of some twelve planets or two-hundred - and -twelve planets lying outside our solar system".

Scientists the world over, regard the Bible as pure fantasy, the brainchild of out-dated views held by unsophisticated people in disregard to scientific proof, which is lacking any fundamental logic.

Yet, those, very same persons back a proposed invention like the space whirl or the wormhole, which is more far fetched than any myth connected to any pagan religion since the recording of time. The Bible says you can move a mountain in faith, this science does not accept. However, to move two Proton-stars (Black Holes) through a SPACE WHIRL and in the process defy time by means of mathematical equations and astrophysics they find, seemingly, so plausibly possible it takes all of the members following the Academics in Mainstream Science and the World-of-Science by storm. Every one with a calculator and a PhD or two or three PhD, suddenly have to urge in challenging the calculations (not test the plausibility), and with such a beastly urge then rushes to calculate and measure the most impractical declaration that came ever since the time that Einstein found the Universe to be a Bantamweight and found that the Universe in total did not hold all that much that that is required in the mass department to survive its fight with time.

This book, whether the reader likes it or not, reveals the lies science accept, (which I prefer to call Xepted science to enable the sane to distinguish madness from common sense) and uncover the fantasy of the **Academics responsible for official opinions**, referred to as **Academics with official status**.

If one can still accept **Academics opinions** at the completion of this book, that person is able to believe Red Riding Hood without raising any questions. Education, at present, is little more than mind control and brainwashing. Should you not believe me, read about it in this book, as you compare myths taught to children, to facts I present and ask yourself why you never raised questions about such obvious ridicules. Then prove me wrong on this statement!

Science always finds a way of advancing and in this case the brilliance of science in its super advancing form, does not advance to anywhere and there comes the misguiding to the surface! Science will not commit to reality while applauding the Wormhole! A short while after, another scientist with equal prominence does comes with criticism about what he conceives as facts about the Wormhole being unachievable. The criticism he holds against the space whirl concept is that the energy requirements needed to establish such a double space whirl, would amount to a value equal to what a neutron star can generate. This is raping the sublime. It is extending what is ridiculous to absolute insanity. To try to put this in perspective I shall try to establish some form of understanding by the following explanation.

Let's say one Super Brain-Powered Academic claims that he found a method to revive a pharaoh. According to his calculations and in the event where you should wish to revive the pharaoh that is dead and buried in the tomb, inside the pyramid, for the last six thousand years, you should move the pharaoh's being inside his Pyramid complete with the pyramid undisturbed and completely in tack, precisely as it is with dust covering it and all other facts included, with him inside his royal chamber, all nicely kept and undisturbed, a thousand miles. Nothing may move but the entire Pyramid as a unit. Then when this is achieved, you must line it up in a location as to place it where you can place another pyramid just as big and undisturbed, in tack as the first one, moved under the same circumstances as the first pyramid that you also have to drag with horses another thousand kilometres. With all the dragging, both the pyramids must remain intact to the last dust grain. The placing of the pyramid should be such that after dragging both pyramids a thousand miles each, you must be able to dig a tunnel that will enable fans in by creating a funnel that can blow with such ferocity, it will blow breath into the lungs of the pharaoh.

Then another wisecrack wizard runs through all the calculations and with him being as smart a scientist as the previous one delivers a highly accurate and intensely calculated verdict afterwards. After calculating the whole process in detail, the second wizard comes up with the conclusion. He finds that that there is not enough fodder in Egypt to serve the animals. According to the second wizards findings when attempting that then in such an event, the storage facility in all of Egypt is unsatisfactory to store the fodder such as you may require. The fodder requirements will out way the horses feeding needs when contemplating the feeding needs of the horses that will do the dragging. The whole concept is bizarre, beyond any description and to top that, another one comes up with the idea that the fodder they will harvest in the time requirement will not feed the horses. It is either I that am small and simple minded or ells they…well think for yourself…

The fact that space is the shape of a dome or a bubble is not a big coincidence but comes as a result from singularity bringing about dimensions. These dimensions are there whether science chooses to ignore them or recognise their powers. Space is a liquid and as all liquids do, space depends on specific densities. With the specific densities, borders come about. The explaining I manage with very simple sketches indicating the principles. I explain with ongoing proof through out the book indicating to the reader how I came to realise that **gravity is not about particles pulling** on each other with some inexplicable force of magic holding an effect on matter or matter pulling matter. Motion brings about the demise of space and that includes space holding material as well as space not holding material. This may sound simple in concept but understanding requires concentration and that makes this book not being a novel to read as it may come across.

What Newton missed in Kepler's work is Kepler's introducing of gravity as a formulated fact because Kepler said gravity is $a^3 = T^2k$, which is the space a^3, that forms through the moving T^2k thereof giving the space a^3 independence by a relevancy coming from T^2k releasing individual singularity from surrounding space. Gravity is space moving in a circle holding space including, that which is in space at a distance from a centre where that is applying the conditions of space in time. In outer space six dimensions secure and stabilize floating objects. Only when such conditions are broken does space fall away and particles come crashing down to Earth? It is the breaking of the balance that leads to objects falling to Earth and not the Earths gravity pulling. This falling comes from a lack of motion by the particle in space in relation to the distance from the centre. Again it is a^3 that becomes too little to support the floating of the object when the spin T^2 reduces in the presence of a certain value applying to k. A little science experiment such as the Coanda effect disproves the gravity-grabbing-on theory. Gravity is about matter concentrating space through the spin of the proton and the reducing space by accelerating the movement of space between the two objects. A good example is a blowing fan. By establishing motion of the air with the spinning fan, volumetric occupation reduces in favour of airspeed coming about from motion. By removing the space, the particles come automatically closer, and that is the principle of gravity in operation. But the diminishing of the space goes about by applying very specific rules and in the adhering of those rules is applied to everything in the cosmos.

Gravity comes about as space a^3 applies motion T^2 and from singularity k gravity is as much part of space as the motion of space is part of gravity. Do not be fooled by the simplicity. Mass is the result of applying gravity by reducing space. Gravity is not the result of mass. Gravity dismisses space and by doing that the stronger gravity can place more particles fitting into less space occupied where that reducing of the space is bringing about extensive mass increases into the volumetric occupied confining more material into less space. That produces monsters like a Black hole having enormous mass and being without space. If gravity was not about reducing of space bringing about the increase of mass because more material will fit into less space, the Black hole does not make any sense. Stars reduce their volumetric size as they gain in mass by creating enormous gravity applying the sphere of influence of such a star. Gravity is the increase of heat in occupied space by the reducing of space in a spherical unit. From the offset of the Big Bang to the process the Universe went through development up to the very point where it is at the present, the process was about converting heat to space and space to heat. I named the expanding of material through the process of overheating antigravity. Where gravity is the contracting we find present in our daily life, the Hubble Constant and expansion through heat applying is the other part of the driving engine applying antigravity with expansion where the heat is transforming to space. Space shifts as heat releases space and converts the Universe in one direction bringing about expanding into more space but less dense space. Remember how the heat came down from 10^{34} to 0 K at present? The density of heat in space surely diminished considerably since then to now. But while the heat was diminishing, space grew as a consequence of the heat diminishing. Gravity on the other hand is exchanging heat through the concentration by removing space, bringing about space loss with increased density of particles and therefore heat concentration. In the centre of all spheres, which all stars are, it is hot. In fact the heat in the centre of the star is the product of the space it concentrates to form heat and in that we can read the gravity the star is able to produce. The ability to secure heat by reducing space becomes the measure of the star. Momentum is the second form of gravity symbolised by Kepler, as **k**. The Big Bang is the result of heat expanding into the forming of space. Gravity, on the other hand is about concentrating space back to heat, and take recouped heat through to material, acting out a balance of expanding while contracting. This way gravity is applying the onset of the Big Crunch by destroying space while space is converting heat to material occupying space. The Big Crunch is coming about because the Universe is expanding where the two processes are one principle.

Where the manuscript presented came from an actual letter I sent to specific Universities, my aim with that effort was trying to keep the printing cost as low as possible. The motive with the manuscript was an attempt to assemble the most basics of many of my ideas in my books in an effort to place as much emphasis on explaining the combination of my work while keeping the printing cost as low as possible. My Main effort driving me to publish the book is my wish to introduce the theory of not special relevancy but introducing absolute cosmic interaction and integrated relevancy as the drive of gravity. Where at first the book was a letter I wrote in an attempt to promote my theory at many different Universities outside South Africa the second attempt is to get it published and let people read it as widely as possible. I realise I have to get the ideas out in the open to as wide an audience reading the work as I can achieve and have people read as much of the information as the effort will allow. I have to show others how I came about finding the most basic aspects of my theory in the most basic form and show how that follows a line as growth progressed and developed. It is an attempt to convey the thoughts in formulating the concept on which I based all my other books. But with this book the motto was keeping it so simple children must understand what I try to convey about my theory and let the other books carry the slightly more complicated arguments. If I could manage to keep the concept that simple and keep a well established link with what mainstream science knows but without compromising facts, I know I could find the key to the accepting and understanding of what I have to say. With the costs as low as possible it will be possible to find a publisher that will publish and sell the book through Amazon Dot COM and Barns and Noble as well as other major book outlets. Without the academics accepting and underwriting my views it is a chance one takes. Another large factor is that only I believe in my views, as being correct because never once were any of the principles vaguely explained, whereas I explain, define and underwrite such explaining double and triple.

$$F = \frac{M_1 M}{r^2} G$$

Let's put the mathematical formula into a practical context.

What Newtonians share with Newtonians, apart from covering up Newton's fraud with absolute conviction is that the two have explicit detailed arrogance about the insight they so blissfully share? NEWTON AND NEWTONIANS ALIKE WERE AND ARE INCAPABLE OF READING KEPLER'S COSMIC WISDOM.

As explained, there is some discrepancies about calculating the force of gravity, because gravity would apply as nicely as it does if it was the perfect balance, a balance exists in space of equal measure bringing about equal seasonal time.
What would drive the Universe if mass does not do the inspiring? The answer is obvious.

For instance Mainstream science has the theory that matter and antimatter developed and some matter formed as particles, which were nicely wrapped in containers we call atoms and stayed on as material or matter where the rest formed antimatter that disappeared. The anti material also formed atoms but then chose to just plainly vanish, as did singularity and other cosmic factors to be nowhere. Singularity is a mathematical position falling outside the detection of the observable Universe in any case so in the case of singularity there is a possible excuse for the disappearing…or is there? We know that singularity produced space-time and space-time produced gravity but then space-time went away leaving us with gravity. Singularity is a single dimensional entity that is not material, holds no space, holds no time and in our 3D view can only be found outside the human visual spectrum. Mainstream science is of the opinion that singularity disappeared after the Big Bang process came about, whereas I am of the opinion that everything must remain part of the cosmos once it was in and was part of the Universe simply because there is nowhere for it to go. The same applies to space-time and antimatter. If it was part of the cosmos in the beginning, it has to be in the cosmos until the very end only and simply because there is no other place available to be or to move to. The Universe is the container that is the only container leaving no other container to contain what ever is in need of containing.

If it is or if it was, it still is in the Universe! Our task is to find the place it went to or find what it changed into. While we then are on such a hunt, we might just as well find the cosmic principles not understood but which is there all the same. Take for instance the Bode principle where all nine planets show a relation with the Sun in precisely the same manner and using the very same method of spacing, yet science brush the Bode principle off as a coincidence. It might be a coincidence when the space between one or two planets shows these phenomena but when all nine planets plus even the fragmented structures adhere to the very same principle, no person can be of the opinion that it came about as a coincidence and still pretend to be serious or professional about cosmology. One then must simply find the proof lacking in our understanding

Mainstream science knows about gravity, the Bode principal, the Roche limit, the Coanda affect, the Lagrangian system and the sound barrier but cannot explain any of the phenomena all though the presence of these phenomena is without dispute. It is the explanations about what causes the phenomena that should be part of the dispute but in science the way they defend Newton scientists go overboard by disputing the phenomena and the phenomena as a principle existing in cosmology or not, becomes disputed. Science fails to give acceptable explaining of such occurrences, therefore disputes the validity of the phenomena and this failing to explain the presence, and becomes disputing the presence thereof. In such a light Scientists must somehow realise they are barking up the wrong tree with the information they have to use to do some explaining. They cannot refuse the phenomena and not realise they must have the cat by the tail as far as cosmology goes. Please remember that with this I am referring to cosmology and not general physics. There is an Earth versus a Universe of difference between the two concepts but Newtonians fail to see that because Newtonians cannot appreciate the differences, thus blurring the understanding of gravity. If there are that many phenomena (it represents all there is in cosmology) to explain and such little ability to explain (science fails to explain even one) by using the information Mainstream science is using to explain the cosmos, then someone somewhere has to realise there is something drastically wrong in the way they present the knowledge they claim to have.

One cannot be serious about science but defend your view by dismissing the validity of all unknown indicating factors presented as such. There then is some gross incorrectness in the way Mainstream science reasons. The Roche limit is there and no denouncing thereof can remove it from the cosmos. They may refer to evidence received from the Hubble telescope as "the star is blowing bubbles" for the lack of explaining what is occurring, but occurring, it does. One cannot say it is some unknown gesture presented on occasions because not explaining the pictures presents the presence of certain foolishness. For fifty years they lost many pilots but still has no idea what brings the sound barrier about, or find the link gravity holds in the process we call the sound barrier. Instead they try to interpret some effect established almost two centuries ago with steam trains back then travelling at the same speed that horses run. No further investigation with the science in hand brought them closer to new facts! It should be a sign telling them they are going about incorrectly but it does not because Newton said so. It may sound as if I am anti Newton but I am not. But there has to be more than Newton with so many pieces of the cosmic puzzle still missing. Science should not serve only Newton but science should serve the seeking of the truth. When I first came upon the unknown, it stirred a sense of disbelief and I decided to respond.

Some twenty-seven years ago I decided to start an investigating quest on my own to see where I could go with my private research. It came as a result of my frustration when I realised the discrepancies there are in theories presented about cosmology and all the unexplained factors no one ever makes any effort to explain. Later I found that no-other person than Newton in person was to blame for the mistake that was made but now I am jumping the gun. If you are a Newtonian and feel a repulsing urge to throw down the letter you will do so at your peril. I say this because I have seen Newtonians get fits in the past when I say what I just said. Whenever I make this very claim the entire science community rejects me immediately without any reservation to any person. When I speak out against incorrectness I see science as a unit and an entire structure reject all further statements that I make without excluding any body active in the field of science. To science Newton is reserved as a god and they placed Newton beyond criticism. When they listen to me criticizing Newton they switch off their mental lights literally. They immediately go blank and I have witnessed it every time. I can visually see their eyes go dim. No Academic Newtonian priest will spend another second to listen to more of my views. Yet they remain unable to use Newton to explain the cosmic phenomena as the Bode law, the Roche principle, the Coanda affect, the Lagrangian system or the sound barrier. But what they do not realise is the mistake was a resulted not about what Newton presented because it came in what he admitted that he could not present. The mistake came as a later presentation of a concept he admitted he could not underwrite by scientific explanations. When he introduced gravity as a concept he admitted he did not know the origins of the force. Newton admitted that he could not explain which he subsequently did not explain later became an institutionalised claim presented later as if he did explain it.

The error is in the facts not yet ever explained and in that which Newton admitted he did not understand. It is Newton's incorrect suggestions he made about gravity that was later accepted as explained and proven science that went on to become institutionalised facts. It absolutely came in the way Newton changed Kepler's formula. Newton admitted he did not know what gravity was and left it at that. He did not offer any more insight than reducing gravity as a force and a force it stayed all the time. Three hundred and fifty years on, science still do not know what gravity is and is still leaving it at being a force. Without knowing what gravity is, all other concepts in cosmology does not make sense because all the phenomena I mentioned a minute ago are sides of gravity and perform (each to its own but still) as another principle where the totality forms one concept we call gravity. If the World does not know what gravity, is then the phenomena coming from gravity will remain unknown to all. With that in mind, Mainstream science still takes a very dim view on my criticizing Newton! In all this time Kepler explained exactly, precisely and unequivocally what gravity is! The unbelievable part is that we all missed Kepler's announcing of gravity and in the "we" I use that 'we' include even Isaac Newton and Albert Einstein in persons and by names.

An Open Letter On Gravity Part 2

All principles I use in the theory I introduce with the publishing of this book are part of nature. I base my theory on heat stabilizing through space using motion to produce cooling. That is gravity. But although this may sound basic, Mainstream science is also most guilty of their usual departing from this basic principle through the employing of terminology and such terminology has the tendency to cover many of the basic meaning behind the principles in nature. For example the one principle I do not applaud is a principle Mainstream science underwrites in the sense that matter in the beginning was coming about and anti matter came to destroy the matter by consuming it. This translates into a packman computer game that has no correlation to cosmology. It is moreover the disappearing from the Universe of that which came as the result between the two opposing materials that I strongly reject. Anything part of the cosmos at any stage before during or after Creation, remains in and part of the cosmos and cannot leave the cosmos because there is simply no other place for what ever there was to go. Leaving the cosmos just is no option there is.

In any test performed today by creating friction through motion the discrepancy there are between objects in motion will bring friction that will produce heat and the heat will result in space forming. In such destruction of matter, space and heat comes about and the net result eventually is space created where no space was before. The cracks showing is space created in the cooled material that was heated to a glowing red-hot always afterwards leaves cracks. The cracks are space not filled but was filled during the heat coming about as a result from the overheating. After the cooling the cracks present new space where there were no space before the heating took place. The heating process started forming and filling space but afterwards when the cooling set in it reduced the filling of the space. The cooling did not destroy the newly formed space. The cracks represent the space not filled. The material in the cool state cannot fill the void that came as a result of the cold material contracting and reducing the space filled when the material was overheated. But the space remains although not filled any longer. That we can see as evidence with material having a heat building up when motion difference brings on friction and such friction brings on heat. I do not share the view that when matter and antimatter came into conflict the product that came from this just disappeared from the Universe without a trace of any sorts. I believe the evidence is present and I think I know where that evidence is. I believe I can show that it is a motion discrepancy that produced matter and anti matter and we do not have to go and look for non-exiting positrons.

A positron must produce a negative proton and such a performing sub atomic structure cannot be functional. By changing legions it must then produce a product where it performs as gravity by rejecting material and pushing away other cosmic components. That will lead to an exploding Universe! I say that discarded material became heat that became space that became outer space in the Universe. I go to lengths to make persons see that space cannot be nothing. This is a factor that science has to accept if Mainstream physics have the will to find solutions about the Big Bang. I say the motion between particles in a cramped space as the case was during the initiating of the Big Bang, would have brought on friction in space between particles present that we couldn't even calculate. The result is that some of the matter particles produced a means of self-sustaining by applying gravity and the demise of the other particle that became destroyed resulted in plasma forming on the one side and material on the other side. I believe even to today and throughout the rest of the Universal motion through space in time, the plasma is transforming to material forming particle growth in space through the motion we named gravity. This was how the cosmos came about and this is the manner the cosmos will conclude.

I believe some of Creation remained as some particles formed by applying gravity in motion and the lack of gravity turned the other particles of lesser motion into heat. In this the destroying of singularity contact came, which formed a product we realise as light, which again I believe (within reason) I do prove (although it is not in this specific book but in "***An Open Letter Announcing Gravity's Formula***"). I believe heat is the destructed form of material and this information the atomic thermo explosions give us. By releasing the heat that is sealed in an atom, such release thereof produces heat, light that can liquefy the eye and most important the unexplained nuclear winds that destroy so much. But to realise that we must beforehand find what space is and accept that space is made of something. We have to see what forms space and why space can be the absolute basic container through which gravity can relay the influence it carries.

I have shown that when particles expand into a more confined are, the average temperature will rise. However I have also indicated that it does not seem to apply in the case of stars because stars freeze hydrogen into a liquid substance and the only way that can be possible is to decrease the average heat levels. This shows that the star is in contra action to what the Big bang heat was. The heat levels back then was zero and we should rather think of the heat growing through time while the stars are maintaining the cold, If the temperatures was without space. Then space came about and heat levels declined. If the temperatures was 10^{34} K in the shade then what was the temperatures inside the atom spinning at and beyond C. From our vantage point we no have about the cosmos the temperature that was 10^{34} K in the shade is now minus -273^0 C or then 0^0 Kelvin also in the shade because it still seems rather dark in that location.

As I said previously by compressing space with the aid of an external piston into a cylinder, such confining of space will increase the heat levels dramatically. This is accomplished by the piston effort to establish a reduction of the space inside the cylinder in comparison to the heat levels outside surrounding the cylinder. The heat coming about has no relevance to particles colliding because compressor cylinders cool down with time and not necessarily with the loss or release of particles. Up to the point where science took humanity to the second millennium, the proof needed the sustain contraction and which we all know rings true, was failing considerably because it seems that Einstein's Critical Density is the biggest fabrication of misleading the public by altering facts. Their inability to face up to the truth about Newtonian science and corrupting facts to try and secure such required evidence, defies the most basic logic. In spite of such deliberate malice on the part of science, it seems all new evidence we receive from outer space is disputing all the findings of Newton concerning cosmology, which also includes that I disproves Einstein's Critical Density as the answer as having even the slightest basis it uses as truth. The Universe will not reach a point of contracting, not withstanding whatever dark matter astronomers try to locate in the vast space. This ridiculous attempt to divert truth from reality, has set science in a frenzy to compensate for the dishonesty coming to light again and now that rush is to find the black matter to force the cosmos to retract to Newton and contract so that it can go back to where it came from.

I don't know if it is naively unintentional or deliberately criminal in all the cases but there is more than strong evidence that point to flat out covering up of the truth about the official view of the cosmos. Why all the denial of the truth about scientific views being incorrect and those incorrect facts reaches across to infiltrate our views that we have at present about the Sun? Where I prove that contraction is at present going on but in a far different way than the magical forces Newton claimed, I show that such contraction is as much part of the cosmos as is the expanding we focus our attention on is. It is our culture we carry from generation to the next generation, which leaves the human view obscured in admitting the truth. If my skin feels the heat of the Sun and I feel the heat every time the sun rises but in spite of my senses there is some idiot that comes along and says the Sun is a frozen ball of liquid ice then for sure my culture will demand the stoning of that mental maniac on the spot. In every case no one will listen further to any debate about the Sun being a freezing ball of liquid hydrogen and every time my mental health becomes suspect before any one even tries to listen to my arguments. This reaction is not resulting from intellectual reasoning but comes from a culture of accepting culture without arguments.

Being the beings, which we know we are, and which we came to evolve to, we have very specific mindsets. We know from experience that only those with a highly distinguishing academic background are worthy of notoriety – in the same manner that we know hydrogen forms gas and gold is as a solid as quick silver is a fluid. Our mind sets a case that originates from a mental state we accept to know as being culture, and culture is what we take from generations from the past as accumulative knowledge. That is the term we used to develop from skin bearing barbarians to the sophisticated persons we evolved into. In further recent development Academics came to reason that only those with strong mathematical skills are worth having an opinion of note, but to those having that attitude, I must also quickly remind that mathematics is merely a tool. If the tool is applied in a concept proving a theory, but the theory has a considerable flaw, mathematics will lend proof to such a misconception but does not necessarily prove the flaw, especially when the mathematician is dead set on not discovering such a flaw. What would happen if the mathematics used, have a cultural flaw and through such a flaw, the picture brought about by culture running for thousands of years carried the crack no one ever noticed? How would that influence the accuracy of recognised theories at present? What would be the difference between mathematics practised and arithmetic practised? I came to the conclusion the difference between the two is zero. But the zero is the number and not the value linked to the concept.

An explanation about the growth of outer space such as the above serves as a typical picture matching every logic view we all have about the Universe, but does Science really provide the answers matching our modern logic, or are we filling in and compensating for science's shortfalls. Does reality really match the logic of official science? Does the Hubble Telescope's pictures match the explanations science provide about how Creation all started... where it is heading...and where it will end? What is motivating the expansion and the moving of Creation? Why in creation would it reduce again, and to where will it reduce...said with the mindset in which we are supposing THAT WHEN again REDUCING it is reducing to the centre of the Universe, but to find such an answer we have to locate such a centre and where would the centre of the Universe be???

$F=G (M_1 \times m_2)/r^2$ Is this truly the answer...?

The fact that I wish to make by my writing of this specific book is that if the Universe can be compressed back to the size it had at the point of 10^{-38} seconds after the Big Bang temperatures of 10^{27} K will come about once more. In normal thought we might think that the expansion was the result of the insufficient space that was prevailing at the time. If the Universe was in a vacuum filled with a soup of heat so thick it was the same as material. From that thick soup that protons froze into substance. However what was available then was as big as it is big now. What were available then are available now and the heat that formed time then is time forming space at present. Whatever is available now will be being available any place in the future and that was what was available when Creation came about.

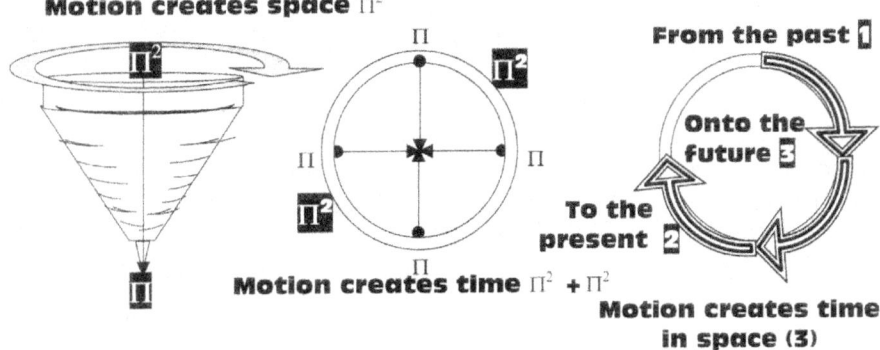

With the Big Bang motion the atom came about as $(\Pi^2+\Pi^2)(\Pi^2\Pi)3.=$**1836**. This forced the atom to establish a six dimensional outer space $\Pi(\Pi^2+\Pi^2+\Pi^2+\Pi+3)=$ **112**. When Creation came about time was tied up in heat and back then what was the temperature of the time was, was also undeveloped space and the soup that time then was, was as empty or as full as it now is. It is the spin or the motion while time is in progress that brought a difference to what was before material filled what material now fills. Then I presume the heat that was present is now in space, which is developed and is in material that developed from movement of space in relation to space allowing material to move.

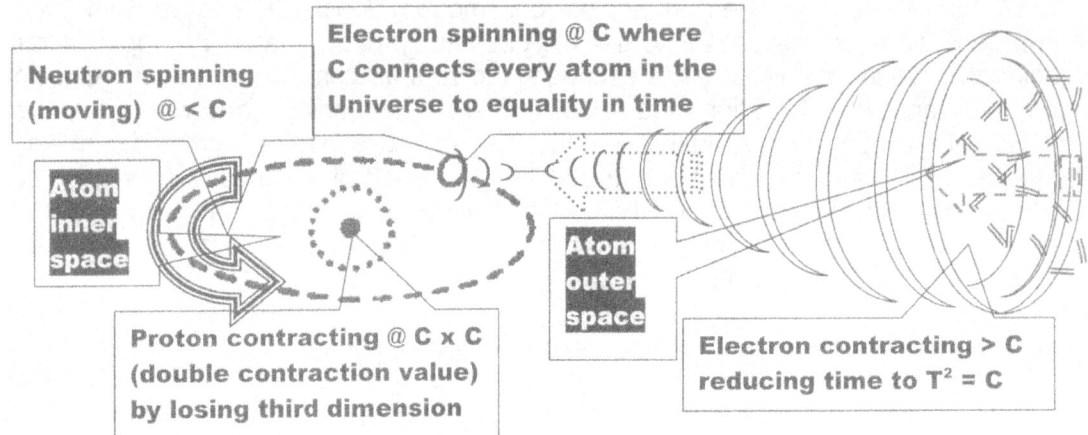

What was is still in the present and only motion of space in time as time implicated any difference to what was and is. The Newtonians will have it that what space the Universe then employed as the space of say one atom, the impression comes through that from edge to edge and from Universal border to border space occupied was the same as one atom will claim as personal space in our present day and age. That which is presented in that way were everything there presently is and the border from edge to edge holds eternity with no end to eternity without any end. It is still what is eternal at present and only the relevancies of what is smaller than then came about.

In their super minds they lost track of one small concept which is the one issue overriding all other facts. As **k** develops so does time expand, so does space explodes and so does time within the Universe progress. $k = a^3 / T^2$ Kepler said space a^3 will grow by dimension of **k** while time T^2 will decline $k^{-1} = T^2 / a^3$. To reverse time every point holding **k** has to reduce k^{-1} back to k^0.

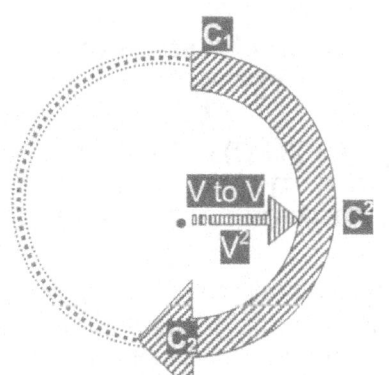

The given formula used to calculate time at the speed of light is $t = \sqrt{1 - (V^2 \div C^2)}$ where as V = matter in motion; C = speed of light; one = the fractionate proportion of the value of t in relation to motion. The square is the implication of Newton's second law (forces apply equally in both directions). Let's revisit this formula and se what implication ass far as mass there is on this formula.

We have seen that this formula is valid as far as the motion in the atom applies to the Universe, but only in regard to matter in motion and not to the whole of the Universe. The rest of the Universe's time duration will not come to a halt just because some fragment of matter is travelling at the speed of light. If C was the measure of time, then it will take no time or time in the instant for light to leave in place of origin and progress to the end of time. That it does not do because it takes light lots of time to reach across space and flow from where light starts to where light is viewed. Light can take almost the lifetime of the Universe to cross all the space and show what happened when it left where it came from as it arrived in my vision. If light was equal to time it is not possible for light to take say 12×10^9 years to show the very first dawn of time because then time was one and one was over before it became valued. But because it took the light to come across from where it started and end in my view so much time, light has to adhere to time and the growth that time in space dictates. Time at one point has to be one but that is far from the point in time we vision and therefore that part of the argument I leave to another book in time.

Far from me to try and reinvent Newtonian formula conscripts I will argue this formula and then stick with it in the manner Newtonians present it as time, notwithstanding how ridiculous that may seem to be as a means to use the formula. In the formula $t = \sqrt{1-(V^2 \div C^2)}$ we have V in a square that indicate motion in a straight line by duplication making the straight line a square. Then also in $t = \sqrt{1-(V^2 \div C^2)}$ we find C going square notwithstanding that it is accepted in science that there is no movement of any structure that can exceed C and moreover reach the square of C. That is just more Newtonian inconsequent arguments and drunken double talk. Light moves at C and light is no able to move at anything faster or greater than C. C is the very limit to what can move, but let's leave this statement hanging in the Newtonian closet. The formula $t = \sqrt{1-(V^2 \div C^2)}$ is moreover an indication of how the atom moves in time and through space than indicate how time presents the Universe. The formula is proof of space-time and not time.

In the centre there is a nucleus from where the entire atom moves its position and then there is an electron that defines the allocated position of the atom in terms of where it moves.

The centre movement is indicted by using a symbol V and the defining movement is done at the speed of light that is indicated as (C).

But let's stick to $t = \sqrt{1-(V^2 \div C^2)}$ and see in which argument this formula do fit.

When anything is moving and is holding atoms, the movement of the structure of light. Time is not the more intense than the speed in a line (**k**) while the that will be square (**T²**). In space (**a³**) is confirmed by

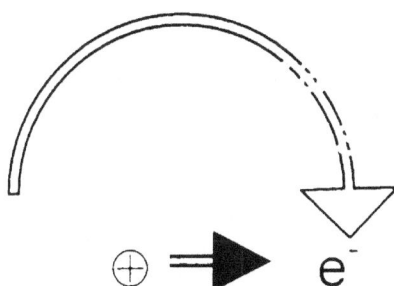

stands in relation to the spewed speed of light since time is a lot of light. The atom has to move electron moves in a circle and this movement the atom in both movements (**k T²**)

It is true that there can't be a C^2 but there can be a V^2. Time will always be the square (V^2) or (T^2) or Π^2, but time will always move from the past (V) or (T) or (Π) through the present (V^0) or (T^0) or (Π^0), and move on to the future (V) or (T) or (Π). This we can deduct from Galileo's time chronograph machine made specially to indicate time (and then Newtonians still say they don't know what time is, but it is because Newton couldn't think that far and they can't think further than Newton and his mass could think.)

Pendulum arm swinging through time while indicating time

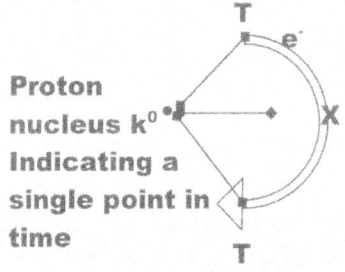

Proton nucleus k^0 Indicating a single point in time

Electron moving in relation to the nucleus k^0 Indicating a reference to space in time

Whether Newtonians ever thought about the fact and whether Newtonians recognise the fact and whether it its in the Newtonian ability to mentally grasp, that is of no concern but a fact more established that Newton's creation of mass is what Galileo brought to the world the time chronograph machine before Newton's fraud in creating mass. But looking at the atom carefully we see that Galileo copied thee atom before the human mind ever knew about the atom. The atom is the cosmic time-measuring-devise as the pendulum device is a time-measuring-devise. The electron moves by duplicating in relation to a specific centre and such movement is time going from the past (T) through the present (T^0) and to the

future (T). Then the question arises whether the electron by moving can go C^2. It can't notwithstanding that in motion the electron should go at C^2, we know it only goes at C which proves that $C^2 = C$ and that brings us to time in the Universe being at T^3 or the space (a^3) is equal to the movement of time in the electron at C or then a linear connection (k) where the electron moves producing the motion factor of (V^2) or then (T^2). That brings us to Kepler's formula where the cosmos and no less than the cosmos through mathematics presented its value to Kepler as $a^3 = T^2 k$ and Newton had the audacity to change what the cosmos delivered to Kepler! The cosmos is time in motion and time in motion is gravity. Therefore the formula concerning time should read that the cosmos forming space-time a^3 is = the movement T^2 in relation to a centre **k** and not $t = \sqrt{1-\left(V^2 \div C^2\right)}$, but be that as it may, let's still proceed with the formula as Newtonians devised it being $t = \sqrt{1-\left(V^2 \div C^2\right)}$. The relevancy in space should be $k = \sqrt{\left(V^2 \div C\right)}$ because $C = C^2 = C^3$ and space should be $a^3 = V^2 \times C$ and that amounts to Kepler's call on gravity being $a^3 = T^2 \times k$ as the formula.

The movement of the atom according to gravity is in regard to the movement of the electron that is spinning round about the moving atom centre.

⊕ Positive space-time   e⁻ negative space-time

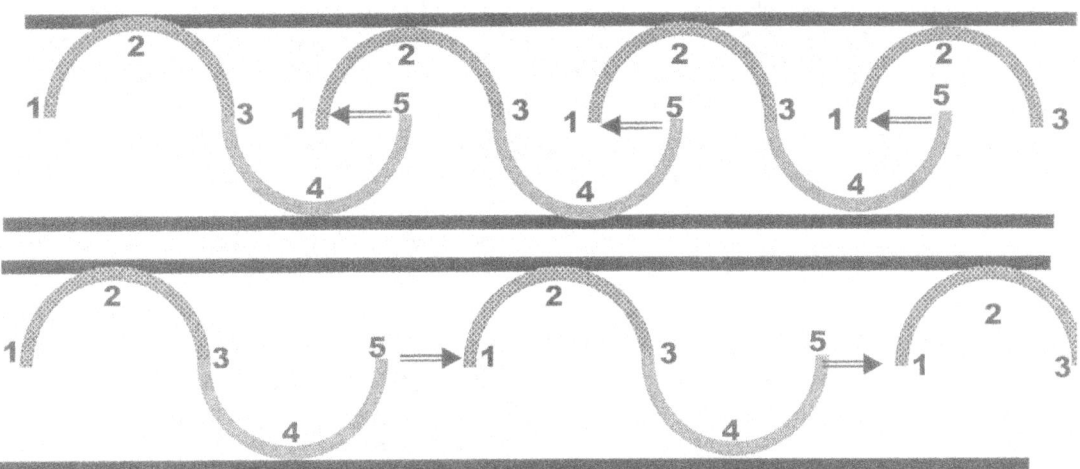

I gave the two different movements two names which has no indication on their status in charge but only applies to the manner I see them translate their positions. Since the electron is always moving away or trying to place space between the position it has in relation to the centre and is always trying to expand, I thought the behaviour indicated a negative attitude. It is not negative and it is not trying to flee and it is not trying to expand but correctly or not I thought it would be easier to remember that action to be somewhat negative in displacing space-time away from the proton.

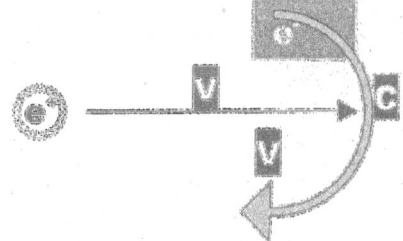

Then in regard to the proton that was always trying to gather or to contract or to displace space-time towards the centre I thought such behaviour shows signs of having a positive attitude in gathering by contracting.

When the atom moves both these factors show the discipline of gravity where the relocation attributes to the space the electron gathers in the relation such re-allocating would allow. As the atom moves in a specific direction (and the atom has to move) since the movement is the state of gravity and not the mass that is an effort to put a disadvantage to independent gravity or to prevent independent motion and is an effort to put claim to space belonging to an independent object. Every point of relocation will find the electron holding a different relevance in relation to the position the centre has

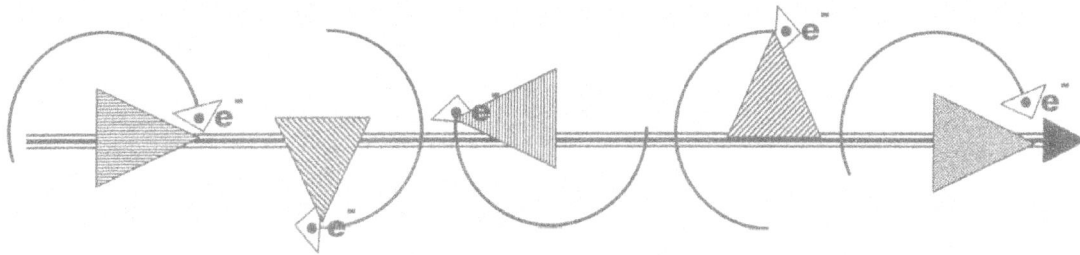

We have the atom forming a circle as the electron encircles the space allocated to the atom in the time instant. Since the electron is going at the speed of light and the speed of light goes above space in a state of form we have the electron encircling the proton at a defined potion of C and in relation to the centre translating time in a new allocated poison in terms of V^2. The space the atom holds is $a^3 = V^2C$, which reminds of Kepler's statement that the cosmos told him (Kepler) that the Universe was $a^3 = T^2k$

As the atom moves in relation to time, so does the electron move in relation to the centre of the atom. While the atom moves with time, or if you wish to use a Newtonian name, they call it gravity, there is a link between the speed of light going at C and gravity going at Π^2 what ever the relevancy will allow gravity at Π^2 to be.

The movement of light is a point where the Universe burst into form and that I explain by proving my statement mathematically in Part 2. Since all atoms are linked by electrons determining the equilibrium time spinning at C, this connects electrons and therefore atoms.

With such an obvious connection we find Kepler's formula giving a relevancy that can and that does apply throughout the Universe since the Big Bang and that equilibrium in relevancy is the speed of light. But the speed of light is not time.

The repositioning of the proton indicates the next location that the electron will be and the relocation of the proton find 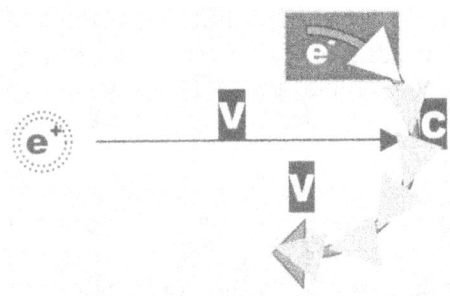 a value by the speed in which it travels and the speed in which it travels finds a value in the gravity applying. The gravity of the atom is the motion thereof and the distance it moves per time Unit in relation to the allocated circle the electron forms, which puts the 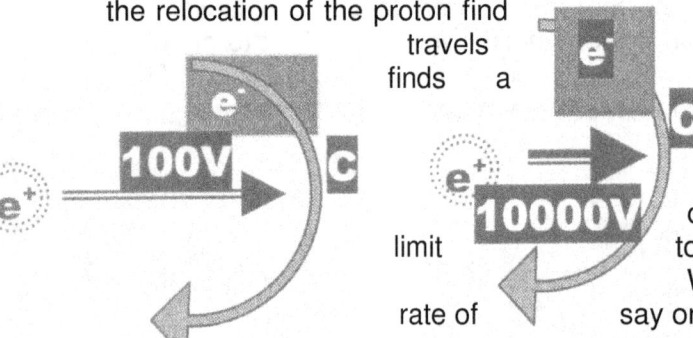 to the space, the atom holds.

When one atom moves at a comparable limit say one hundred Units of whatever length the rate of measurement takes, the electron would hold a much different circle in space that what the distance would be when the atom moves at a rate of say 10000of the same Unit measuring length or when the atom moves at say

10^{10000}. It is the movement in a specific direction, which is the factor that we call gravity and the speed of such movement establishes the size of the atom. That means as easy as Newtonians wish to create their Universe with placing mass in every hole they

have and filling every concept they don't understand with the vacuum filler called mass, a constant Universal time there is just not. That is gravity. It is the positioning of the space the atom secures in relation to the position the Universe allows the atom to secure. The gravity has a direct implication on the intensity of movement in relation to the size that the atom has.

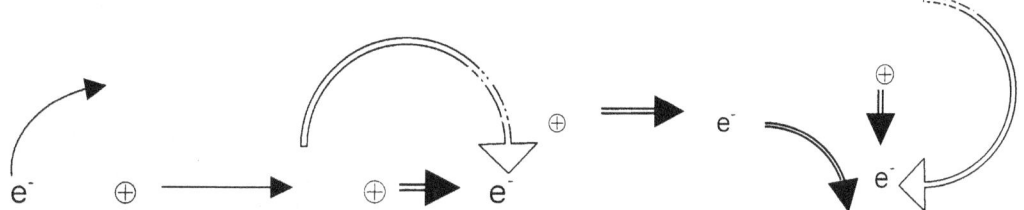

The electron that is moving at C or the speed of light connects the Universe just by doing so and while it is doing so the electron is forming the Universe as a woven structure, just as it had during the GUD era, but not in terms of equality in duration of time lapsing but rather in terms of time development forming a string of time. Every point standing in regard to singularity applies a different ratio to time connecting and there is no constant in time except for the constant being that there is no constant.

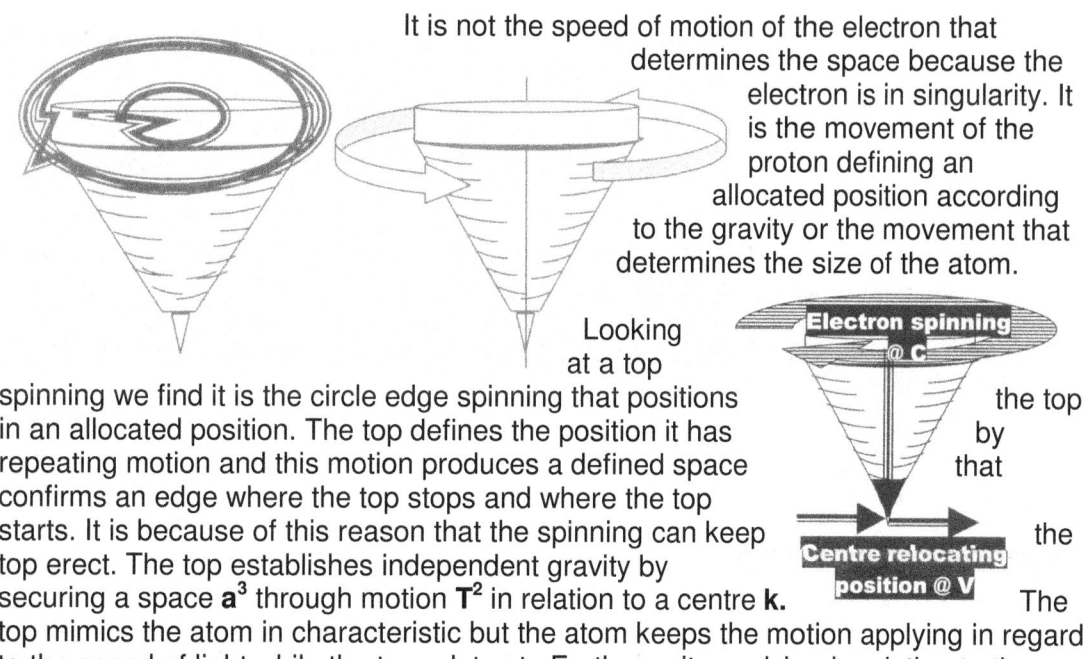

It is not the speed of motion of the electron that determines the space because the electron is in singularity. It is the movement of the proton defining an allocated position according to the gravity or the movement that determines the size of the atom.

Looking at a top spinning we find it is the circle edge spinning that positions in an allocated position. The top defines the position it has by repeating motion and this motion produces a defined space that confirms an edge where the top stops and where the top starts. It is because of this reason that the spinning can keep the top erect. The top establishes independent gravity by securing a space a^3 through motion T^2 in relation to a centre k. The top mimics the atom in characteristic but the atom keeps the motion applying in regard to the speed of light while the top relates to Earth gravity applying in relation to the energy life bestowed on the top.

Movement humans apply is only gravity in access. Yea, sure Newtonians call it amongst many other things by many names but Newtonians are famous for naming when they have no bloody clue what principles apply and to hide their incapability's they use elaborate names as to confuse every one to the maximum and then hide the Newtonian lack of understanding science. Gravity is the movement of material in relation to a centre securing a hold on the Universe, which is the very same as the role time fore fills.

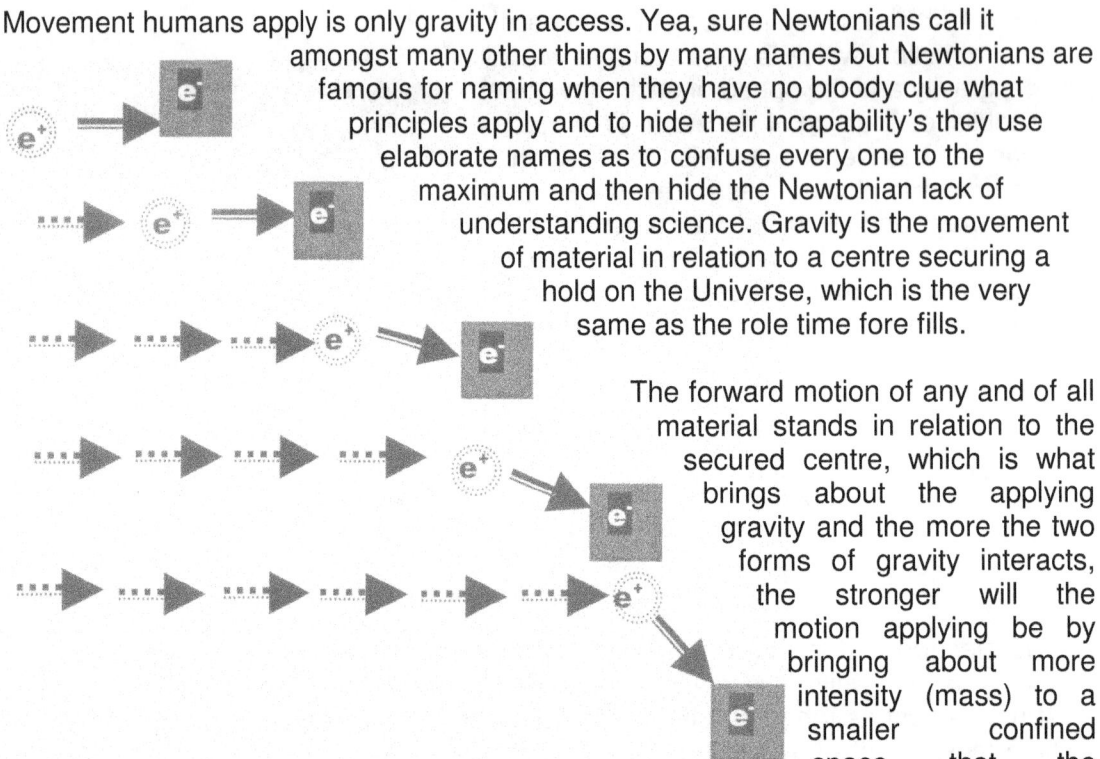

The forward motion of any and of all material stands in relation to the secured centre, which is what brings about the applying gravity and the more the two forms of gravity interacts, the stronger will the motion applying be by bringing about more intensity (mass) to a smaller confined space that the electron provides (structure size). Mass is the result of gravity and gravity is not the result of mass and everything in the Universe is connected by time and therefore are relative to time where time varies from position in location to position in location while never being the same but is still forms and provides the overall connecting of everything to everything else.

⊕   Positive space-time   e⁻   negative space-time

Only the electron can move at the speed of light and that places the movement of the electron in relation to the movement of the atom in the Universe $t = \sqrt{1-\left(V^2 \div C^2\right)}$. The electron moves at C from C_1 to C_2 and this movement is C, simply because not even the electron can move at C^2.

But moving at C has no implication on the movement as such because moving at V^2 shows gravity in relation to gravity measuring the worth of the atomic structure and that C^2 is no the speed of light but refers to the gravity that established the atom during the period the grand Unified Theory envisages. That was when the entire whatever was, was parted by an electron and all gravity then had the compactness of the electron. At that stage the Universe was C in relevance with C^2 in gravity placing the Universe C^3 in the atom. That simply means the Universe were apart by electron and had then the size of what puts the atom in space-time, which is the neutron. It all still just proves the correctness of the Big Bang concept. At that point what ever did travelling between atoms did it at the speed of light as it still does it but the concept of the speed of light in terms of light holding space-time changed considerably. If time was equal to light travelling then light must take time in the instant as well as in an eternity which then will be equal to cross the entire Universe because time is both the smallest instant no one can ever fathom and eternity, which is totally beyond the realm of realising the concept to the mortal man.

The truth of the matter is that nothing will come to a halt if everything in the Universe is travelling at the speed of light. If that was the case, we should still be experiencing the Big Bang, because everything was travelling at the speed of light, not only in relation to one another, but also to the completely geodesic time structure. If it did not apply then, it does not apply now. Even if the Universe with everything in it will travel at C, the Universe will still expand, only because it did so in the past. Nothing stagnated after the Big Bang, but everything developed through time. If Prof. Einstein was correct and time is C, time it proved him wrong, because everyone can witness the progress that occurred since then. Those friends of mine made some fundamental flaws to the argument. Firstly, if the graph's perspective is from a mathematical point of view, then C never touches the Y-axis but remains a fraction of one. This implies that there is no zero ever reached on the X-value, so no mirror image can be made. Secondly, matter never acquires a state of motionlessness, because of the Hubble Constant. As the Hubble Constant is forever in progress, matter could never be in a state of zero inertia.

I decided should show you the reader (again) where to find singularity. It is so incredibly easy, but if singularity is not located at the proper place and singularity is not found and correctly valued, using the cosmic given value then none of the other phenomena makes any sense. Without taking guidance from Kepler and using what he gave to enable me to trace singularity, my effort would have been fruitless. By not finding singularity such not finding of singularity will then not uncover the other phenomena such as the Roche limit, the gravity statement behind the Coanda principle, the true value of the Titius Bode law and how the Titius Bode law bring about gravity, the location and identity of space-time in its true meaning, the working principals behind gravity, the reason why the five apply to the Lagrangian system in relation to a centre where singularity is secured and what causes the sound to break the atmosphere.

When Kepler's direction is followed it is so incredibly easy to find what ever you wish to find.

IN THE MIDDLE OF EVERY ROTATING OBJECT NOTWITHSTANDING SIZE, RUNS A LINE THROUGH THE OBJECT WHERE THAT LINE DOES NOT ROTATE BECAUSE THAT LINE HAS NO SPCE.

Every aspect that can rotate does rotate but the rotating is around the centre line that is standing very still and is without space to be in. It is a mathematical calculated line that does not exist except that it takes charge of the entire spinning process and it covers every detail the spinning process maps. Without being in our Universe it is an indicating point where what is on the left hand side meet what is on the right hand side. It is the point taking charge of where lines start and where lines end indicating the very end of all lines on either end of every line. This it does without reserving space from its positioning or taking away space-time in any manner from either the right hand side or the left hand side of whatever is spinning. We find this spot in the place where if any of the sides will continue to reduce, it is at the very point where such continuing of the reducing of the line would end being on this one side and the very next instant it will be going from that side to where it will be landing on the other side at the very next point of the line running inwards that is forming the divide. If such a line will run to the inside of anything rotating it is able to shift to the other side of whatever is rotating without being in the centre singularity point and still being on the other side of the spin. It is crossing the absolute divide where the motion that is coming from the back to the front on the one side of the spin then suddenly swaps ends and run from the front to the back. That is where one locates the presence of singularity.

In such a line singularity is found. It is the centre of everything that rotates and Kepler said everything rotates because space is equal to the motion of the space in a space.

$$\text{Space } a^3 = T^2 k \text{ Time}$$

In the centre of all spinning objects there is a specific point that has to stand still and such a centre point is eternally motionless as much as it is space less. Such a point divides left-hand spinning from right-hand spinning and even as the direction changes, such changing do not influence that specific centre since that centre spot is in the single dimension of forming 1. As the single dimension it holds no space and without motion or space it has to stand very still. In considering the **spinning motion** in the **fraction of time** in the **detailed instant** we find that every aspect of rotation will turn because the incentive of the motion is going straight. In the motion-instant of moving in a straight line the line will alter direction in **every instant of change in time**. The motionlessness of the space less ness where one would locate such a line running and forming, from such a point the line has the inability to move whatever it controls and such control it exerts, changes the moving of the straight line to become an eventual circle motion. When the line changes from going straight to rotation the line changes alliances by motion. Although **the points spinning** had the **same characteristics** only micro split seconds before, they oppose the characteristics it had just before and just after the very micro split second in which they are and to which they relate by similar points also in rotation. This is because by going straight it rotates and by rotating the motion phases a totally different Universe than what it intended to move towards. This change of direction by a centre incentive removes the adding of Π to form a circle because the line can be nothing but a circle when rotating around the space less motionless centre. The point indicates as much as divides the rotating object in equal sectors forming harmony as well as precise contrast in every sector.

In the **precise middle** of all **objects in rotation** is a precise centre dividing the object in sectors that will **start the spinning initiation** from that centre point. But the spinning object **will have a middle point**, a very specific **centre point that does not** spin and only holds k^0 as a specific value. One value such a line **cannot have is zero** because **zero does not start any** line and therefore the **value of the line must be infinite**, just as described in **accordance** and by **the definition of singularity. That point** albeit hypothetical, is also as much **a reality none the less** and singularity is found where that point runs because everything in that specific line **must be standing still** as every line **running from that point is directly** in **opposing directions in relation to all other points** and are also **in opposing directional spin to each other.**

Now test your personal understanding of the most basic and simplistic of mathematical expressions $a^3 = T^2 k$ when it is converted to English.

By using mathematic rules and laws correctly and investigating the formula that Kepler introduced intensely but without Newton's interfering the world has unwittingly been aware of gravity for four centuries. Now is the time where we have to take on Kepler and use Kepler on Kepler's word without forcing Kepler to read as science wishes to have Kepler read in terms of Newton. What he (Kepler) has found was so far ahead into the future science is still trailing in Newton's advice but by investigating nature science beginning to catch what Kepler introduced. It is time that science see Kepler instead of allowing Newton to tell the world what Kepler saw and science now have start and look at what he (Kepler) found. Then modern science may find how much they have missed while they were influenced by Newton' view about Kepler's inferior mathematical skills compared to that of Newton. They would see what Newton missed when Newton said he does not know what gravity is and all the while Kepler saw what gravity is. He (Kepler) said that the cosmos said $a^3 = T^2 k$. The space is held in check by motion from a centre and that is gravity. It becomes more than clear that space a^3 is time by dimension T^2 and time is space a^3 involving dimension k. Gravity is a^3 / k but k is an addition of motion T^2 when motion reproduce space.

If it is that simple why does everyone make singularity and everything concerning the locating of the centre of the Universe so complicated?

$H_o \neq 0$. This is only as far as the mathematical principle is applied, but it still proves nothing about space, time, space-time and its relativity to each other.

This proves yet again Kepler's findings and it proves the formula correct which Kepler got from the highest authority on cosmology which is the Universe in person. The atom is the Universe and every atom is another Universe where time holds every atom as a Universe in space-time. The atom a^3 stands related to the electron circling at C which substitutes in this case **k** and as C are square (C^2) as well as the cube (C^3) (a detail I plan to explain later in this book) the movement of the atom in relation to its relocation where that is in terms of the (electron) which connects the Universe to the value of C we find that it again substantiates the correctness of the GUT universal theory. We have the atom as $a^3 = V^2 C$ which is confirming Kepler's formula $a^3 = T^2 k$.

However mathematically that places singularity (any value that is exponentially having zero as an indicator in example $1^0 =1$, $1000^0 =1$, $a^0=1$ $k^0 = 1$ and so on. That is the value of singularity because that is the only value singularity can have in spite of Newtonian brilliance flair and genial wisdom in mathematical forecasting. Looking at Kepler's formula if we have $a^3 = T^2 k$ then we have to have $k^0 = a^3 / T^2 k = 1$

The point of **k** connecting k^0 is indicating a differentiate space-time value in every case where **k** performs the task of applying a relevance factor that indicates by what margin does singularity relate to time ($a^3 = k \times T^2$). The point of **k** connecting k^0 is indicating time growth from its origin where all originated at singularity and since singularity throughout and everywhere is equal $1^0 = 1$. The fact is that $k^0 = 1^0$ no matter what is involved. Getting the growth of any space growing away from the point of origin is creating time distinction away from where all sides meet in singularity and singularity is the place where there is no space dividing the Universe without showing division but is still dividing. It is where there are no boundaries and yet every aspect of k^0 within every individual particle commits a boundary in relation to the singularity that is maintained. As we are part of the 3D and on top of that locked in time motion it will therefore be very wise for us not to try and understand the fist dimension or singularity at this point. In the second dimension there are sides committing motion to the third dimension but in the first dimension all is alike with separation coming about merely from being a unit. With every singularity progressing by **k** from the centre of the galactica since the Big Bang, **k** became the measure of progress. It progressed from a Universal governing singularity charging an accumulative control that responds to the growth of individual singularity forming and from that more individual atom singularity comes about. As singularity finds heat through gravity it progresses by establishing less support in the atom's individuality and by passing support onto the governing singularity.

The atoms are at task to remove heat from space thereby eventually they remove the atom's individual independent singularity to favour the stars governing centre singularity. That is why atoms grow when heated and reduce when cooled. If the atom did not get larger when objects heated, the entire principle of expanding when heated and reducing in size when cooled would be untrue. If it applies in any lab or in any workshop then sure as hell it's applying across the entire cosmos. Hubble's law indicate the expansion to which I am referring. The expansion that the Hubble law indicate is about galactica and then the most typical response comes from Newtonians on this matter. If there is an expansion, then that expansion is there where over there is billions of light years away and is so far we have no intellect to comprehend what there is at such distances because here by us god Newton said we are contracting and if god Newton said we are contracting the Universe will have to comply1 No, the Newtonians favour the believe that the Hubble's expanding Universe only concern galactica at a far distance and not here in our neighbourhood where even God has to comply with Newton's standards.

Then the reader may as what is it that Hubble's law indicate and how does it apply when we disengage Newtonian madness and find reality favouring our thought patterns. The following is my view about the Hubble constant as well as the Hubble law. **Hubble's law** is the relationship between a galaxy's distance and its speed of recession, announced in 1929 by Edwin Hubble. Hubble found that distant galaxies were receding at speeds which increased in proportion to their distance: the farther the galaxy, the faster its recession. This result, termed Hubble's law, shows that the Universe is expanding. Like spots on an inflating balloon, every galaxy moves away from every other, and no galaxy is at the centre of expansion; therefore, the expansion of the Universe appears everywhere the same. It follows that the rate at which two galaxies move apart is proportional to their separation – galaxies twice as far apart recede at twice the speed. Hubble's law received an immediate explanation in the theory of relativity, which regards the expansion of the Universe as a uniform expansion of space itself. The actual rate of expansion is given by Hubble's constant.

Hold this in mind and see how the Newtonians cheat science out of this predicament. First they put the movement of such expanding at a constant. We know that gravity is the result of atoms and if expanding is present, then expanding also would result in atomic interaction. That says that where more atoms are present more expansion will take place and where less atoms are present resulting in an area of space with less density the expansion must be slower. If there is an expansion then such expansion is not only contradicting Newtonian magical gravity by contraction but it is contrasting and counteracting Newtonian magical gravity by contraction. In the case where certain galactica prevails there are phenomenal movement, but Newtonians marginalise this by putting all such motion under one blanket and giving it a constant Universal undetectable movement that doesn't bother any one because it is happening afar and at places we cant even see. If it was on our doorstep it was blatantly intervening with Newton but as it is far away we have oodles of time to find how it will correct and convert to reality as the Universe than will become all over equal to Newton's prediction.

An Open Letter On Gravity Part 2

Hubble's constant is a measure of the rate at which the Universe is expanding. Hubble's Law shows that a galaxy's speed of recession depends on it distance; Hubble's constant is the figure that relates velocity to distance, and is determined by observation. Some figures put Hubble's constant at 10 miles per second per million light-years (55 km/sec/mega parsec). Galaxies, in other words recede at a rate of 10 miles per second for every million light-years of their distance. Some theories of cosmology predict that the rate of expansion of the Universe may change over time. A rough measure of any such change can be found by comparing the local expansion rate with the rate at very distant regions of the Universe. Because of the time the light from these distant regions has taken to reach us, we see them as they appeared many billions of year ago. Current results suggest scarcely any measurable change, which seems to indicate that the expansion to the Universe will continue indefinitely. If the expansion of the Universe has been constant over time, then the inverse of Hubble's constant yields the age of the Universe – the is, the time since expansion began from the Big Bang. I do not agree with the measure accepted as the Hubble's constant and I wish to share my reasons for rejecting of these figures

The expanding is happening between every atom in the Universe. A galactica is a lot of stars. A star is a lot of atoms. If the galactica expands the stars are expanding and if the stars are expanding the atoms within the stars are expanding. If the atoms within the stars are expanding then singularity defines the time between any and all atoms throughout the entire Universe notwithstanding any place in particular. If the expanding happens all over then sure as hell it is happening at our doorstep and then it is happening in our bodies and we call such a process aging. But only Newtonians would have the ignorant audacity stemming from an idiotic belief in magic to seek methods of all sorts to circumvent aging. The process we think of as aging and having wrinkles and the entire collapse of the body coming with age where this age is the manner in which the body never stops growing and it grows itself out all of proportions in the end. Every atom in the body remains growing while the body can only house that size and therefore to accommodate bigger atoms, the body has to reduce the number of atoms forming the body. This aging Newtonian madness wishes to abolish is the healing process we depend to repair damage and the growing of nails and hair is all the same thing…it is all because of Hubble's law applying anywhere as much as everywhere. This puts the Hubble law inside every atom in the Universe.

What would be the most important issue our Newtonian Brainy Bunch completely misses in all their arrogance and ignorance, the Hubble law is time. The Hubble law is the movement of time in its entirety. If you alter time then start with the Universe at large because singularity connects everything to all things by putting singularity in control of whatever is controlled. Singularity is everywhere and everywhere singularity is not equal but singularity is the same. Singularity is $1^0 = 1^1 = 1^2 = A^0 = C^0 = X^0 = B^0 = \Pi^0$ where every one is identical to the next one notwithstanding where anyone is. Changing one specific condition relating to singularity has to alter every aspect throughout the entirety if singularity everywhere because time is the alternating of singularity in relation to singularity remaining unaltered. The Hubble constant is the proof of this where everything in the Universe grows away from one single point remaining still. The factor time T^2 provides the duration of time but k indicates time development and if our time travellers wish to return back to the past (to do what only they would know what they wish to do in the past) then they have to reduce **k** by taking time and **k** back as far as they need taking **k** back. They have to take the value of **k** in as far as the entirety of singularity will permit in order to establish a Universe that had whatever they require available at a specific point in time. Should they wish to see the Grandma again arising from the grave and turn the clock back hundred and fifty years as to ask her about a hidden fortune, then they not only have to pull the Moon closer (without the aid of magical gravity this time) but they also have to redevelop relevancies (**k**) throughout the entire Universe because it is the changing in relevancies (**k** = **a**^{3} / **T**^{2}) or as we experience the Hubble law (**a**^{3} = **k** **T**^{2}) that determines time in progress. Time is not just a direction but time is progress of stars, galactica and atoms forming countless Universes enrolled into countless Universes. Remember the least singularity is equal to the ultimate original singularity by virtue of a value being 1^0 and that makes whatever point there is holding singularity as much a Universe as the entire Universe forms a Universe. Those whishing to travel through time must accept the reducing of space (of all space known and visible as much as unknown and invisible) and the increase of heat that accompanies the journey. Suicide by means of shoot into Jupiter will be a lot easier to alter time. By the reducing of the relevancy of (**k**) it will reduce their personal **k** which will alter their duration of time spent on Earth the person they think they are than to establish a duration extension of time. By the way, those that has the inclination to travel down a Black Hole and create a space whirl should first visit Jupiter (or something we know is much less space redundant than the nearest Black Hole) and see how they survive Jupiter's shrinking before they attempt to convert the nearest Black Hole into an amusement park roller coaster. Everything is departing and moving away notwithstanding Academic fraud and Notwithstanding Newton's personal corruption. There was a time when the Sun had **k** as relevance being the very same as size (holding the same space) as Jupiter now holds and Jupiter then was presumably even more frozen and much less of an independent cosmic reality than Pluto is at the present. Being frozen back then adhered to a much different meaning than being frozen now. Coming to this conclusion is simple. All one needs to do is study Galactica and see what the galactica lectures when they say in light what they say at night.

It is said that the expanding is coming from everywhere. That means the expanding is directing the relevancy **k** from every centre in terms of all other centres.

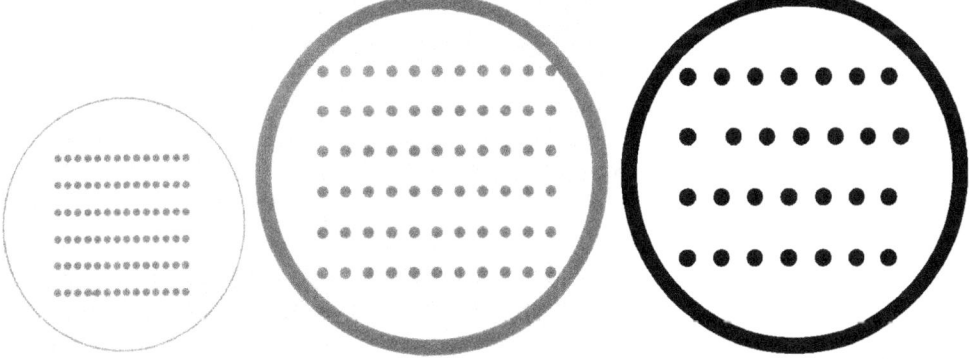

The Hubble law states very explicit that the expanding developed from every point there is outward not favouring any particular direction but grew evenly and similarly. This can only be the case if every star in the galactica grew evenly as the growth is pushing development outwards in all directions equally. If every star grew in this manner from every direction in which it maintained growth, then the effect must result from every atom growing in every star. This means that the relevancy to singularity (k^0 going to k or $\Pi^0 \to \Pi$) pushes the space-time of every atom outwards and such pushing then repositions every other atom accordingly to the new relevancy in spin applying. This is how the Big Bang developed through the Hubble law from the very first moment of Creation.

It is not only the relevancies of heat and time putting the Big Bang Universe much different from our perspective but the relevancies produced a completely different Universe altogether. It is not shocking that during the time and while experiencing the Big Bang the Universe was a nice average temperature of 10^{34} K because back then with conditions applying, it was just a normal day during the Big Bang. It was another Universe, one being in the same space that a neutron holds today. That fraction of space in time had to accommodate an entire Universe potentially filled with everything as we now see.

The Universe then had completely other rules than we have today. It is all coupled to the relevancy we find that singularity holds whereby gravity developed space-time. Everything changed as k extended and that proves that the factor k is the determining factor of space a^3 and time T^2. The progress of k unlocked all other factors including layers in galactica, which in turn unlocked stars from the galactica cradle. The stars in turn had layers developing by measure of k and with that the Sun also developed because all other stars develop in this manner and so does the cosmos develop using this manner.

Then again there was a time when the sun was the same frozen state as Pluto now is, but freezing then meant a temperature of what we now consider to be 6500°C throughout the entire outer space. Outer space was a pretty cool 6500°K keeping the surface of the sun nice and icy. It is all in relevancy because what is freezing to one is glowing to the other. At a point 6500°K was many times colder than was outer space, but outer space represented one end and matter indicated another end of the relevancy there was and lying between what we measure as hot and cold. It is apparent that in the sun hydrogen freezes heat to a liquid state at 6500°K and with oxygen further down towards the colder centre parts of the sun, it is any person's guess at what temperature oxygen will freeze to a solid state in the sun. We also know that deep inside the sun where things really gets cold at 18×10^6K it gets so cold, a hydrogen proton freezes to anything including an iron cluster. But with oxygen freezing to a point of using the condition, is considerably different with the governing singularity setting other rules as tow hat we, with life, can understand. The variations in conditions are in relevancy with the Sun's governing singularity and is setting grounds, which apply as the Sun sets different conditions and in no manner may the conditions the Sun's singularity dictate be compared to conditions favouring life to develop and to that conditions in which we are that we find to suit us, living on Earth. I do not wish to make presumptuous statements, but such rules apply very similar factors as the rules that bond compounds where singularity locks the space-time of different elements in a relation set free the elements. In that manner stars will regulate layer conditions but not by creating compounds ... no, only by producing similar types of rules. This means not only does the Big Bang look different, but also everything about the Universe was different then from what is now applying.

As sure as the Sun shines today so will Jupiter one day also shine as another Sun and then Jupiter's moons will be planets as large as the Sun's planets presently are, but by that time the Sun will be something awesome and awful? With the ratio fitting this tidally, the time duration and NOT TIME AS SUCH but the time it takes time to tick becomes infinitely shorter as it is coming from the eternally longer. In that is found also just a ratio. This goes totally against Newtonian religion and I am about to explain why that is in a minute. At what temperature will water boil on Jupiter and the answer will most definitely not be equal to a temperature it boils on Earth. Even on Earth the temperature of boiling water runs along a spectrum covering many scenarios, but the temperature of water boiling on Mount Everest is not equal to the boiling point of water at the Dead Sea level. Ignoring such indicators only brought scientific miscalculations and scientific mistakes. If water does not boil evenly at all levels on Earth how science could establish constants in outer space? It is known that even on Jupiter the freezing point connected to hydrogen varies as one continues down the atmospheric layers of the Micro star. At one point the equilibrium between the temperature of the Sun core and that of outer space matched setting the singularity within the core free and allowing the Sun to have a free inner core that then began applying gravity individually. We must presume that the outer regions of the Sun then were as frozen as Pluto is at present. We may think it was hotter but it was just less spacious on the outside. The cosmos was a lot smaller but that only made it more concentrated. This statement I shall retract but in order not to get ahead of myself we will keep using this statement for the time being. As space grew and reduced at the same time the growing in space by reducing in heat intensity came as a result of a changing **k** factor that represented both the demise of time as well as the increase of space. This Kepler introduced while no one took any notice until Hubble came. By that time every one forgot to look at Kepler again. Claimed space increases their space while time demise as the relevancy re-applies in favour of materials. Materials holding space a^3 is inseparably linked to time in the square $k = a^3 / T^2$ as structures in orbit apply duplicating motion and $k^{-1} = T^2 / a^3$ as contraction recoups time that brings about motion. As space increases time demises. This I say full willingly knowing such a view does not represents the Newtonian and therefore the overall human outlook because they focus on what is present in what they can calculate by using constants.

We measure a distance by the length of a meter, which is a distance we take from where a line begins to where a line stops during specific time duration, that we then convert to a cube, which we attach to a mass by the thousand that we connect to the distance the Earth travels while rotating around the sun as well as applying the gravity. Our measure of $\Pi^3 = \Pi^2\Pi$ is completely different to the same measure we find in Jupiter or in the case of the sun. Every cosmic structure will have gravity at Π^2 but the gravity will be gravity aligning with Π in relation to the governing Π^0 we find applying in that Universe. That is why the sun can freeze hydrogen at 6500°K and Jupiter can turn hydrogen into liquid at −150°C. Jupiter takes less cooling to also have hydrogen in a semi-liquid state. Gravity on Jupiter will be Π^2 no matter what, because the Jupiter space Π^3 has a built-in seclusion from the rest of the Universe. Again the space Π^3 is directly in relation to the relevancy Π, which we find that Jupiter has, which stands in relation with Π that is the extension of Jupiter in singularity Π^0. That makes Jupiter a nice other world and our measure by meter per second and weighing in kilograms will be very different on Jupiter then we have on Earth. A six-foot man will most probably be some four-foot down at Jupiter's sea level where hydrogen becomes a sticky metallic thick liquid, but the mans would not be an inch smaller. The man cannot be shorter because the man remains the same man as the man was on Earth. The man did not change fore he only came into respecting different relevancies. The foot in distance has changed and the six in number could have changed but the man stayed the same.

Our tunnel vision comes from our stance where we see the cosmos we wish to see because we also only fit into a small slot. Our slot is blocking our view but don't tell the Newtonians that! As **k** develops **k** has to develop in all aspects of the Universe that is if we are to believe in the Hubble constant (another nice little constant with no applicable use anywhere). That means the layers in a star and in a galactica is the prone ones to changes brought about by the rules that the Hubble shift adheres to and the layers brings growth to singularity as singularity in governing stars come to life. They take charge as the atom singularity progress and as the atomic singularity shifts the dominance to support the governing singularity in the star centre. Then **k** progressed further and with that the **k** within the Sun and indeed all the solar structures progressed in equal terms but not alikeness since every one has in place the charging of its space-time its coming as a result of another and most different singularity. It is this relation we have with the Earth and with Earthly relevancy we hold so dear in our occupying space on Earth that the Earth **k** holds every aspect including our minds and our thinking under control. The Earth is our Universe and that we can see from all the constants and ultimate limits we place on the Universe in order to bring the Universe in line with our Universe which is the Earth and all standards going according to the Earth.

We tend to see the Universe from our perspective we have where we are filling the centre of the Universe. With all the absolute phenomenal achievements science accomplished and more so the past sixty years, by going to the moon and splitting atoms and visiting planets and... was there ever one that took the time to find the centre of the Universe? How can anyone tell how much mass gravity attracts in the entire Universe in relation to the critical density of the entire Universe when you are incapable of knowing where to look for the centre of the Universe? The Universe with gravity's attracting action must be pulling us to the centre of the Universe and because you can't judge the direction we are going, you don't even know where to find the centre of the Universe. When any person is standing on any place anywhere, while viewing the Universe, that person is filling the centre of the Universe. That is not only applying to Americans in particular, but to all persons that were born through childbirth. This however does no apply to animals but that is in another book reserved for another day.

Let's get more personal. We tend to think of our position as having the position only the most important person in the Universe can have because every one thinks of himself or herself as the most important individual that is holding the centre of the Universe in the entire Universe. Professor Einstein said gravity is where the Universe draws flat, and the Universe can only draw flat where the centre of the Universe is because only gravity can draw the Universe flat. Consider this while looking at the night sky outside where light pollution has not destroyed the view... All the light that come across and travelled all of the vacant space from any and all possible positions in space runs directly towards your position using a straight line towards you where you are filling the centre of the Universe. With you being able to draw the entire Universe flat so that all the light through out the entire Universe come together to meet you in person in the position you hold, you must therefore have the most intense gravity by your effort of drawing the Universe so flat, in order to have all light running directly to you. Not allowing even excluding the effort of one photon, all light is heading to meet you where you are in that centre spot and not one photon will pass you by. Not one photon dare miss you because if they do they miss the effort that all light has to accomplish and that is to locate you as the person filling the centre of the Universe. Should you decide to shift your position to any other place in the Universe, you will shift the centre of the Universe to that location as well because the light will track you down in your new position. If you install a camera on Mars, the light is obliged to acknowledge your relocating the centre of the Universe at your will to reposition you're taking control of that centre of the Universe.

All the light that ever left its destination crossing the vast spaces of the Universe, excluding no particular light, travelled all the way just to find you filling the centre of the Universe, right where you are. By you're standing anywhere, you fill the centre of the Universe, and the entire Universe admits to that because all the light comes to meet you there. If you shift from the North Pole to the South Pole you will shift the centre of the Universe because all the light travelling throughout the Universe will find you where you then moved the centre of the Universe. The light left its destination billion years ago as it travelled through space at the speed of light so anxious it is to acknowledge you're being in the very centre of the Universe. No photon will be able to pass you by where you are in the centre of the Universe because all light is heading your way from their starting positions. No wonder every person born has the idea they were born to fill the centre of the Universe, which we do fill. The Universe is spinning around you or I, which is filling a centre where all motion is connected. It implicates gravity as wide as can be... Some things mathematics is able to explain but other explaining goes beyond mathematics.

Some aspects of the Universe go beyond mathematics and some even go beyond words. It is our task to find space, to find time and moreover it is our optimal task to find the Universe. This line of thought is in concept a joke but as much as a joke it may be it is the truth as no other. It also is the scientific Newtonian inspired approach to science that brings the thought pattern of truth in all people to mind. Because we use the Earth and on the Earth a meter is a meter in a second, we believe that meter per second will also be a second in measure of all meters everywhere and the distance duration will be the duration of all seconds per meter everywhere. Let's see how much this joke is a reality in the minds of Newtonians.

Being in the centre of the Universe is frightfully Newtonian. It is very clear how Professor Einstein and his compatriots followed their genius in their arguments when they argued about the critical density and the manner in justifying an attempt that is as much a cover up for a scandal that has no political proportional rival to date by correcting Newton's obvious misconception. However fortune favours the brave. They went about shouting about not enough mass being the culprit for the expanding that Hubble uncovered, which was contradicting the compressing theory Newtonians so stubbornly cling to. Nevertheless the occasion presented the man and the location supported the deed in rectifying the error of directional flow of what the Universe was up to. To this day no one came up with a reason what besotted the Universe to go in contradicting the direction of development to what Newton declared. To count the mass they were fortunate enough to be in the centre of the Universe because only from such a vantage point could they see and measure the entire Universe. From where Newtonians stand even to this day the Newtonians have the fortune of seeing the entire Universe from edge to edge to edge and so forth…and that too applied to Professor Einstein. Professor Einstein could see all the edges and Professor Einstein also new that beyond the edges was nothing more than just the edge of the Universe. Although we know the Universe is without an edge, Newtonians being in the centre of the Universe can see the edge that cannot be there but to Newtonians it is being there because by magic Newtonians create gravity. They see the edge of the Universe every day from every possible telescope and to hell with those saying the Universe are eternal without borders and edges! So there is an edge where no edge can be because they are Newtonian and Newtonians fill the centre of the Universe.

However it is not for us to criticize so therefore let's journey back and see what Professor Einstein saw…and remember we are back in time so tenses becomes an issue. When Professor Einstein and his fellow Professors look outside at night they see the edge of the Universe to the left of them. They can see where the Universe ends in that direction. Looking to the right they see the end of the Universe to the right of where they are looking because there to the right where they look, they see the Universe ending at the edge. Looking to the front the same happens and to the back the same happens with edges in all directions. Even when looking up into the night sky they can clearly see where the edge of the Universe defines the end of the Universe in that direction. It is to their fortune that they are where they are because by being where they are they are filling the location holding the centre of the Universe in the most magnificent place they could ever choose to be. They are in America and we all know that Americas is very much the centre of the Universe. More so is the fact that they are in an institution called Harvard, which puts them in the Academic centre of the Universe. By them being part of the physics department of Harvard brings them in line with the astrophysics Academic centre of the Universe.

Things are going from good to better to best to excellent, because with then being in Professor Einstein's office they are in the brains centre of the Universe and with Professor Einstein being in their midst they are placed by his presence and intellect smack in the centre of the Universe. Their place in Harvard's physics department right inside Professor Einstein's office standing next to Professor Einstein puts them smack in the centre of the one half of the Universe. Now they do not have to worry about finding the bottom half of the Universe because America having Harvard with a physics department having an office where Professor Einstein presides takes cover of the bottom half of the entire Universe and puts their bottom half smack in the centre of the Universe. Professor Einstein is the living presentation of everything that is not stupid as Professor Einstein just has to walk outside and look at the light coming directly to him. One spin of 360^0 would ensure him that being Professor Einstein and all... he then must be in the centre of the Universe because he can see the edge of the Universe in every direction possible! If he wasn't the centre of the Universe there was no way he could measure all the mass in the entire Universe because his allocated position of not being the centre of the Universe would bring obstruction to part of the view required for the measuring task in hand.

That means being in America and more so within America's Harvard Physics office that takes part of the bottom centre of the Universe and Professor Einstein in person is then taking care of the top half of the Universe finding the location that presents the centre of the Universe where the entire Universe aligns at that point. I know every one has sleepless nights wondering why they gave the problem of measuring the entire Universe form edge to edge to edge to edge to determine the critical density calculations to a person such as Professor Einstein. Wonder no more! There is a possibility that it has something to do with his mathematical abilities but that would not count for much if he was not able to see the Universe from edge to edge to edge to edge to... But with Professor Einstein in the place where he is, he can see all the stars sending light directly to him and telling him how big and how far they are. If he were in the incorrect place in the Universe his measurement from not being in the centre of the Universe would no have been trustworthy at all.

With Professor Einstein being Professor Einstein, he knew he was the most important set of brains America could present to the Universe. By the Universe realizing this fact and acknowledging the fact while sending all the light to them at the centre of the Universe and without causing delays, the Universe responded by sending all the light at the speed of light to Professor Einstein. The light came from near as it came from far. It came as much from the very edge of the Universe to the right hand side of Professor Einstein as much as it came from the very edge of the Universe to the left hand side of Professor Einstein. Then the light came from the front as far away as it came from the back of Professor Einstein. From the top as well, all the light travelled as far as light can travel and travelled as fast as only light can travel just to acknowledge and support Professor Einstein in his task to calculate all the mass in the entire Universe because he has to prove Newtonian incorrectness automatically correcting by self-preserving determination.

Of course with him being in America and at Harvard's physics department and moreover in Professor Einstein's office placed the bottom half just as accurately in the centre of the Universe as Professor Einstein found the top half to be aligned. If it was not for America and if it was not for Harvard's physics department and Professor r Einstein's personal office the bottom half of the Universe might have mismatched its effort to align with the top centre and then the lot was not in the centre of the Universe from where they could see every possible edge of the Universe. But with the fortune of things being as they are the top and the bottom halves of the Universe matched and in that Professor Einstein could now fill the entire centre of the Universe on top, at the left, at the right, to his back and to his front as well as the bottom half of the Universe. If that was not the case then what a tragedy that would have been because only from being in the centre of the Universe could Professor Einstein view the entire Universe and see what there is in the form of mass to calculate and measure every star there is in the entire Universe.

Fortunately for mankind the world had a person such as Professor Einstein to fill such an important position from where he was then able to see and measure the entire Universe. How gratefully we should be to the Academics of America for allowing us to share America's centre position in the Universe. More than thankful we must be for the Academics that allowed us in sharing the physics department of Harvard's central position in the entire Universe because from there it was a hop to get into the office of Professor Einstein and share his position of seeing all the light from every corner the Universe has to share with him and use such a marvellous position in the entire Universe to gauge and measure all the mass we can see…and if you say Newtonians Academics don't put them and their position they have from where they see them being and filling in the centre of the Universe …then please do a rethink! The theory they present puts conditions we find on Earth used by the entire Universe in the entire Universe. Being on Earth we can see that water freezes at zero Celsius and we can see that it is one bar of air pressure that we find at sea level. We have the element table with solids and gasses nicely arranged for us by nature in the table in column order and we know that absolute zero is absolute zero because absolute zero is what we measure when we measure absolute zero. How can the Universe have the tenacity to have absolute zero anything else than what absolute zero should be where we measure absolute zero.

The Universe has not the capability to change anywhere because if it did dare to change we will see such change as we fill the centre of the Universe! From the Newtonian stance the Universe grew from the size of a Neutron, however its official Newtonian policy that the Sun was the same since time began and the atom was always what we measure the atom to be. With Newtonians filling the centre of the Universe we know the Universe can grow and expand but that feat is quite impossible for the Sun and the planets to achieve. It is completely anti Newtonian to think that the solar system is getting bigger just because the Universe is growing bigger. All Newtonians would recognise change the instant change occurs because from filling the centre of the Universe Newtonians will notice change immediately and after all one has to consider that Newton said the lot is contracting. Saying anything to the opposite will be quite sacrilegious to Newtonian religiosity. The reality is that the Universe grew and the Universe still grows even in our part of the Universe. Amidst all this evidence Newtonians have a constant speed of light, a constant time since time began, a constant gravitational force, a constant expansion and every other aspect that will bring along nice and easy calculations, so the Universe will have its constants just to keep life a little simpler for the Brainy mathematicians.

Go ask every Newtonian and that Newtonian will tell you the Universe has no edge because it is limitless, but being as important as only Newtonians may be they fill the position where they are able to see all the edges the Universe cannot have. However I should warn any one that listens, don't tell Newtonians of their double standards. When I confronted professors in the past and accuse the Academics of limiting the Universe to benefit their views about claims they make or to support Newtonian claims, I am compromised being the one referred to as the incoherent, as the raving idiot because they say Newtonian science will never do such a thing.

Then the next minute, I see Academics limit the Universe to having an edge, which they find in very clear telescopes. They see a boundary where the Universe ends. To prove my case I present I challenge anyone to visit the web site and go and visit the many such web pages carrying this very claim. A red giant was found on the edge of the Universe or some galactica conjunct with some other galactica at the very edge of the Universe. It is such normal every day Newtonian practice to use a double forked tongue. It states clearly that science caught big bright stars on the edge of the Universe. Any one can glance at this on the condition they have excess to the web page and can use the web page.

This clearly shows the lack of understanding on affairs Newtonians claim to be knowledgeable of. Those Academics (the lot of them do advocate that the Universe does not end) are those giving the Universe an edge…to do what with. What ends at the edge or the border and what bring the edge about, forms the wall that allow no more of the Universe to continue. However I am sure that there must be some persons in education that will share my view that Newtonian science did not yet leave the shores as Columbus' sailors did when they found no edge of the world. But Newtonians science did manage to take science much further than science was before. Newtonians managed to shift the edge of the world so far it became the edge of the Universe and gave the Universe with no end, an end. It is about time that the entire philosophy of cosmology is overhauled and is revised from the backwardness of five hundred years ago to a more fitting approach in the five hundred years that time went on.

Then they claim the Universe is expanding…expanding where too. Where can anything go that covers the lot and has nowhere to go? How can anything get bigger when such a thing is as big as anything will ever get. How can the Universe get bigger when it is limitless? Notwithstanding what logic needed they will tell you the Hubble constant is about the Universe expanding that which is not expandable and going where there is no going too because (I guess) the edges that are not there is shifting further out to nowhere. The Universe is getting bigger but where is the Universe getting bigger too because wherever it is going the Universe is surely already there! There can be no place without the Universe already being there so where is the new claimed territories that is gained by a growth that it is gaining from what because it clearly already claims all there is. I realise that Newtonians being in the centre of the Universe can claim to see what we others with less intellectual means are unable to see for we do not have the grant in privilege to see the Universe from the centre of the Universe.

It is rather comical to think that Vasco Da Gama and his sailors would not set sail on a voyage of discovery in fear of confronting the edge of the world while currently our cosmic sailors in waiting desires to get going on such a ridiculous voyage and reach the edge of the Universe because they can see the edge of the Universe. How much did things change just to remain the same will you not think? Today the schooled Newtonian opinion about cosmic science is that we just have to hop into some craft, blast off to the unknown into the unknown at the speed of light and send a post card back home when we get to the edge of the Universe. ...And all the while Newtonians have no clue how light travels. They show some mat-like surface with graded blocks that should represent space and time but it puts space and time in some single dimension holding a square of some sorts and present that as the travelling road light supposedly takes as it journey all the way to them where they are in centre of the Universe.

Whenever I am presented with this explanation I so dearly wish to ask the person 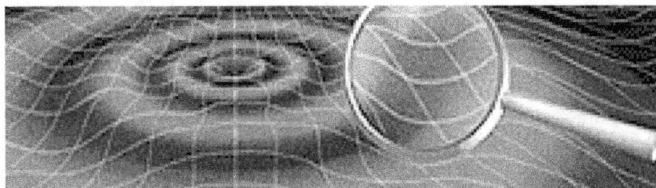 presenting the argument what he did with the rest of the Universe? Where did that person find a place to go and hide the other five sides? Only nothing can vanish and the cosmos is definitely not nothing. He took the three dimensional six sided Universe away, put it somewhere I don't know and then left only one of the six sides where I am suppose to see only one of the six sides...and how am I suppose to know where he put the rest of the Universe! Here he is so smartly showing a stupid bloke such as me what happens when a Black Hole comes about.

That is fine and that much I truly understand even with me being as stupid as I am. What I do not understand is what he did with the other five sides of the Universe. He is showing one side of the Universe that has gone flat, but what did he do with the rest of the six-sided Universe. How did it disappear and where is it gone to...how long will it stay there and how is it coming back...who is strong enough to bring it back...it is all viable questions asked in sincerity...therefore even in my stupidity I deserve an answer, after all it is my Universe too. Just look what did they do to my Universe...and who is going to repair it! How did the Universe get that flat because gravity is not in outer space...gravity is in the atom no, moreover it is in the proton. It is the proton pulling space-time flat and not outer space pulling flat.

They show a centre and they show a flat Universe with light of all things travelling about a flat Universe. How light which is the very focus of the three dimensional purpose of the 3D Universe can go flat and travel in a flat Universe along where no travelling space is available, but the light travelling is just the thing that is establishing the three dimensional space we see. How light under those circumstances can travel flat through no space is beyond that which I shall ever be able to understand. Fortunately, I am the stupid one around and they are highly educated. If this is true then the electron should pull the atom flat because the electron is the epitome of light and what density light can ultimately achieve.

It is not only the relevance applying from the centre to the electron but it is also that the electron is holding an allegiance with the centre in the precise manner as planets do in relation with the Sun. It is because of the dual relevancy that the electron stubbornly clings to the newly elected centre before being overpowered by the Earth providing the controlling centre. It is all about relevancies attaching centres that formed when creation came about. To break those relevancies we have to take time back to before those relevancies because of the gravity gluing the relevancies in a unit.

Einstein's General Theory of Relativity
The General Theory of Relativity is an expansion of the Special Theory to include gravity as a property of space. Let us start the debate by looking at this Gravity Tutorial.

The Equivalence Principle
The Theory of Special Relativity has as its basic premise that light moves at a uniform speed, c = 300,000 km/s, in all frames of reference. This results in setting the speed of light as the absolute speed limit in the Universe and also produced the famous relationship between mass and energy, $E = mC^2$ The foundation of Einstein's General Theory is the Equivalence Principle, which states the equivalence between inertial mass and gravitational mass.

I firstly wish to challenge any and all Newtonians to convince me there is something such as mass by proving to me why when I am in outer space I have no mass. I have maximum gravity since I move at a tempo taking me beyond the Earth gravity, but I have no mass since I am floating with no containing of my position any where.

When I am on the Sun my mass of 100 lbs or approximately say 44.5 kg rockets to a one ton or 2200lb value as a factor indicating mass. My space I hold reduced on the Sun while my mass increased since the Sun holds more material in less space. On may consult the picture to the top to see how much mass increase with the space reducing taking place as the Star grows smaller in space but bigger in applying gravity. It is the movement T^2 that increases which reduces the atomic relevance **k** while increasing the density of the material increase exponentially by reducing the space the material holds a^3 as a unit. It is the relevancy that the material has in terms of the gravity of the location that changed that increases and there is no hint of mass unless I wish to contaminate the truth as Newton did and as Newtonians still corrupt science by falsely supplying mass to conceal their lack of understanding the cosmos.

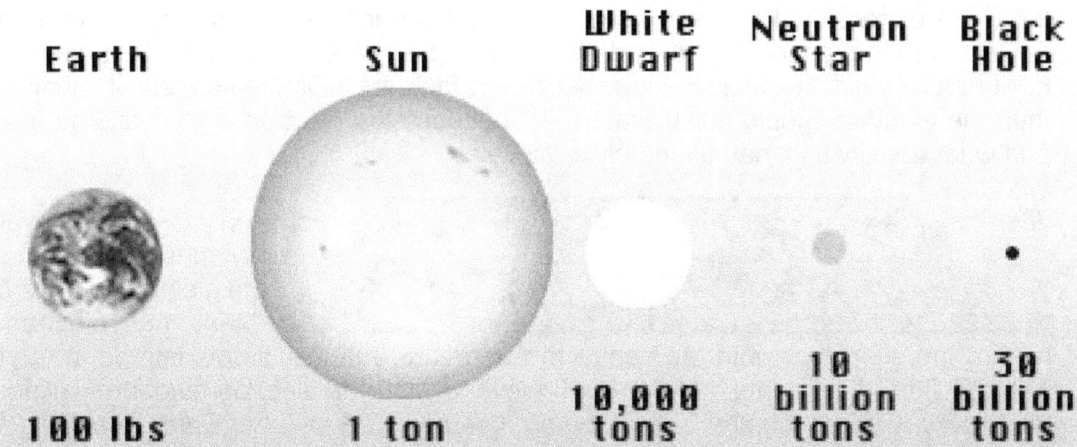

Inertial Mass is the quantity that determines how difficult it is to alter the motion of an object. It is the mass in Newton's Second Law:

This is their saying so please don't frown on me when the argument sounds a little drunk. **F = ma** *This imply that gravitational mass is the mass which determines how strongly two objects attract each other by gravity, e.g. the attraction of the earth:*
It is the apparent equivalence of these two types of mass, which results in the uniformity of gravitational acceleration -- Galileo's result that all objects fall at the same rate independent of mass: $F_{grav} = \dfrac{GM_\oplus m}{R_\oplus^2}$

Galileo and Newton accepted this as a happy coincidence, but Einstein turned it into a fundamental principle. Another way of stating the equivalence principle is that gravitational acceleration is indistinguishable from other forms of acceleration.
$g = \dfrac{GM_\oplus}{R_\oplus^2} = 9.8/8^2$
According to this view a student in a closed room could not tell the difference between experiencing the gravitational pull of the earth at the earth's surface and being in a rocket ship in space accelerating with a = 9.8 m/s².

Can the corruption I try to point out be more obvious than it is in this statement? I would like to challenge any Newtonian to show where Galileo in his entire life confirmed his views on Newton's mass extravaganza. Newtonians are so corrupt they are unable to distinguish between fact and fallacy and they make up facts consistent with fallacy while pretending it is unblemished truth. Galileo never pointed to mass and in fact he said all things fall equal which destroys any credibility that objects fall with mass. Newton and Galileo said the very opposite but to the Newtonian blurred contaminated vision Newton and Galileo said the very same thing. Sometimes I think Newtonians have gone mad with misrepresenting the truth. There is no evidence of mass and there is no evidence that mass creates gravity and there is no evidence that the cosmos apply mass in any way or form. But that is not where Newtonian corruption ends. They relay get bizarre when they get stuck into the unreal world of Newtonian cosmology and create a Universe that can draw flat from edge to edge without ever explaining where the edges of drawing flat starts and stops.

So we have heard all the hype and the brilliant mathematical expression of the impossible where the star goes bang and the gravity goes mad and it implodes (how ever that may be achieved) and the whole cosmos goes bananas because a star has gone lost for one eternity and now has died a tragic death. That is so complicated that there is little available to explain.

I do not think I have to convey my disgust in their fraud they commit. Let any Newtonian Brainy Boy wonder show how it is done when one takes the mass of the Earth and multiply that with the mass of any or all objects while multiplying that figure with the gravitational constant and divide this lot with the radius of any object standing solidly on Earth as the formula $F_{grav} = \dfrac{GM_\oplus m}{R_\oplus^2}$ $g = \dfrac{GM_\oplus}{R_\oplus^2} = 9.8/8^2$ will indicate we find a value of gravity and then have this lot resurface in

In order to get to a value of 9.8 m/sec.
It is disgustingly corrupt.

The answer to my explaining is so simple it is laughable. A star is about fusing atoms together and on that am I correct or incorrect? Then what happens when all the atoms fused together in the space of one atom and what gravity will such an atom produce? If only a handful or maybe just one atom remains which is the final result in fusion of what ever contraction finalized all the possible fusion between the atoms in the star, and one atom ends up with all the gravity the star had which was initially delivered by all the atoms in the star that produces gravity and such gravity is now within the space that one atom holds, then the gravity will be devastating. The fusion has to end somewhere because by fusion the star is heading in a direction that will eventually combine all the atoms in the star.

Curved Space-time

The figure above represents a two-dimensional slice through three-dimensional space showing the curvature of space produced by a spherical object, perhaps the sun. Einstein's view is that the planets follow the curvature of space around the sun (and produce a tiny amount of curvature themselves). That again is crossing monkeys with watermelons to bread marble. They picture a two dimensional (flat bedded) mat like surface that light uses to travel by. According to the Educated Wise, gravity will pull the Universe as flat as the topside of a mat. However, some of the content within the Universe escape the fait the Universe has because some things in the Universe does not draw flat although the entire Universe supposedly draws flat.

The Curvature of Space caused by a Massive Object.
By allowing the object to have a hole is returning the object back to 3D from the single dimensional surface it is portrayed to be when being in such enormous gravity. However there are certain requirements and there are certain rules that has to apply in such conditions. The conditions will apply to all and will not exclude any object in the vicinity. One aspect of gravity is that it applies to everything equally in the space in which the gravity is acting. It is very obvious that the claim of the curvature of space-time is made without any support or backing about reasons why such curvature is there in the first place. Why would such curvature appear and why a circle forming curvature?

Professor Einstein's view is of course completely Newtonian which means at best, it is totally rubbish. The curvature of space-time is the result coming from Kepler's formula stating that any space a^3 will become secluded a^3 by any straight line **k** eventually forming a circle T^2. The straight line will end up in a circle and the circle will define the area seclude by the object moving around any centre object. That is why everything in the Universe is some circle in form.

The second fundamental principle of General Relativity is that the presence of *matter curves space*. In this view, gravity is not a force, as described by Newton, but a curvature in the fabric of space, and objects respond to gravity by following the curvature of space in the vicinity of a massive object. The description of the curvature of 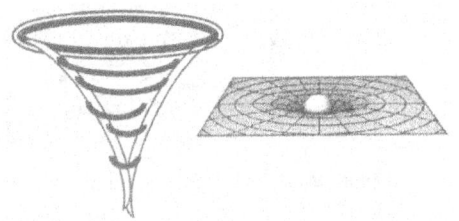 space is the mathematically complicated part of general relativity involving "metrics," which describe the way that matter curves space, and tensor calculus.

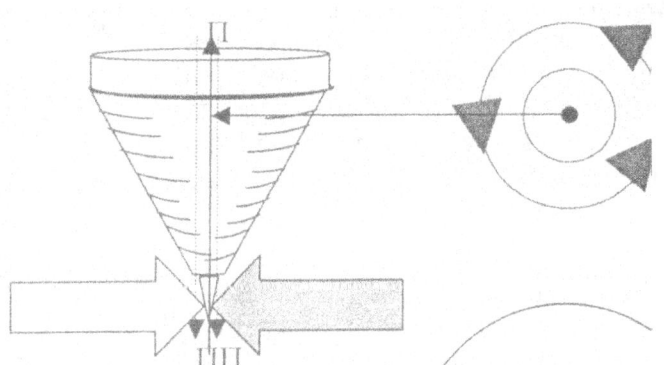

The curvature of space-time comes as a result of singularity extending from the single dimension into the circular sphere. The cosmos is a sphere and that I prove in Part 2 but any person with a little insight do not have to weight that long Any person can see that in spinning the top spins straight while going straight results in circling the centre or singularity position.

I have very disturbing news for Professor Einstein and his followers, those I call the Brainy Bunch. It is the fact that space goes flat. However, they are looking at time and presume it is time that goes flat. For space to be space the space has to have three dimensions where three are opposing three more sides. That is space a^3. Where space becomes flat the space is moving with time and such space then has motion in relation to an ever-changing position and location endorsed by the square of time T^2. However, it is the motion of the three-dimensional space that holds the square value and not the three dimensional space is going square by losing one dimension. Time or motion of space has the square which gravity also has being Π^2 The square apply to the moving of space Π^3 in the third dimension but it is very unrealistic to place space Π^3 by measure of losing one dimension. Moreover it is confusing and more so to the Academic preaching that as a fact in physics. It confuses the rest in believing what the rest of what the person has to say. If such proposal is made, one require a reason why it went away because gravity is not space therefore gravity is the motion of space. One cannot say gravity pulled space flat. How flat is flat and what is flat? Singularity is flat but singularity has no dimensions. Every atom is a Universe because the atom holds space a^3 by the motion of time T^2 relating to the relevancy k that singularity k^0 puts in place in relation to defining that specific atom-Universe- space. The drawing flat of the Universe is the proton conferring heat with singularity. We are never going to experience the time between the Moon and the Sun drawing flat because then we will be roast meat in an instant. If the entire Universe draws flat, to which point does it go while it draws flat and from where does it return once it bounces back into form? Is the entire Universe drawing fat from a Newtonian edge to the other Newtonian edge or does the drawing flat entail some segments where the one drawing flat leaves a universe departed from other Universe drawing flat. Newtonians always love to leave so many questions unanswered when they break their backs to confirm their distinctive stile of stupidity.

Then in this carpet that represents the flat two dimensions there appears a hole in the centre. The presence of this hole returns the whole picture back to being a three-dimensional surface because the whole produces a third dimension to the point where gravity is pulling everything into a flat two dimensions, but the hole is adding one dimension to the bargain of the flat square. How did gravity produce a flat two

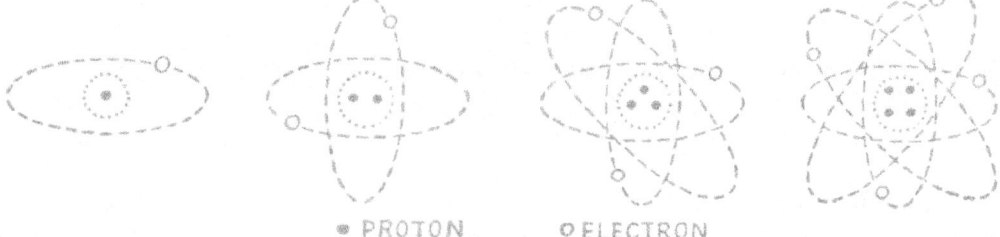

• PROTON ○ ELECTRON

dimensions while equally at the most intense point, gravity produces a third dimension in the location where gravity is supposedly lurking to give is our flat Universe. Then comes injury to the insult already inflicted on our weak minds; they place a three dimensional ball into a flat Universe to show us a three-dimensional whole in the flat Universe. If that is not done by the magic of Newtonian gravity, then there can be no other acceptable explanation about the whole affair. They have no valid reason for the Universe to go flat except blame it on gravity. However, gravity is the strongest where space is the least. That would then allow gravity to curve the space-time surrounding the sphere from the centre, evenly, in all directions equally. It would put in place the seven degrees the circle of the sphere insists to have. It would use Π by many dimension placing Π in relation to the centre where gravity is the strongest. With this in mind there is no need for all the hype about the curvature of space-time, we may just refer to the sphere being in place.

Our Super-Educated first puts on the table a flat surface. No one ever try to bring across any explanation or reason when the surface went flat or is flat except that gravity is making it flat. They left the guessing to us to fill in what produces gravity in outer space where bodies float around centre objects because such explaining might just mesmerize their theories while they wish to mesmerise our brains. The Universe does draw flat. I admit, but what is the Universe that goes flat?

As any symbolic picture of the strongest possible gravitational force will show as in example a picture of a Black Hole, the gravity deforming the surrounding Universe is where no space can be located.

In the centre of a sphere that is holding the sphere in form as well as the surrounding space attached to the sphere is gravity. It is forming the surrounding space from a point inside the sphere that has no sides and no space other than merely form in which it puts all space attached to the gravity. In one picture they indicate gravity pulling flat whatever gravity can pull flat. Then mysteriously where gravity is pulling flat, gravity is pulling a hole in the flatness. How that comes into the realms of the possible is impossible to explain. It is either a fact that gravity pulls the lot flat without any part of the picture able to fill a three dimensional stance, or gravity is unable to pull anything flat. One cannot depict gravity as being selective to prove the thinkers thoughts. If the ball in the hole is three-dimensional the hole is three-dimensional making the picture with the light portraying the picture three-dimensional. Or everything is flat without having a hole to fit a ball. If gravity pulls the Universe flat then gravity is quite unable to curve the Universe by the same margin. Again we arrive at a Newtonian fork tongue. This can only happen when one put mathematicians in charge of theory. It is still the same double standards to give the Newtonian view validity. Gravity is having the Universe go flat or so did Einstein portray our gravity stricken Universe. Gravity also produces gravitational lenses...but they never mentioning whether it is while or before it is going flat because there is a choice to be made. It is either going flat or it is having a lenses but it cannot have a flat lens.

$g = \dfrac{GM_\oplus}{R_\oplus^2} = 9.8/8^2$ In a picture of outer space Newtonians place a gravitational constant. That means there is a form of gravity keeping order to the vastness of outer space. In such vastness light is known to travel in a straight line between points. Gravity is pulling objects to a centre where mass is concentrated. In this vast region without any end they claim a specific gravity and that gravity has a specific value with a specific name being the gravitational constant. If it is gravity being out there in the blackness, then it is pulling to a centre. The biggest Newtonian question then is: Where is that centre and what produces the centre whereto the pulling is going. Where is the centre of the region thought of as outer space? If outer space had gravity there has to be a centre to which the gravity is pulling or moving objects. The only centre there is, is the centre material form. It is the centre of the sun, or it is the centre of a galactica. That is material and can hardly qualify as outer space because such space is openly controlled by material moving about in such a space. Other than that there is complete lack of gravitational evidence.

Curved Space-time

The figure forming the spiral represents a two-dimensional slice through three-dimensional space showing the curvature of space produced by a spherical object, which is perhaps the Sun. Einstein's view is that the planets follow the curvature of space around the Sun (and produce a tiny amount of curvature themselves). That again is crossing monkeys with watermelons to bread marble. They picture a two dimensional (flat bedded) mat like surface that light uses to travel by. According to the Educated Wise, gravity will pull the Universe as flat as the topside of a mat. However, some of the content within the Universe escape the fait the Universe has because some things in the Universe does not draw flat although the entire Universe supposedly draws flat.

However, it is the motion of the three-dimensional space that holds the square value and not the three dimensional space, going square by losing one dimension. Time or motion of space has the square which gravity also has being Π^2 The square apply to the moving of space Π^3 in the third dimension but it is very unrealistic to place space Π^3 by measure of losing one dimension. Moreover it is confusing and more so to the Academic preaching that as a fact in physics. It confuses the rest in believing what the rest of what the person has to say. If such proposal is made, one require a reason why five dimensions went away because gravity is not space therefore gravity is the motion of space. One cannot say gravity pulled space flat. How flat is flat and what is flat? Singularity is flat but singularity has no dimensions. That, I can prove because I can show where singularity is because of singularity not being there.

Then in this carpet that represents the flat two dimensions there appears a hole in the centre. The presence of this hole returns the whole picture back to being a three-dimensional surface because the whole produces a third dimension to the point where gravity is pulling everything into a flat two dimensional state, but the hole is adding one dimension to the bargain of the flat square. How did gravity produce a flat two dimensions while equally at the most intense point, gravity produces a third dimension in the location where gravity is supposedly lurking to give is our flat Universe.

Then comes injury to the insult already inflicted on our weak minds; they place a three dimensional ball into a flat Universe to show us a three-dimensional hole in the flat Universe. If that is not done by the magic of Newtonian gravity, then there can be no other acceptable explanation about the whole affair. They have no valid reason for the Universe to go flat except blame it on gravity. However, gravity is the strongest where space is the least. That would then allow gravity to curve the space-time surrounding the sphere from the centre, evenly, in all directions equally. It would put in place the seven degrees the circle that the sphere in form insists to have. It would use Π by many dimension placing Π in relation to the centre where gravity is the strongest. With this in mind there is no need for all the hype about the curvature of space-time, we may just refer to the sphere being in place.

One cannot depict gravity as being selective to prove the thinkers thoughts. If the ball in the hole is three-dimensional the hole is three-dimensional making the picture with the light portraying the picture three-dimensional. Or otherwise everything is flat without having a hole to fit a ball. If gravity pulls the Universe flat then gravity is quite unable to curve the Universe by the same margin. Again we arrive at a Newtonian fork tongue. This can only happen when one put mathematicians in charge of theory. It is still the same double standards to give the Newtonian view validity. Gravity is having the Universe go flat or so did Einstein portray our gravity stricken Universe. Gravity also produces gravitational lenses...but they never mentioning whether it is while or before it is going flat because there is a choice to be made. It is either going flat or it is having a lenses but it cannot have a flat lens.

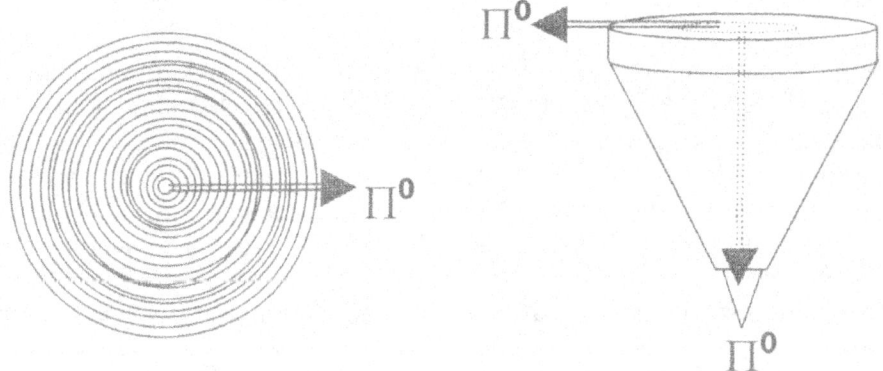

That is space a^3. Where space becomes flat the space is moving with time and such space then has motion in relation to an ever-changing position and location endorsed by the square of time T^2. However, it is the motion of the three-dimensional space that holds the square value and not the three dimensional space, going square by losing one dimension. Time or motion of space has the square which gravity also has being Π^2 The square apply to the moving of space Π^3 in the third dimension but it is very unrealistic to place space Π^3 by measure of losing one dimension. Moreover it is confusing and more so to the Academic preaching that as a fact in physics. It confuses the rest in believing what the rest of what the person has to say. If such proposal is made, one require a reason why five dimensions went away because gravity is not space therefore gravity is the motion of space. One cannot say gravity pulled space flat. How flat is flat and what is flat? Singularity is flat but singularity has no dimensions. That, I can prove because I can show where singularity is because of singularity not being there.

They have no valid reason for the Universe to go flat except blame it on gravity. However, gravity is the strongest where space is the least. That would then allow gravity to curve the space-time surrounding the sphere from the centre, evenly, in all directions equally. It would put in place the seven degrees the circle that the sphere in form insists to have. It would use Π by many dimension placing Π in relation to the centre where gravity is the strongest. With this in mind there is no need for all the hype about the curvature of space-time, we may just refer to the sphere being in place. If the ball in the hole is three-dimensional the hole is three-dimensional making the picture with the light portraying the picture three-dimensional. Or otherwise everything is flat without having a hole to fit a ball. If gravity pulls the Universe flat then gravity is quite unable to curve the Universe by the same margin. Again we arrive at a Newtonian fork tongue. This can only happen when one put mathematicians in charge of theory. It is still the same double standards to give the Newtonian view validity. Gravity is having the Universe go flat or so did Einstein portray our gravity stricken Universe. Gravity also produces gravitational lenses…but they never mentioning whether it is while or before it is going flat because there is a choice to be made. It is either going flat or it is having a lenses but it cannot have a flat lens.

$R_{grav} = \dfrac{2GM}{c^2}$ In a picture of outer space Newtonians place a gravitational constant.

That means there is a form of gravity keeping order to the vastness of outer space. In such vastness light is known to travel in a straight line between points. Gravity is pulling objects to a centre where mass is concentrated. In this vast region without any end they claim a specific gravity and that gravity has a specific value with a specific name being the gravitational constant. If it is gravity being out there in the blackness, then it is pulling to a centre. The biggest Newtonian question then is: Where is that centre and what produces the centre whereto the pulling is going. Where is the centre of the region thought of as outer space? If outer space had gravity there has to be a centre to which the gravity is pulling or moving objects. The only centre there is, is the centre material form. It is the centre of the Sun, or it is the centre of a galactica. That is material and can hardly qualify as outer space because such space is openly controlled by material moving about in such a space. Other than that there is complete lack of gravitational evidence.

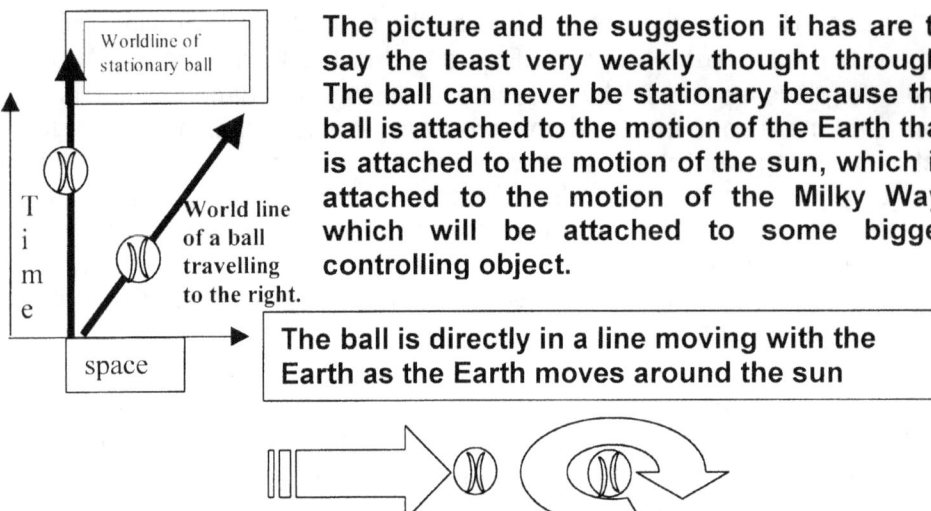

The picture and the suggestion it has are to say the least very weakly thought through. The ball can never be stationary because the ball is attached to the motion of the Earth that is attached to the motion of the sun, which is attached to the motion of the Milky Way, which will be attached to some bigger controlling object.

The ball is directly in a line moving with the Earth as the Earth moves around the sun

At the same time the ball is spinning with the Earth around the axis of the Earth. At no time can the ball ever be stationary in relation to the rest of the Universe and that relation in motion is time.

Just above is a sketch. The sketch represents a ball thrown through space and through time. The ball is representing the movement of time. See how equal space and time is? See how space and time form a mathematical vector. See how easy it is to draw a line that represents space and represents time in precise proportions. Does that not make the life of all mathematicians much easier? Can any one argue that space and time is represented even handed by equal proportions just to make life much simpler to calculate the Universe? This (I suppose) must be where gravity draws space flat and time flat and space multiplying by time gives some square from which a vector then forms. How flat can time get and what happens when space goes flat but more still, how does the flat then multiply the flat to form a flat square. There is no mention of dimensions although we all realise dimensions is the issue concerning the Universe.

Please note how the Newtonians suggest that time and space is both the same value and is equal in partners being one and the same of value although being far apart in differentiation. If that was the case then the Universe was one because the one in space would cancel the one in time making space eternal and time infinite, and then they claim to be the highly taught. How they can bring space in as a flat and the same as time although in division of time (space dividing or per time or space-time) notwithstanding Einstein declarations goes beyond any normal logic. How they can put time and space as equal partners and still find that space can move through time also is beyond what silly old me can explain. Fortunately for every one concerned I am the incoherent one. Time is always in control of space by motion of space. Time positions the location of space in the relation time where time places all in relation of all space to the rest of the Universal spaces. If science wishes to put space in a motionless stance science should produce evidence where space is motionless or bring proof where time can find the ability to stand still.

In the use of the formula there is a deliberate admitting to space-time being used. In the formula they put 9.8, which are the square of pi to a unit of meters (space) divided by seconds (time) to the square. However that is applicable to the Sun as much as all other objects. Then the formula is taken to apply to Black Holes where the concept science then put foreword changes totally. It becomes Using the formula in the Black Hole suddenly calls for the square of the speed of light to come into the formula as a factor. The square of light comes in to support the diameter of the star. How can they get the diameter of the star married off to the speed of light?

The one concept is a length a distance between points of compacted material and the other is the flow of liquid space through time. The one concept is measuring distance and putting that into and as an additional factor, which by any standard, is inexplicably unrelated to the distance of the diameter, but still comes to support the square of the speed of light. Motion supports the diameter of material by the square. $E = mc^2$ Light cannot go into a square because the speed of light is the epitome of motion in space through space. There cannot be a speed of 1.005 times the speed of light or be twice times the speed of light. The speed of light is the ultimate and the optimal. It is where velocity ends and no increase can come into furthering the calculation. Yet the Brainy Bunch has this unexplained novelty of squaring off the speed of light where we locate a diameter to the measure of the third dimension as r^3. I do not say that doing that is incorrect, but the motive and the explaining why in could be corrects totally lacks in science and even their realising the incorrectness passed all of the Brainy Bunch by. I also do not say going square by motion in conjunction of the diameter is incorrect, but they do it by using all the wrong reasons which they never mention. If that is true and they can use C^2 in place of a diameter to perform as gravity, then they have to admit to my statement that gravity is no force but pure motion. I say gravity is the difference there is in motion between particles having unequal relevancies to a specific centre controlled by singularity and the gravity is the extending thereof. Still gravity is the motion of space in time through time.

The time that Newton froze on paper in his establishing of a single t that is representative of time is effective in remembering the viewer of an event but that cannot be the event that is part of the present any longer. If we reduce the moment to a snapshot, the picture we focus on can only be what we see the image to be. The image remains in ink on paper, which froze not time but an image of time. It holds an image that was part of time for a very short instant and then forms that which was how the event occurred during the time from where the camera shutter opened T_1 to where the camera shutter closed T_2 and the time frame T^2 was then during the open period of the camera shutter. As soon as the shutter shuts, time moved on and another T^2 formed leaving the image taken as time serving an image never to repeat again.

Afterwards the image we see is not time. It is an event that occurred during the flow of time at a specific stage in the flow of time. It did not freeze time but took an image of time distorted forming space as the picture represented **t** at that moment of **T²**. When looking at the picture, the looking at the picture also became an event. That event happened during a specific **T²** that went from where one is taking the first look to where one is looking away from the paper. The event lasted while carrying the first dimensional image of an event gone by. That is at that stage a representation of **t** in another milieu of **a³** = **T² k**. It is not time standing still. If you show the picture to a horse, the horse will try to eat the picture because the horse will be unable to recognise the image in ink on paper. The last thing the horse will experience is a freezing of the moment. The **t** in the single is when mathematically presented as only **t** indicating a mathematical single flat dimensional view of time being part of paper by means of ink. The image we recollect is in our minds and not in time. It is an image that is then correctly applied because it represents a reminder of a four-dimensional event **a³** = **T² k** that went single dimensional because the moment in the fourth dimension which was then frozen in a single dimension on paper. With the paper being part of space-time while the fourth dimension **a³** = **T² k** soldiered on and time will always be representing **T²** as Kepler stated.

But I showed that **k** = **a³** / **T²** and Newton's claim is that **a³** / **T²** = G (m + m$_p$) / 4 π²

Π^0

Time is in the square, and that is allocated to space having a cube. Kepler said gravity is **a³** = **T² k** at a time even before gravity got a name. But reducing the dimension of time to a single **t** one will find the ability to mathematically design the paper on which the photo image will be printed in time **T²** using space **a³** in the third dimension to apply the ink in the third dimension. Printed on the paper is an image that is not part of space-time while the ink used is space-time and the paper is space-time. The ingredient of ink on paper all holds different values since the image has a value we as humans grant the image to carry such a value. The image, the ink and the paper all hold different relations but the image only relate to thought in our mind. The image has no **k** indicating only references with and to what forms part of a realistic different singularity coming from the Earth centre and connecting to individual atom groups, forming individual as well as group space-time. We are all so painfully aware of Newton's claims about a spinning wheel standing still because of the fact that Newton could get time to stand still and then when time was standing still he would divide the one in the other and best of all he would get zero.

If Newton's vision about the wheel not producing works where the work does that electricity do come from. The generating is done by a "wheel" turning and the "wheel" contracts space-time by measure of heart (yes electricity is just heat and because it is just heat it give burn wounds) If Newton is correct, no driveline can form because by driving a line in infinity (1⁰) takes charge of the motion and drives what is in need of driving. I am aware of this hornet's nest I (again) disturb and was marched from campus about this on previous occasions but I cannot agree with what id disagreeable to me. The whole idea of time standing still is wrong.

If gravity was mass induced then what explains the fluctuating radius between the Earth and the Sun because the radius does fluctuate. That would prove either that the Earth grows larger at some stage during the year or on other occasions it shrinks way. If that argument is valid then me living in the south must be on a different planet from those living in the north. The sun and the Earth remain the same size which brings this mass idea Newton had into serious decongestion because with mass pulling and the radius shrinking and stretching there has t be another as well as a much better thought through idea about this concept of gravity managing the solar system. It brings the whole Newton concept into serious doubt.

$$\frac{dJ}{dt} = 0$$

This disputes mathematics. DJ / dt can have any number from eternity to infinity, only excluding one; it cannot be 0. By placing the one in division of the other, you bring in relevance. You cannot then say there is no relevance. By doing such, you proclaim that one of the factors is non-existent.

$$\frac{dJ}{0} = dt \text{ or } \frac{0}{dt} = dJ$$

In both cases, one of the factors then does not exist. Such a claim is incoherent, because you proclaim that a circle has no radius, or a radius has no circle. When calculating a circle, you multiply either the square of the radius by Π, or the quarter of the diameter at a square by Π.

$\frac{dJ}{dt} = 0$ constitutes a circle and is also therefore $\Pi \times r^2$ = CIRCLE

I do not, for one second, deny or dispute the revelation. What I do encourage is place the event into its correct context. It was merely, and simply an apple that fell from its branch to its roots. The apple did not pretend to be a meteorite that fell from the heavens. If it were a meteorite, I am sure, with the man's genius, science would be somewhat different at this stage. However, as a young man, being very impressionable, as all young men are, and with the attention this brought about in the world of science, the matter overshadowed the fact.

I am not disputing Newton; I am disputing the relevance of Newton's scientific breakthrough. It was not two objects of cosmic proportions, colliding in a show of spectacular. What he concluded the day the apple fell was not an event of cosmic proportions. It was, after all, only an apple falling from a tree. With this miracle revealed Newton found he was competent to improve on the work of Kepler and if I may dare say this, there must have been some political agenda behind this act and the accepting of it for Kepler was a German and what German can ever teach any Brit. The very same politics are still the order of the day forming international rivalry on all fronts.

Newton, and science, made one enormous blunder, from this stance. They took the radius of a wheel not to have any influence on the wheel. In doing that, they removed the very fact that keeps the universal attachment together. They put two objects in an attaching relevancy and then announced no relevancy. Doing that is breaking the most fundamental mathematical principle.

$$\frac{dJ}{dt} = 0$$

This disputes mathematics. DJ / dt can have any number from eternity to infinity, only excluding one; it cannot be 0. By placing the one in division of the other, you bring in relevance. You cannot then say there is no relevance. By doing such, you proclaim that one of the factors is non-existent.

$$\frac{dJ}{0} = dt \quad \text{or} \quad \frac{0}{dt} = dJ$$

In both cases, one of the factors then does not exist. Such a claim is incoherent, because you proclaim that a circle has no radius, or a radius has no circle. When calculating a circle, you multiply either the square of the radius by Π, or the quarter of the diameter at a square by Π.

$$\frac{dJ}{dt} = 0$$ constitutes a circle and is also therefore $\Pi \times r^2$ = CIRCLE

If you remove r it then is $\Pi \times r^2 / r^2$ = CIRCLE.

You cannot then say $r^2/r^2 = 0$ and therefore $\Pi \times 0 = 0$. That is nonsense. $\Pi r^2/r^2$ will always be $\Pi \times 1$, and that is the eternal circle.
When looking at any rotating object, there has to be a point of no rotation and no rotation means "no rotation", not no existence. No rotation means a factor of 1, not zero. That then is singularity. The eternal Π, the Π that may not have significance but still it is a Π of value.

If it is that dJ / dt = 1 then that is exactly what Kepler said when he said $k^0 = a^3 / T^2 k$. He also said the motion brings about the filling of singularity $k^0 = 1^0 = 1$ when he said that the space is filled by the matter in the motion through the time period.
($\Pi=\Pi^3/\Pi^2$) or in Kepler's terms $k = a^3 / T^2$ presents the duplication of space while

($\Pi^2=\Pi^3/\Pi$) or in Kepler's terms $T^2 = a^3 / k$ presents the destruction of space and

An Open Letter On Gravity Part 2

($\Pi^0 / \Pi = \Pi^2 / \Pi^3$) or in Kepler's terms $k^0 / k = T^2 / a^3$ presenting the final act or the demise of space. $k^0 / k = T^2 / a^3$ becomes totally dominating when there are a displacing ratio of **6** (materials Number) in the square of space **10** which then presumes the value of **60**. Space will remain duplicating as long as there is space available to convert to heat and as long as there is heat that can be converted to material and material available to transform to singularity. Nowhere is they're having a free ride coming to any part or dimension of Creation. There is built in only a lot of hard work and a dear price to pay for every inch gained or lost in growth.

In this mentioned ratio between the dismissing and the duplicating of space through motion stars form by accumulating material in a giant sphere and keeping the atoms secluded from outer space. All the atoms within the star that are forming the space within the star that is forming the star are as much the star as the star is all the atoms combined. Since all the atoms in the star works towards a mutual goal as to provide the star with the required security to provide the maintenance that brings about survival to components in the star is as much one atom in cosmic space as all the atoms are individually cosmic structures. When the flow of space is exceeded by a certain specific number of proton abilities to dismiss the space the dimensional walls keeping the space in form no longer can sustain the flow of space by substituting the demise with a flow of space. We have to remember that the initial motion was equal to the initial expanding that was in turn equal to the initial space a^3 that developed. The distance **k** that came about was the same value as space a^3 and the motion of the space T^2. The Universe divided innumerably as it remained one structure. Relevancies came about that excluded no possibilities and whatever one may think of being in the Universe came into place through relevancies between innumerable factors all acting in groups that remained one. There was no space but the space created in that cycle motion. There was the motion that provided the expanding in the straight line that was precisely the same value in the half circle and that brought about the triangle also to the identical value as the other two and was securing the space that was precisely the same as the other two factors. Reflecting on matters we still find this very same trend applying to light at the present time we live in. In the development space grew because the diminishing was only half the growth and size rather than direction became a major influence. The direction is the foundation and came in as a bases in the very first instant. The second repeating instant changed the scenario. With time progressing the distance k^1 will tend to lag behind as the rotation T^2 has to compromise for more space a^3 involved. With more space coming about the circle that had to produce the rotation was at the start equal to the expansion distance and there fore $T^2 = k = a^3$. Since then the ratio changed to $a^3 = T^2 k$. That is how space/time relates according to the original calculations Kepler introduced.

$1/k = T^2/a^3$ We know that there is a demise of space relating to the growth in space proving that when distance **k** reduces space a^3 will do the same and

$k = a^3 / T^2$ when **k** expands it will produce space in relation to motion.

$T^2 = a^3/k$ When **k** demise the growth in distance **k** expand time T^2 will increase by the square but distance **k** will diminish by the single therefore time T^2 will grow faster than space a^3, which is the result of **k** will diminish.

At the beginning a trend was set that apply throughout Creation ever since. As the space increased the time ratio decreased since the distance in relevancy reduced in relation to the available space and that is the relevancy I simulate in the atoms ratio of space-time being $(\Pi^2+\Pi^2)(\Pi\Pi^2)(3) = 1836$. But as the star takes time in space back towards the earlier scenario conditions applying to the atom will reduce the space it holds and as such the time will bring the compromising aspect to the changing ratio. It is most important note that when the comet liquefied it was all the atoms within the comet that was liquefied. It was not a case of the comet destruct but the atoms within remained preserved. The comet is as all cosmic objects are, made up by all about the atoms within the structure and also the way the elements arrange their various positions to assimilate the sphere to prefect detail where the entire group of atoms as a group manifest in the form of a sphere spinning about an elected axis. As the group of atoms work in the way an ant ness will where the group form a unit, so does the atoms form a unit representing all the atoms as a group in the group where the group retained the position as one atom with one elected singularity in a charged centre. As the Universe formed by dividing innumerable times and yet remained a unit throughout every aspect, this leads on to groups forming galactica and stars where all the atoms within the star becomes the star and the star becomes anther single atom. The atomic relevancy apply as that quantity in the outer space and with Earth not being that much better developed the atomic relevancy in outer space and in the Earth centre is $(\Pi^2+\Pi^2)(\Pi\Pi^2)(3) = 1836$ meaning the proton displaces space-time at a relevant value of 1836 times that of the electron, or the other way around is that the proton is 1836 times more intense going on route on to singularity than heat is going liquid through the strainer we call the electron. We can see that the atom formed in accordance with Kepler's prediction and followed the route Kepler introduced.

As far as ordinary physics go, nothing changes at all. Every aspect of physics remains the same, except the way science views the cosmos. The formulae I show, has NO CALCULATION ABILITY. The only value in the exercise is proving what no person ever proved before, AND THAT IS THE INFLUENCE THE ATOM HOLDS ON THE UNIVERSE, AS THE ATOM'S INFLUENCE EXTENDS BEYOND ALL COSMIC BOUNDARIES.

To me everything makes perfect sense and while saying this I do admit full heartedly that I am not a Master such as yourself with the knowledge you possess. In that light, should you feel there are aspects I do not explain to a sufficient standard, I am willing to work on it.

An Open Letter On Gravity Part 2

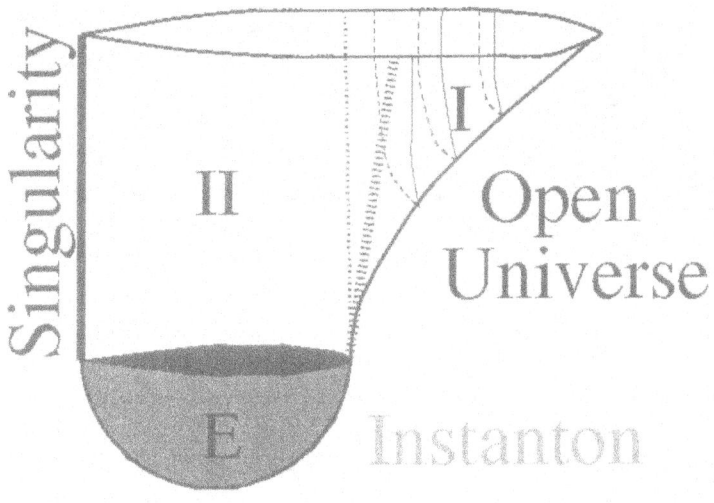

The only concept an illustration such as this represents is to show how little Newtonians truly understand real physics, as nature represents physics. There is supposedly a Universe that holds the basic Universe (◌) and then there is an adding to the basic Universe (◌). As to where the adding is coming from I have no Idea. In all my investigating I have not come across any person that can show how the Universe can grow, since no one ever detected a "White Hole" that is supposed to fill the losses or refill the Universe.

What is the Universe? This is such a simple question everyone gets wrong because of the relevancy we humans place on the universe and the relevancy what the universe truly is. SUPER–EDUCATED–WIZARDS really get tide in knots with making all about nothing so complicated it absorbs everything holding back nothing.

The universe is a compliment of matter, ranging from as little as a photon and radiation to as large as a Black Hole and Galactica; still it is a compliment of matter, holding space. There is no open universe and there is no singularity running free in the wider universe looking for some space to be within. There is only one universe holding many, many, many compliments of parts in the form of atoms to one space that does not even exist. This means every particle holding singularity is the universe and the complimenting total of that is the universe. Light is only part of the universe and not time as such. Light in as much as the photon is space in the form of heat that we can see and space in the form of heat is light we cannot see because it diverted from our direct line to singularity. Light is the smallest particle that maintains singularity while running away from a mutual singularity amongst photons to join a singularity within larger matter. The darkness of space is light we cannot see. Light is the darkness of space we can see. There are three components in the universe, one is time, the other is space and the third is matter. Looking through the looking glass there is only matter that matters because we can feel it and matter that so far did not matter because we cannot see it. Then Einstein brought in light as pure energy and with that he was correct up to a point. Einstein led us to believe that energy was some fluid magic where "gravity" on the other hand was some solid magic. If that is the case, then what is gas magic? If there is two of the three forms identified, the third form also must be somewhere. By introducing singularity as the third, magic is surely not on. He introduced singularity but never identified singularity. That brought about that singularity was also some witches brew of magic and ghosts and all the rest. There is not even a universe at large because take away matter occupied or unoccupied and you are left with nothing. That is precisely what space is. Space is a virtual thought. I have a huge problem with the fact that Einstein saw singularity, but never could place singularity, while the position of singularity is so very obvious. I AM VERY AWARE OF THE FACT THAT ALL NEWTONIANS BRAVE ENOUGH TO ENDURE MY VIEWS UP TO HERE ARE STEAMING WITH ANGER AND I AM GRATEFUL NOT TO BE PRESENT IN THEIR MIDST. BY NOW EVERY NEWTONIAN PURE AT HEART WILL INSIST ON MY EXECUSION. At this point I have Newtonians stretched to all limits with my going on with the second largest Newtonian that ever walked the earth. Newton was of course the biggest Newtonian of all, and the rest of science filling all the other positions from top to bottom according to personal importance and devotion. I truly consider the world of science a conspiring collaborating Brotherhood. Blame me if you wish, but the Newtonian religion and devotion to Newton has no bounds. Everyone is supporting the other while covering for the group as a whole. If they wish to castrate me for saying that, then answer some simple questions. No Newtonian knows time or space by definition or place, yet they want to bend, stretch and tie it in knots. I should think they must first locate the objects before they intend to torture it with such tenacity. What is the universe and where does it end? Those are questions that one should address before going on to the more significant major issues like time travel and galactica infesting. The Universe begins inside every atom and ends where the atom's influence on space ends. The Universe runs from atom to atom and the lot holds the universe in line with individual singularity. The universe is the total of all heat holding space occupied or unoccupied in whatever form heat may be. The universe ends where one will find the last of matter occupied or unoccupied wherever that may be and that will be nowhere because of the vastness of the universe. The centre of the universe is in the centre of every proton, where singularity is. There is no collective universal centre because the universe is holding together atoms, which is holding together space in different dimensions.

Einstein declared time to be the same as the speed of light $t = \sqrt{(1 - (C^2 - V^2))}$ AND THIS MEANS WHEN $V = C$, TIME IS ONE.

It is a great mistake Einstein made. He confused time and space. To this reason, what you see is not what you get. The influence of time duration to light displacement is of vital importance, when assessing the universe.

According to Einstein, the universe in time is at the speed of light, with "gravity" stretched to the limit going into a single dimension where time moves from the past, through the present into the future. In answering such a huge statement one should firstly find the location of gravity and what generates gravity. Gravity is within each proton, whether it is in a cluster or on its own in a single atom. After that location one should accept that that is not too far from singularity. However, it is not in the general cosmos out there somewhere. What will generate the GRAND OLD GRAVITY CONSTANT, if not for the number of protons holding that space to occupation in divisions of space unoccupied?

That is the logic of the second most notorious Newtonian on record. To think of the BIG BRAINS arguing in such a manner truly scares me from the position that I am in being one of the small brains. I admit I know very little compared to the TRULY LARGE AND WELL ESTABLISHED NEWTONIANS, but from where I stand I can see a tower of strength on its way to collapse. Then one wonders, if these are the issues too small for them to waste their energy on, how far the mistakes goes to alter the issues that are so large it keeps them very busy. If everyone knows that the proton generates all gravity, why go on a manhunt in outer space? Is that not blowing the issue large enough so that the Big Brains can find Big Issues to keep them busy? What will ever produce singularity in outer space, if one discounts the genies, witches and evil spirits that we all by now know is not there? No sooner did man rid himself of a flat earth, than Einstein introduced a flat universe. I should say that is going slightly in the wrong direction. Let us consider Einstein's FLAT UNIVERSE, another topic every Newtonian wants to explain to me, because I am too stupid to understand. As with Newton's gravity, I truly cannot see what there is not to see. Time in progress makes the universe go flat! It is the senselessness about it I cannot see through.

By applying slight logic, the argument becomes senseless. This argument can only be true, when there is a total dark non-existing universe. This we know is not true, because every seeing person can observe by means of light transmitted through space-time that nothing is going flat, even by the speed of light. If only this part of my reasoning is true then Einstein is the one going flat. It takes time to fill space. Only magic is instantaneous and excluding magic, such an occurrence is unnatural and unreal to scientific evidence. Time as a factor can't ever be lost because time as everything else in the Universe can never depart once it entered the Universe. Time becomes space as time moves through space forming gravity or motion. Time is never lost because when the time is going on the move it takes the Universe into more space as the time in every instant fills the Universe with space and that cannot go flat.

Newtonians tend to take the Mickey out of the Flat Earth Society while those same Newtonians dump us all in the centre of the flat Universe concept. I am not against this flat Universe but Einstein was not incorrect about the Universe going flat, he was incorrect about what he thought the Universe was that was going flat. If He had the idea time, space and space-time all three was going flat then he deserves the laugh and ignorance this remark would bring about and what stupidity this should reveal. The Newtonians form the FLAT UNIVERSE SOCIETY but they are just as wrong as their counterpart: the FLAT EARTH SOCIETY. Later on I shall show how only half the universe goes flat and therefore only half the Earth goes flat because only half the universe goes flat at a time.

The universe cannot be in two places at the same instant because that is time maintaining space implementing the double proton. The universe is in every atom and there is no universe at large but the compliment of atoms claiming space outside singularity towards time in motion and that is what we wish to call gravity.

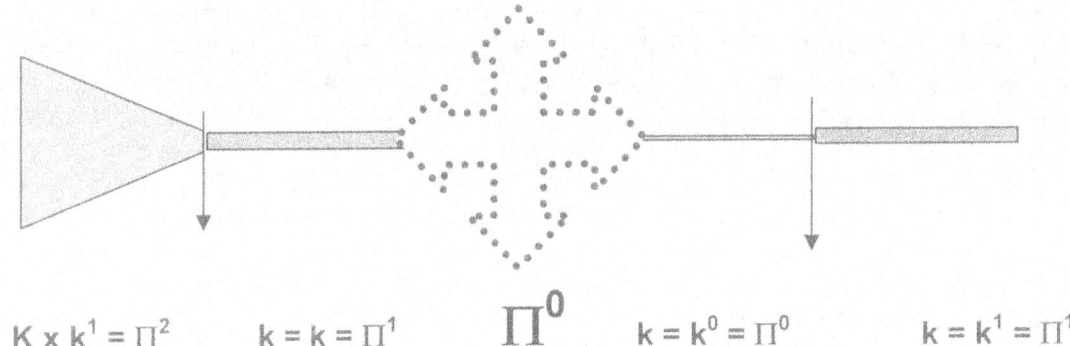

$$K \times k^1 = \Pi^2 \qquad k = k = \Pi^1 \qquad \Pi^0 \qquad k = k^0 = \Pi^0 \qquad k = k^1 = \Pi^1$$

That is the function of the double proton and that is the reason nothing in the universe can be in two places in one instant. While the one half of the universe goes flat (inside the atom) the other side of the universe holds space (inside the atom).

If the photon was travelling at the speed of time, the photon has to remain in one place. The photon, and therefore light as such, has to stand still in time.

In order to have this book reach some conclusion which we may use in part two as a starting point, I wish to end with the last thought on light because the Universe is light and that I prove in another book.

Have you as you sit reading this part at this minute sat back and gave a thought about the light enabling you to read? Such a thought brings to mind the most simplistic answer one can imagine. The light hits the page bounces from the page and makes contact with the lens of your eye where the lens conveys the photons becoming electricity to a part of the brain that translates the electricity to an understandable message and that makes one read. It is as simple as that! Ever gave a broader thought about light streaming across the night sky, coming from the ends of the Universe, we do not even realise it is there? How do the photons manage to convey one complete picture coming from as far apart and as wide an area as it does? With a few photons connecting the eye or lens no one ever noticed the wonder of light. The photons reflect a view that seems as if coming from all the billions upon billions of stars, but most are coming from darkness covering an area no man can measure. Yet how many photons can actually connect to the lens of the camera or to the eye? Still a few photons coming from a single direction directly ahead eventually tell the entire storey. It is very simple to take the process of seeing by means of photon conducting very lightly and I have never heard one of the Brainy Bunch really in sincerity dissect the process to its potential. It is impossible that light from such an array of assorted sources can simply come together at the eye lens and show a picture of objects spanning across a universe as wide as our mind can receive where the objects they reflect is beyond human measurement and the quantity is inconceivable many.

I have so often heard the argument that we on Earth holding life can't ever be alone in the Universe because there are so much out there to fill and why would we be alone with that much out there to fill. We just can't be the only life in the Universe. Well, if ever I heard an egocentric Newtonian argument it must be that. The Newtonian finds a position filling the centre of the Universe. This is substantiated every night while the Newtonian looks at the sky! Going along with this argument would be the next part when we ask why the rest of the Universe is created as we are here and no where else. That is taking centre stage in the Universe. That is seeing how the entire Universe comes to secure my place in the centre of my Universe. The argument that defines this is more a human attitude we find in the way we see us with all the light through out the Universe coming to acknowledge our position in the centre of the Universe and if we as life did not fill that centre position in the Universe then why the hell is there a Universe to start with. This is not God's Universe but it is my Universe because if not for me why did God Create a Universe to start with? That is putting life at a centre stage in the Universe while life is not even a decent attachment, but is much more a thought in the Universe that came after the fact. We are not some conclusion as to why there was a Universe to begin with, but we are an after thought that was placed to vanish in a very short while afterwards, leaving the Universe to be for ever and unmarked by human life. The following that I wish to say is a personal feeling and is not scientific fact but I get the impression we might be some experiment God embarked on and we are failing the expectations quite substantially. We might think modern man in the twentieth century and onwards are very successful but the only success we had this far is to harvest and use energy and with that we are going to destroy the Earth with Nuclear wars that is coming. A Nuclear bomb takes Creation to where it started, and it started with light. Light is much more than the medium science makes it to be. Light connects the Universe in a way we cannot contemplate. Light being far apart originating from regions not in the same time or universal space connects in a way that present us with a picture holding the universe in an understandable content. From the point we stand and we watch the universe, the significance of what we see surpasses the sense of understanding of what we are experiencing. How can the few photons that our lenses catch coming from such an area which is covered by the night sky transmit the complete picture of what we see? Take a few seconds and glance at the picture of the night sky then rethink the picture applying the full content in the picture to what the size of your eyes are. Think how big the picture is that your eyes take in and translate that area to the size of your eyeball in an effort to determine a ratio. One will be forgiven if one thinks of the ratio as eternal to nothing. Yet a few pages back I showed that according to mathematics there couldn't be anything as nothing. Consider the path the light followed from the source connecting to light from all other sources where all particles of the other light may come from and bringing a full picture to the lens one use to look through. In your mind, connect a line from every atom producing light and connect the lines to your eyeball and see how you can manage to fit all the lines, as small as the lines may be.

Scientists think of outer space as geodesic zero, with nothing in outer space but space. Geodesic zero means the light travels in a straight line from where it originates unhindered all across space to where the light connects the eye. Such an idea by itself is outrages because the stream of photons are reduced in space to such a minute quantity that taken the area the photons travel and the space in vastness it covers, the chances of one photon coming across many hundreds of light years through billions upon trillions of cubic kilometres of space and selecting my eye to convey the electricity to, is less than infinite. Yet such conveying takes place every second of every minute. The position of the location of the second singularity, which is the precise duplication of the first singularity but in a diminished capacity, is obvious to miss when one is not applying a detective mentality, as one should in scrutinizing the cosmos. Culture will have us believe that when one sees a colour shining from an object, the colour is associated with the object. Logic tells a different storey. A yellow dot is all the colours in the spectrum but yellow because it is disassociating with the yellow. That goes for red, blue and all other colours we may visualise. I think the norm accepts this as scientific fact with very little argument or substantiating proof about that required.

If light came as individual streams of photon flurries our visage would translate that as such shown in the fragmented picture above. It would be a picture unconnected bringing across some photons in the manner where every object stands apart not being related in any way and that will be what we see, if it is anything that we see. That we know is not the case but that means geodesic zero is as much rubbish as anything Scientists regard with simplicity and with careless thought. Geodesic zero means nothing and how can I see nothing as darkness because "nothing" is not darkness, nothing is "nothing" and the darkness I see is darkness showing the darkness as something.

What then about colours that are technically not colours as is the case with black and white? White is simple. By spinning all the colours in the spectrum, the colour white shines through. Black is quite another matter. A friend of mine whom is one of the best painters I have ever come across told me that one couldn't paint black but have to make black a dark blue to show shade on the canvass. That apparently is his success in achieving the realism.

He also went on to explain how many variations of dark blue form the shadows in one simple tree. This remark set my mind in motion. One cannot see black because black has no colour to show, but black is the colour most prevalent in the universe. One can see only by colour and since black is not a colour we should not see black, but we do.

If the darkness was the representation of "nothing", then that should be exactly what we must see, nothing but the stars. Taken from the top picture some stars and leaving the rest to nothing is what we see in the picture below. A blind person sees nothing but when we look at space, we see something that we think nothing of as we see as space. One cannot have the ability of sight and see nothing except by closing your eyelids and then you see nothing. But in that case you do not see "nothing" in contrast to "something", you see "nothing" without it contrasting to "something".

Nothing is all about not being and "not seeing". By the ability to see the darkness, renders the darkness something other than nothing and that changes the acquired value of the darkness from nothing to something. There is an eternal difference between something in infinity and nothing.

The arguments introduced up to this part of the introduction prologue only touches the most basic aspects of my work and by no means can such an introduction secure an opinion. Yet, not once through all my long investigation in the past thirty or more years have I found any other person claiming such views that I have brought about even in this skimpy way as I do in the prologue. As it applies with all things, so it does in this case as well that when delving deeper into any issue, the complexity of the issues truly come to the foreground when analysed in more detail. I wish to advise the reader to treat the seven books as seven different works and in that light I have separated each work in volumes of seven separate books with individual I.S.B.N. numbers with adding one part, the one you are reading, with one sole purpose and that is to bring about an academic introduction to clarify a quick perspective. Then the next three parts being of a general introductory nature there are overlapping in some sense but each highlighting issues in a different manner as to clarify facts used in the last three parts bringing conclusion to different cosmic perspectives. Yet the work is seven parts of one thesis and as such it serves.

The speed of light was little understood and Einstein's explaining did not make it easier to understand. Einstein first said that light travels in waves of photons which do not look that complicated to understand but get severely confusing when dissecting the integrations or the web. One must recognise Einstein's calculations that came with the era it was in. The speed of light does not transmit through a vacuum in the sense that there is nothing to transmit the light. Light is the smallest space-time permitted by singularity.

UNIVERSAL TIME (UT)
A worldwide standard time-scale, the same as Greenwich Mean Time. How perfectly simple is the Newtonian Universe! How absolutely uncomplicated did the Newtonians made their life! How simple can the Newtonians constant standardised time scale be in order to make the Universe manageable for the Newtonian intellect? Universal Time is the mean solar time on the meridian of Greenwich. It is defined as the Greenwich hour angle of the mean sun plus 12 hours, so that the day begins at midnight rather than noon. It is closely linked to Greenwich Mean Sidereal Time (GMST), since the mean sidereal day is a precisely known fraction of the mean solar day. In practice, UT is determined by a formula from GMST, which in turn is derived directly from such observations of the meridian transits of stars. The version of UT derived directly form such observations is designated UTO, which is slightly dependent on the observing site. When UTO is corrected for the variation in longitude due to the Chandler wobble, a version of Universal Time, UT1, is derived which has genuine worldwide application. When UT1 is compared with International Atomic Time (TAI), it is found to be losing approximately a second a year against TAI. Broadcast time signals use the time-scale known as Coordinated Universal time (UTC). This is TAI with an offset of a whole number of seconds. The offset is adjusted when necessary by the introduction of a leap second, and UTC is always kept within 0.9 s of UT1. On this issue there is much more to explore than the meagrely mentioned. Time stands related to the position an object holds to a centre such an object refers to while in rotation. Kepler found for instance that T^2, which holds the orbit to a rotation specific, is directly dependent on **k** to value the space a^3.

By contracting the Universe is expanding and everything is based on gravity providing both actions. The universe rides on a balance and we have to locate such a balance. To prove my theory I firstly had to locate the centre of the universe. Even admitting to such a notion sounds like madness or in the least a tasteless joke, but please give me a chance to explain in more detail. I realised that my effort to locate the point holding singularity only stood any chance of success if the reducing of the line enabled me to backtrack the exploding universe to its origins. By applying some basic effort I have located the position from where all movement came and the direction it took moving forward in time...and yes, while all of that took place, I was also finding the centre of the Universe which I might add I even located at the same time. There are two standard mathematical formulas used to calculate a circle. The one uses a "r" to indicate the radius and the other uses a "D" to indicate the diameter, which is double the radius and therefore needs to be divided by a four to eliminate the Newtonian inverse square law amounting to the difference there will be between the two. This has the significance that it implicates time.

The one using the radius is Πr^2 and the other formula using the diameter is $\Pi D^2 / 4$. At the very start of my interest about matters concerning cosmology, it lead to investigate the travel of light through time and in particular what Einstein said about time and light. In my involvement as I progressed in cosmology I arrived at the point where I had to understand what Einstein said about light travelling one year in opposing directions and being one year apart instead of two years apart. From this I made conclusions, which resulted in my forming my personal theory about the cosmos. In cosmology normal mathematical principles do not apply that straightforward as we try to envisage. Please allow me to try and explain myself as follows. My understanding of my future theory started when I was trying to understand Einstein's view on light in motion. When light depart in opposing directions from one point jointly shared by all and the light departed will travel in a straight line 180^0 in direction to each other they are all still relevant as a continuing line.

Starting point

$180°$

I have explained this next bit previously, but in conclusion to this book I wish to elaborate more on the matter. The question that nagged me for many years and on which I spent almost half a year just trying to solve the puzzle is how can light travel for one year in opposing directions and after travelling for one year in opposite directions being at two different points but was one year apart.

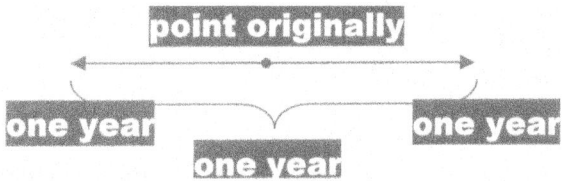

After one year of travel both points still are the same distance apart from each other as that which both points are from the centre. Both points are from the centre just as far as each point is from where the light came from and all the points are at an even distance. The point of origin has to be evenly matching on both sides of the divide where the divide serve centre point from where the two points originated.

Under normal circumstances, when applying normal mathematics the light will be two light years apart. If one could stop the light travelling to the right and have that light that is stopped, standing still, while the light flowing to the left can make a complete turnabout, it will take the light coming back one year to reach the point of origin once more. It will take the light one more year to reach the other point that is having the light that was standing still then at the time for two years running. Einstein proved that the normal way humans use mathematical thinking power is not the way rules apply with the speed of light as it is in the case we find with light. Light travelling in opposing directions for one year will be one year from the source it came and the two lights will be one year apart from each other.

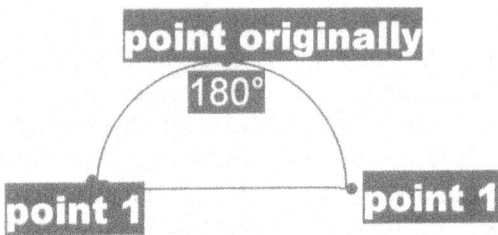

In order to give this argument mathematical logic, is to put the light equal to time but this tests logic even further. Einstein's claim that this comes about because the light is equal to time did not make sense to me. Then I came to the conclusion that light became a cosmic factor before mathematics as we understand mathematics in its fully developed state. It is where mathematics started off with Pythagoras that comes into affect. If the light was time, as Einstein interpreted light in motion to be, then time was no factor to light. In that case light will travel through space in a ratio of one meaning the moment it releases from the source it is on the other side of space notwithstanding the distance the space has. It holds a factor of one with no distance to measure any restraining. That is nonsense because light is restrained by 300 x 10^3 kilometres of space in one second of time. Light is restrained by space-time. Light cannot be time, because light is just a simple speed ratio like any car driving or aircraft flying or spaceship launching. Light was forming distance during time duration and that comes down to being pretty fast, but it still remains speed and speed has just the same relation with time than placing time in relation with space forming distance. One should think that Mainstream Science would see from this that normal mathematics used on Earth does not apply in astrophysics but apparently that slipped past their noticing. Let us again gauge what is happening when light is travelling.

After travelling for one year the light had a distance of C multiplied by the seconds in one year to each side of the source. The points of light travelled on either end of the dividing line, which was parting the points, and securing their independence. Light had that same distance apart taken from the point the light started from. The two markers are just as far apart from each other as they are apart from the original starting point. With this in mind the use of C^2 by science might prove convenient, but it also proves with this mockery how a big farce such innovative calculation is. There is no chance of anything going at C^2 because there is no exceeding of C by light or any other particle. If that were the case, light would not be present in the explosion or antigravity. The speed of light is not a force but it is a speed. It is a ratio putting space in relation to the time in the space density the speed will establish. This whole argument pointed to Kepler holding the straight line in relevance to space and time. From that point I concluded that the link must be the value of a straight line sharing a dimensional value with a half circle and the triangle. If we look at the line supposedly travelling straight we find that the straight flowing is equal to the square relating the triangle. This is completely Kepler indicating gravity being $a^3 = T^2 k$.

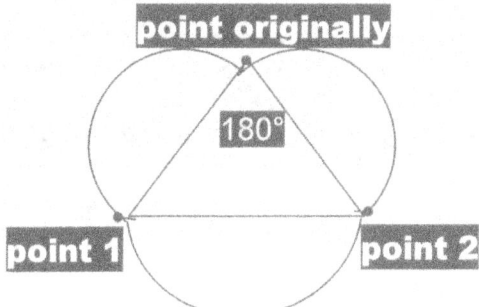

Look at the dimension and not the number implicated. It is $^{1+2+3}$ and transfers that to the line being the ^{1} and the ^{2} being the square being equal to the triangle as ^{3}. But it diverts very much from normal mathematics and that is precisely what Kepler's formula also does. With Kepler $a^3 = T^2 k$ and with mathematics the volumetric size of space must either be according to the measure of normal mathematics if it is a cube then three sides form $a^3 = L \times B \times H$ and in the case of a sphere the measure will be $a^3 = 4/3\Pi r^3$. This was a triangle in relation to the square we find in the half circle standing again related to the half circle. It is not standard mathematics and anyone drawing links between mathematics and the speed of light has no idea about what is involved. With that I have again antagonised millions of the most important people with which I have to share a view. I do share their view on the Cosmos but not their view holding mathematics as a standard fit-all and applied anywhere in the cosmos. It is about lines carrying dimensional properties and with that we have to consider the line once again.

Before coming to the mathematics I would first like to bring your attention to the practical side. I am promoting a theory in which I am able to prove there is as much contraction (moving in the direction of the Big Crunch) taking the cosmic universe back to the size it had during the Big Bang as there is expansion (moving apart by Hubble's Constant) and the contraction is as much part of the expansion.

All the difference we find is seated in the human mind. We humans set differences because we look at the cosmos by placing humans and the life we find on Earth in a pivotal centre in the cosmos instead of placing singularity in the centre and life where it belongs; only found on Earth. Einstein proved mathematically that in the presence of a strong gravity such a strong gravity slows time down. Surprisingly with that evidence being around this long nobody in science since Einstein's discovery took those statements and made any further progress from that. It seems to have been left in some drawer to dry. Science still sticks to the opinion that time did not change, not even slightly, since the beginning of the time and holds the same pace ever since the start of the Big Bang notwithstanding the implications this concept carries. Before the Earth took one year to circle around the sun and even before the sun was there, a year was still the same duration of one year. How odd... don't you think ... that the only aspect in the entire Universe that is beyond change is the aspect of time? With the entire Universe including all the gravity now present and not excluding one Black Hole or dust speck pressed in such an area that was possibly the size of a lepton even then, the gravity extending from that circumstances must have been beyond what words can ever describe.

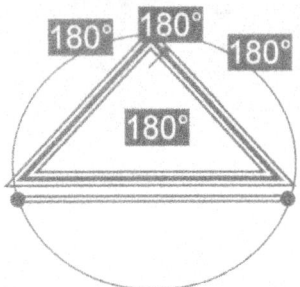

If light cannot be two years apart when travelling in two directions opposing one another but can only maintain a distance of being one year apart, light is unable to exceed C. How can light then reach the square of C?

With light not able to exceed C there is no possibility of

When everything was that small when the Big Bang took charge, the gravity at the time was beyond light, because even today in the Black Hole the gravity is beyond the speed of light. If the gravity was that high and Einstein already proved that strong gravity slows time down, then there is one logical conclusion and that is that time was in fact at the time of the Big Bang standing still. Mathematically it is incorrect to allow gravity to compress the Universe into a spot smaller that an atom and exclude any other factors and relevancies to change.

There is no chance of anything going at C^2 because there is no exceeding of C by light or any other particle. The electron forms the limit of C but after C the space-time breaks ranks with the third dimension and accelerates to π and π^2. Light might equal gravity π^2 in space 3 becoming $3\pi^2$ but it cannot reach any value above C in the third dimension. This is the fundamental fact in cosmology and breaking this concept is reducing cosmology to rubbish. If that were the case light would not be present in the explosion or antigravity. The speed of light is not a force but it is a speed meaning it is a ratio of space over time $k = a^3 / T^2$ where space is a distance a^3 = km, k is a value 300 and time is T^2 = seconds distance of space in relation to time. It is a ratio putting space in relation to the time in the space density that the speed will establish. This whole argument pointed to Kepler holding the straight line 180° in relevance to space and time ($a^3 = 180° T^2 = 180°$).

From that point I concluded that the link must be the value of a straight line sharing a dimensional value with a half circle and the triangle. If we look at the line supposedly travelling straight, we find that the straight flowing is equal to the square relating to the triangle in ratio. The normal manner of mathematically calculating diverts completely from that which Kepler was indicating as gravity being a sphere in motion $a^3 = T^2 k$. Look at the dimension and not the number implicated. It is $^{1+2+3}$ and transfers that to the line being the ^{1} and the ^{2} being the square, which is being equal to the triangle as ^{3}. But it diverts very much from normal mathematics and that is precisely what Kepler's formula also does. The fact that $a^3 = T^2 k$ diverts from the accepted norm of $4/3\pi \times r^3$ It is a clear indication that what Kepler saw does not in any way translate to normal mathematics. What Kepler saw as $\pi^3 = \pi^2 \pi$ is not normal applied mathematics. That, which Kepler saw, predates normal mathematics and it is our duty to investigate why that is instead of changing it to our thinking and our liking. With Kepler $a^3 = T^2 k$ and with mathematics the volumetric size of space must either be according to the measure of normal mathematics if it is a cube then three sides form a^3 = L x B X H and in the case of a sphere the measure will be $a^3 = 4/3\Pi r^3$. This was like comparing a triangle in relation to the square. It predates mathematics to a time when we find in the half circle standing 180° related to the triangle (180°). It is not standard mathematics and anyone drawing links between mathematics and the speed of light has no idea about what it involves. If I take what I unleashed in the past with this statement, the past is telling me again that with that comment I have again antagonised millions of the most important people with which I have to share a view. I need acceptance of my view but if my statement is not well understood I get rejected. It is therefore most important that what I say is understood. Cosmos mathematics is a standard to fit all and to apply anywhere all over the cosmos. Cosmology is about lines carrying dimensional properties and with that we have to consider the line once again because what I try to introduce is not the general perception one finds in the view of science...

Let us find the smallest possible line first. We already have reached the conclusion that by reducing the line, the reduced line will eventually leave all sides on the same spot. Such a spot must be round in form. With the line being the smallest line, such a line will start off as a dot that moved away from a spot. With all possible sides being in precisely the same spot we have all possible sides onto one spot. Mathematically the spot is in the single dimension, where the space is a factor of forming a value of one ($1^0=1^0$, $1^1=1^0$, $1^2=1^0$, $1^3=1^0$, $1^4=1^0$, $1^5=1$) and exponentially zero. There the space moved over to form the dot. We now are reaching into areas only the human mind can venture by understanding and nothing more. The understanding of this concept demands our reaching the point where the mind of the animal cannot reach.

If it starts with a line, that line only represents two sides being one and as such that is rather a flat Universe. The spot is not yet round because being round requires a shape or form and this lies beyond or before a time when any form or shape came into the cosmos scenario. It was in a period where shape and form was a part of the distant future hidden in and beyond eternity. In that time the line must have been so small it had reached a point not yet dividable in any way. If any further dividing took place, such dividing would have brought growth because there then would form space between the sides going in the opposite direction. The dividing brought all there is having all sides literally on the precise same spot, and I have located singularity in just such a spot.

I came to the conclusion that the spot I found had to be singularity purely on the grounds that that spot holds only one side to serve as a start to the starting point of all directions possible. In that side is only one spot where there is only one side applicable and one dimension present. With all the factors given one can only come to one conclusion and that is that there can be only singularity. In such a case more dividing by two will land further positions on the other side of the divide. That point is serving as a position for all possible points and cannot allow further dividing as it is in the smallest line or spot there may ever be. This spot is the result of a most basic process of reduction as the Hubble constant is a most basic process of expanding during a matter of time. By reducing the line constantly, the only value that will eventually remain without dispute from any party arguing about the facts is exponential zero. By only having exponential zero instead of a numerical zero and a radius as one in the square (the radius effectively becomes one holding any and all sides on one point) such a point might become any value of any significant measure implicating anything but zero as the radius. By expanding the line, it will be an evenly spaced structure growing into the most perfect round dot ever possible anywhere at the point when it starts to grow.

The reducing of the line is one dimension in six and although such reducing is representative of two indicators, all the other indicators must still be accounted for. Therefore the ring or circle is the only way to include all six sides in one aspect. In mathematics there is the formula used in calculating the volumetric inside of the sphere: $a^3 = 4/3\Pi r^3$ which holds two major components that will establish final value where as the rest are indicating ratios. In mathematics there is a line being one quantity and the circle indicator Π being the next circle indicator. Reducing the line will erode the value of Π by ratio. That will eventually lead to having a circle ratio of Πr^2 and eventually lead to Πr^0 but that is not the point where the circle ends. That is where the ratio applying factor ends but it cannot exclude the circle. The circle as a concept can still reduce when it abolishes form to the single dimension. It is not the radius that is responsible for the circle but the figure value of pi and by abandoning π only then does all the aspects fall back into the single dimension. The circle can reduce one step more when the circle eliminated r completely, but the elimination of r as the factor reduced the major factor to the single dimension in Π^0. That will not reduce the cosmos to zero it will only eliminate all potential lines r^0 to potential circles $\Pi^0 r^0$ and from there the circle Πr^0 will come about as manifesting as a line but that manifesting can firstly only establish a circle Πr^2.

The only value that singularity can have, although the single dimension may host the entire universe, is Π^0. Pick a number and elevate it to the power of zero and in the process one may have established another point holding all points in singularity but that is not the value of singularity. Only Π^0 can ever be the accurate value of singularity while singularity will then host the rest of all the possibilities in the Universe. The first value there ever was came in the form of π. Where mathematics was still an idea in development the universe granted values of the triangle being 3 circles as π^3, which was 180° and π^2 which was half a circle also with the value of 180° and finally the straight line also being 180°. Mathematics was not yet established, but the most basic came about. Science is not taking the cosmos back as far as possible, science is taking mathematics back as far as they can but mathematics does not go all the way. Mathematics presented as numbers and symbols only became valid (as did all other aspects) later on in development. But the most basic of mathematics was in place when the spot moved on to form the dot by going from π^0 **to** π.

The reactions of those in charge of producing official policies which are responding to my argument is of the opinion that my argument is silly, but should that be your personal opinion too then test where the silly part applies. Bring the zero into the calculation, the zero that science so eagerly place in outer space and see the mathematical result. The forming of densities is once again establishing certain relevancies and when one removes one factor with a zero the density relevancy goes incoherent. By applying the distance one accepts automatically that the figure becomes calculated with a one, since one is a representative of a factor that is having a value and not being without any value because as a factor it represents at least one in being part of the calculating process of the cosmos. The calculation, as all calculations normally are, is in order to calculate something and the something will at least stand in as one in relation to the rest being part of the calculation. When replacing the one with the nothing that science do when they say they are calculating that which is contributing to space, then you can see that nothing is not what you may find in a Universe filled to the point of overflowing. But saying that the factor of one in fact represents the nothing which becomes a name and not a number since nothing is then a factor of one as it is that much the part in the calculation being calculated, then the one has to replace the zero as the fact of the factor of being calculated. You may also think that nothing can connect a half circle, a straight line and a triangle except their sharing of a value, but I try to prove that your granting of nothing is in this case a calculated value being something.

The claim becomes obvious when observing the connection between the half circle, the straight line and the triangle, which could also promote all the qualities lurking behind the pyramid. Consider the connection between 180^0 sharing three different forms all part of mathematics where each is different in form, but equal in value and then one may realise in considering the very basic in mathematics being the Law of Pythagoras on which all mathematics are focused. The triangle stands in for one factor represented by one at a value of 180^0. So does the straight line become a factor of one and the half circle also becomes one where the factor of one equals all 180^0. All three are most seriously part of shapes in the cosmos. Revalue any one form to zero and the rest too must follow and share the same value.

The only manner in which light can move one year apart form one another while each one is staying one year apart from the source it left, is when the straight line that light uses to move holds $180°$ true to the half circle they are apart ($180°$) and connects the two half circles in a triangle $180°$. Only if light exceeds mathematics and becomes part of pre-mathematics can light find validity. $\pi^3 = \pi^{2+1}$. What we find in this sketch is what we find in the law of Pythagoras.

The Law of Pythagoras is about angles in relation to lines and not one angle can represent zero because that will reduce all the lines also to zero. The measure of angles between stars at a distance uses parsec as the indicator, but the parsec between the stars indicating an angle has to represent an angle whereby one may measure distance and such a distance cannot be zero because then the parsec will be equal to zero. Again it is multiplying the factor with the measure but if the measure is about a factor of zero, then the factor too becomes zero. That is as basic mathematics as I can present.

The flow of light transmits as a single straight line from any one point to any other point and enabling such a flow disallows the cosmos the value of holding zero as a factor. To state that the cosmos is a vacuum can only be understood in terms of defining the definition of what a vacuum is. If the vacuum is the absence of material, then one must define what material is.

Material can be all space holding heat on the inside of the electron, securing the space as much as excluding all space to singularity by developing time through spin. However, on the other side the electron also gathers heat securing a specific density in heat by allowing a zone holding heat not occupied by the material, as we understand the term to be.

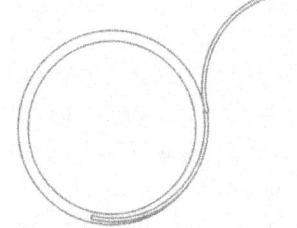

By travelling in a straight line in relation to singularity, the straight line becomes equal to travelling in a circle, since the circle is standing in for the straight line as it is a similar value to the straight line in form and in form it holds an equality we cannot see but we have to accept because when passing through singularity, we are too slow to see the similarity. In Part 2 I explain this graphically and much better. This is where Pythagoras (in the triangle) becomes the corner stone on which the Universe is built. From Pythagoras we can see that the straight line altering direction has the value of the half circle indicating a duplication of the same presence in the two opposing presences of the cosmos.

The cosmos holds four quarters as equal and also as dual from any aspect throughout the cosmos. It is in the relation to singularity not moving while space is moving through the quarter available at that moment. The best example one can find is the Black Hole where the star turns space to suit singularity through the double value if four quarters in space-time. Holding this in mind and seeing how that fits into the moving of light, one can see duplication present in the equality shown by light travelling.

Light travelling away from one centre point will seem to circle away but to our thought the light is conducting the motion as we seem to look at the motion taking place. The light would seem to depart in a line by an immeasurable number of points where every point that the light might travel in will leave another point and the two points will be in opposing directions while each is using a straight line that forms 180^0 in the direction of travel. The motion however is relevant to singularity applying the half circle in motion. From the straight line flows another straight-line with light at an angle of 90^0 to fulfil the requirements of singularity set by singularity on light in motion as a moving part of space-time. That means to singularity it is merely directions applied and not

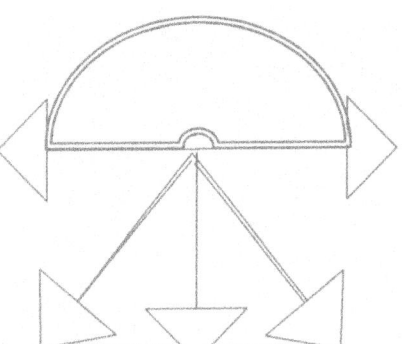

dimensions differences.

One may never look at individual photons because then it becomes senseless, as senseless as the matt-like singularity presenting picture science wants to portray. When light travels on the edge where the photon presents another side of singularity, the straight line is the same as the four quarter circles meeting as they travel. The drawing depicts precisely what should not be done as if by looking at one single photon at any time.

One should rather see the flow of photons as a duplication of many and where one stands in for the other in duplicating to a precise standard as everyone independently presents information in forming a Universal unity.

GRAVITATIONAL REDSHIFT

The redshift of light or other electromagnetic radiation caused by a strong gravitational field; also known as the Einstein shift. It arises because radiation loses energy as it passes out of the gravitational field of the emitting body. Therefore, the frequency of the radiation decreases and its wavelength is shifted to the red end of the spectrum. The redshift at wavelength λ is given by $Gm\lambda/c^2r$, where m is the mass of the body, r is the distance of the emitting region from the centre of mass, c is the speed of light, and G is the universal gravitational constant. A gravitational redshift has been observed in the light from

GRAVITATIONAL WAVE

A wave-like motion in a gravitational field, produced when a mass is accelerated or otherwise disturbed. Gravitational waves travel through space-time at the speed of light, and their amplitude is proportional to the rate of acceleration of the body producing them. The strongest sources are those with the strongest gravitational fields although the waves, like the force of gravity itself, would be very weak. Gravitational waves have not yet been observed directly. However, the decay in the orbital period of the binary pulsar PSR 1913 + 16 is attributed to loss of energy through gravitational waves.

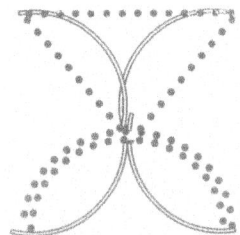

If Einstein is correct, and there is no doubt that he can only be correct, when he declared that the photon travels away from an object holding a space-time position in relation to an exact opposing position where the two directions of light travelling at C will be one light year apart from the source it came from after one year has lapsed, but the light will also be one year apart from each other. The whole definition never made sense but was still accepted by science since the correctness was established mathematically.

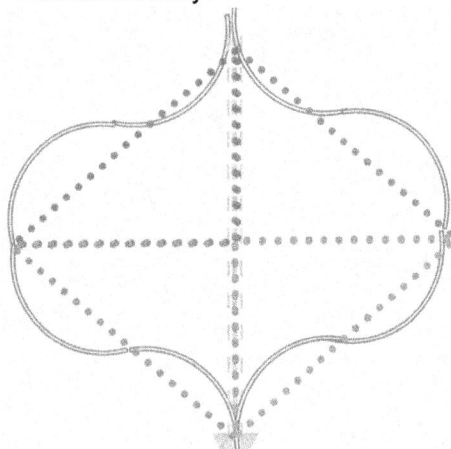

From the human 3D position light travels in a straight line, but seeing light from singularity it holds a much more complex position in space-time converting space-time to motion.

From our position singularity is less than a straight line, flat as it is round with no shape we can ever detect. But the whole concept is covered in relevancy and from singularity the light travelling goes around as half circles in a straight line by the triangle because from singularity there is absolutely no difference in the forms.

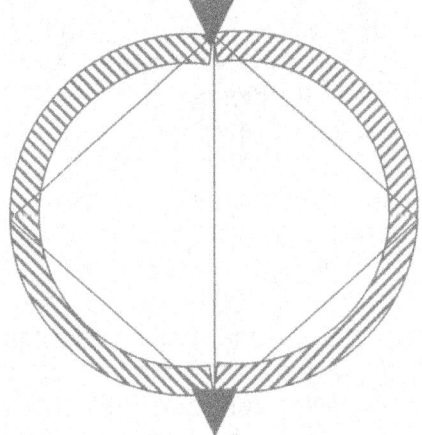

Light falls on the rim of singularity applying in the 3D while representing singularity in 3 D.

An Open Letter On Gravity Part 2

From our position it is a straight line, but following singularity it goes by the curvature of space-time and that goes in a half circle as much as a triangle as much as a straight line.

We take so much light for granted, never thinking for one second how impossible our relation with light truly is. This totally extraordinary relation we have with light must be one of the reasons why we humans put our position we have in the Universe in such a pivotal place. The fact that we persons all with the ability to use our hind legs to walk on Earth, have the idea that the Universe was created especially for us, us being those holding life. Such an idea is absolutely bizarre. It would be the same as if the ant in the park thinks a thousand people maintain Central Park in New York with one purpose and that is to please that one ant. And yet that is happening with the light and us. Every person is standing in a specific position in the Universe and all the light throughout the Universe is directly flowing towards the very point the person is standing. It happens to all of us. The place where I stand or any other individual for that matter is standing is positioned in such a manner that every beam is directly flowing to that very specific spot. From all the corners of the Universe one line of light is especially directed to that location. The light departed from every location following one direction and that is the spot where I am filling that spot. All the light in the Universe is coming to me. Straight to me where I am standing filling one spot on Earth. Go outside and see the vastness the light is coming from. It is coming from all over. It is coming from areas so large not even Einstein can calculate the size and it is rushing towards me specifically. There is not one ray that is going to miss me by fluke or accident. The light has one purpose and that is to meet me at the point I am. Every beam has my name on it and it is coming for my eyes. Can anyone imagine if a person was standing in a location and found all the persons in that city was running towards him where he is occupying that point, how frightening such a person must feel? Yet it is happening to everyone from wherever the vastness of space is situated and is coming across space to that very specific point the viewer is standing.

Even if I shift to another position, the light will change direction and trace me in my new location. Even if my new location is on the other side of the Earth, the light will still get me at that location. The light flows to me from where ever and to top that, it is also flowing to all other persons. That means it is not the Earth that is that important but it is where the location is and that point the observer is using to view from that is that important. If it was only the sun that the light was streaming from that is choosing me as the centre of the Universe it then cannot be that very exclusive. The sun is close and the light is plenty. But it is coming from all over.

That is just one small part of the fantastic affair. Some of the light left the stations they come from some 12×10^9 years ago to meet little old me in this spot I am filling. The light has been travelling 12×10^9 at the speed of light, which I might add is long before my birth, crossing space and time, rushing all the way to meet me at this point. No one ever thinks how was it possible for the light to know I was going to stand at this point and wait for the light to arrive. How did the light know I was going to take centre stage at that moment and fill the specific centre of the entire Universe? I have to be in the centre of the universe because all the light is travelling to this spot filling the centre of the Universe. The light takes two million years only from the closest next galactica to meet me here and after all the time it is meeting me in the centre of the Universe. How important can I ever be? Light is coming across time measured in millions and billons of years through space measured in millions of trillions of kilometres, ignoring all other places it could go to and came to meet me in the centre of the Universe.

To the light on route time means nothing and space even less. Light cannot be more motivated to reach me at this point that I am filling at this moment. Not one ray is by accident missing me except by my choice prevailing. It flows through the Universe in time and in space in the hope I whom is filling this spot, the spot all light is coming to is graceful enough to notice the light. If I were not in the mood to acknowledge the light, the light would have done all the travelling just to be disappointed by my not meeting the light. An effort spanning billions of years and an effort stretching trillions of mega kilometres was all in vain because I neglected to meet the light. From everywhere the light is coming my way and that miracle is passing me by because I am feeling even more important as to acknowledge the total importance I have. The light is tracing me specifically at the location I am occupying just to please me and serve me with all the information about the history of the Universe. I can accept and acknowledge the effort or I can dismiss and ignore the lights efforts. I suppose that will allow me some arrogance and encourage me to think this all was specially created with only me in mind and if I wish to draw a map about the Universe I have all the right in the world to place me in the centre of the universe from where the all of the everything is meeting. After all, all the light is doing it!

We stand on Earth holding our space in the space of the Earth. We cannot have space if we do not fill the space we have on Earth. While being on Earth my position is $a^3 = T^2 k$ where k is because of the mass in movement standing in for k^0 by being k^{-1}. Being k^{-1} we are also T^2 / a^3 which is reducing us in the space we hold our mass as we try to reduce a^3 further to comply with the T^2 the earth is applying and which we have to use. If we move we will produce a larger k factor to the order of at least k^1 to find the ability to move from k^1_1 to k^1_2 which will allow us because we use T^2 to move from k^1_1 to k^1_2. So we have to improve both T^2 as well as k to accomplish motion. But that puts Kepler's formula in question. Using $a^3 = T^2 k$ and producing a larger $T^2 k$ it means a^3 must also improve. That it does by doubling the space it uses during the motion. This is not that uncommon physics. A car holds the space a^3 and is moving by T^2 through the distance of k if the car speeds up the gravity will increase because the distance k will reduce and with the mass or space in motion remain even it is the gravity or the time by the square that increases. That is what gravity is The way gravity is applying is everywhere acting in the same manner but man has subdivided the concept under so many names for each fragment we divided we cannot even find the basic principle any more. Gravity is not a force as Newton suggested but a motion between space occupied and space filled forming a relevancy and this applies throughout the Universe. The only force there is can only be found on Earth in the form of life. Life is the force on Earth and only on Earth with the ability to manipulate space-time under its control by providing motion other that and above the motion the cosmos does provide. It is precisely such a manner that light travels in from singularity to singularity. Because singularity in space, which we consider as dark and therefore invisible, breaks down and rejuvenates space much faster than light in the photon does, the photon can release and join the next singularity in the period of removing space-time and the next in rejuvenating space-time which then will include the photon reassembling with the next singularity forming the space-time of the next singularity. Looking at the issue in this way we can begin to appreciate that light is the duplication of the photon by the singularity charged by the motion that provides the singularity by charging the intensity. It is about duplicating more than dismissing although dismissing does form part when the photon changes singularity. In that way the light loses intensity as it travels and is recharged by the singularity on route to somewhere in the future.

In contras to the duplicating of light is the duplicating of material. The duplication of space filled with material is done by the use of heat in the space surrounding the atom which provides the material the ability to confirm of space. The heat surrounding the atom is there to be used as a stabilizing medium and as it holds quantities it is much more than what singularity will ever require for sustaining heat balance and therefore also much more than that which the particle will ever require to sustain cosmic growth. Singularity in charge of light can generate by duplicating the photon whereas material use the heat the photon provides when clashing with the atom to dismiss space-time by atom singularity making redundancy of space-time, but even more it requires more than the photon can deliver because that is why there are shadows on the dark side.

In the search for time in space, the most obvious place to look for the factor, time as such must be where it is excluded from the space factor and stands alone. Therefore, one should find the place where space is zero leaving time to be eternal. At such a point, it would be impossible to locate and place a value on time. However, there is no such a thing as zero time or zero space.

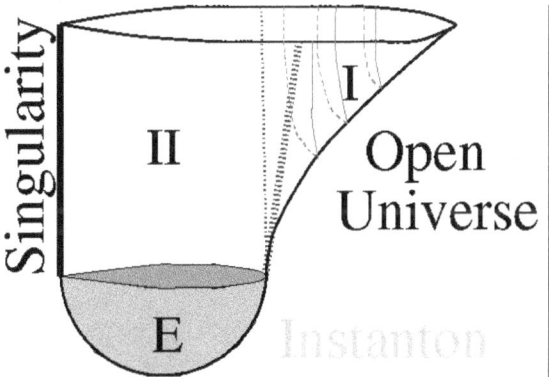

Einstein declared time to be the same as the speed of light because he focused time on $t = \sqrt{1-(V^2 \div C^2)}$ AND THIS MEANS WHEN V = C TIME IS ZERO. Time can't ever be one or be zero because the only thing time is not capable of is being without motion. Time will always move and that is the only evidence we have of time being the fact that everything is forever changing and time is moving everything by changing everything.

As far as the speed of light is concerned we know that putting C^2 in any formula as to indicate that the speed of light is going square is so ludicrous we can only call such a conclusion Newtonian. As even Newtonian believe, I showed that $C = C^2 = C^3$ which means that light doesn't move in space, but light moves on the edge of atomic time. I do now what C^2 is indicating and I explain that in another book, but I am not entering that conundrum in this book yet! Putting $t = \sqrt{1-(V^2 \div C^2)}$ as a time factor is quite silly. The formula $t = \sqrt{1-(V^2 \div C^2)}$ would have us believe that t can go single and that means that t as in time stands still. That, time can't do in our Universe because time is one point (k^0) being in relation with the movement of space (T^2) that then forms space a^3.

It is a great mistake Einstein made. He confused time and space. Because of this reason, what you see as space is not what you get as space but as time. The influence of time duration to light displacement is of vital importance, when assessing the universe.

According to Einstein, the universe in time is at the speed of light, with "gravity" stretched to the limit going into a single dimension where time moves from the past, through the present into the future. By applying slight logic, the argument becomes senseless. This argument can only be true, when there is a total dark universe.

Every person can see plants grow, therefore, where there are plants, there must be water. All persons were unable to match the answer of one plus one because every one was attempting his (or her) form of rain dance to produce water. In order to impress one another they made the rain dance more complicated, until they all seemed silly. Their conclusion that in the desert there is no water is as fatal as the rain dance they performed. In real life, none of them would be alive beyond a few days, because they were silly and stupid. However, I realize it is not my place to call the SUPER-EDUCATED-MASTERS-IN-ABSOLUTE-REASON silly and stupid and I shall never do so, because they all made an extreme effort to show me my place. The fact that there might be water under the sand never crossed their minds, because they hired authors of fantasy books to help them think up more complicated and ridiculous steps to improve their rain dance. The rain dance that is evidently not producing any results from the start.

The fact that time might lay in the micro, and not the macro, never once crossed their minds. With such superior logic, what else could they be than atheists? Once again, I shall never dare to call the SUPER-EDUCATED-MASTERS-OF-THOUGHT mindless, but one question I could not clarify is: Are all mindless animals not atheists? Should you the reader not believe me then convert your cat to the religion you favour and wait for a miracle to happen. There has to be intelligence behind the reason to respond to being converted and I know of not one animal ever converted. The mass of the elements, structure and ratio of the elements within the star produce the positive space-time displacement that is important. This balance between linear displacement and circular displacement is how the star moves in space-time within its structure. With a long stretch of the imagination, one might regard this action to be the star's way of breathing. To think in these terms is only slightly wrong and altogether not right, but it will do, in order to form the basis for a better understanding. In an attempt to find time, let us first exclude time from space-time.

The answer lies in (arguably) Newton's biggest blunder. Instead of discarding Newton's concepts when Roche found the way binary stars act, our SUPER-EDUCATED-MASTERS-IN-REASON discarded the phenomena as "unexplainable", to carry on with the concepts of Newton. By this time, it should be well clear to all why our EDUCATED-HEROES are atheists. It proves a clear lack of understanding basic concepts, but it is far from me to say this, because one must always regard them, with the utmost respect!

As a human is looking at the universe through a strong telescope, the observer is at present in the fourth dimension of time. As this is the case, it is also true that the light the observer is witnessing is part of the third dimension in time. Therefore, the observer is looking at the duration of time and not a picture of an object. Every line represents the relativity of the duration of one time unit extending back into space.

Light in motion

No object can at the present geodesic space-time be in motion at the speed of light or to any related value, therefore no object can move towards the earth or away from the earth at even a fraction of the speed of light. What this indicator reveals is the geodesic space-time value during that specific era when the light was permitted to break free from the time concentration during that point. The higher the geodesic time-space value was, the more time was to space and the less time was committed to matter. There is no space but only time as well as the history of time written in light as space and time forms space in the past where the past is written in light as history.

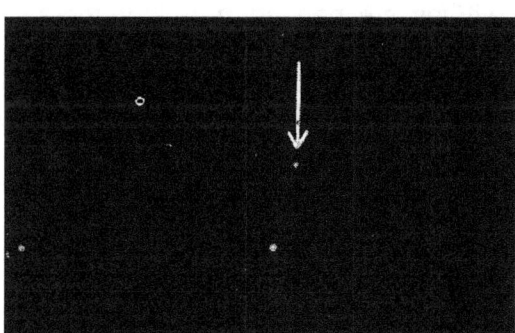
Everything there is also is in a variety of light. Even the concept of darkness is light and all there ever can be is in some form of light. The Universe is what light composes of. Whatever the Universe holds, even in a form such as material, is in another form of light. When material decomposes into fragments, the fragments are intense light (think of a nuclear explosion). The Universe are made of light and light forms the entire Universe as a manifestation of time keeping the history of time in a space capsule by the forming of light. The Universe is light and when God said let there be light, God said let there be a Universe! Let's argue the most complex issue using the simplest reasoning and find where that lands us.

Why would that one spot indicated by an arrow in the picture to the left show light whereas the rest surrounding that point of light show darkness? It is just not possible that the point holding the light has to be something while the rest we can't see is nothing. If I see it, it must be something and when I see it, it must be light because I can only see light. That brings us back to the question about how it is possible to see darkness when the eye can only see light. We can't see nothing because that is one thing that just can't be. There can't be nothing so let's make rubbish of the most admirably astonishingly stupid Newtonian impression that outer space is nothing and that is why we see nothing…it is because it is nothing that we see nothing…can you believe that…and they are the intellectual cream of the human crop! In a ratio of one to ten regarding Newtonian stupidity the reasoning that makes outer space nothing must top the list and being in ratio with the rest forming the order of stupidity this idea of outer space being nothing even surpasses the most rotten effort in stupidity that Newtonians can come up with.

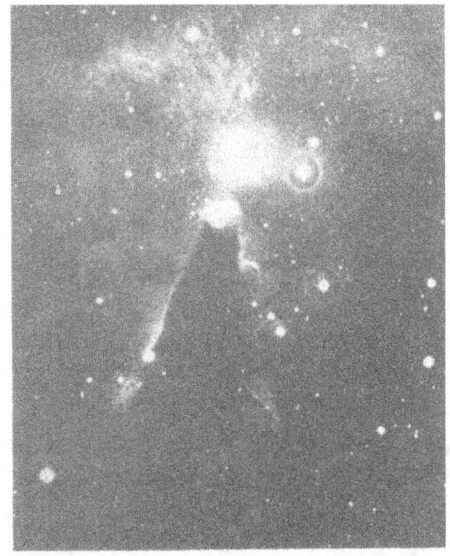

Let's put some sense and clarity to the issue by simple reasoning. We see the light because the space where the light is has a concentration of light. There is a lot more of the medium thought of as light per cubic measure of space as there is in the surrounding area. Then it stands to reason that the darkness we see has a lack of any concentration of light because we see not nothing but a density of the medium we think of as light reduced in the location that is seemingly dark. With the concentration of light in that specific location it is a fact that the light will move away from the density of space holding heat (which is what light essentially is) towards the area I am in. If I se light it is because the light I see is rushing towards me from an area holding less space and therefore more light than the area which I am in. That means the area I am in has less concentration of space overheating which means it has more space because the space it has is expanded more than the area is from where the light is coming whereas the space where the light is, is expanding due to an overheating of light. Areas holding material also holds more heat because the material is a concentration of heat controlled by a proton forcing an electron to move the light in a distinct manner and by such movement the material freezes the heat through applying movement…but more about that in a later book. It seems darker because more light is present and the light is more expanded by movement albeit in material or expanding by the movement of the Hubble shift.

Light moving away will always move from a denser area to a less dense area and that is a cosmic law, not an interpretation or a notion of nothing. The denser area has less space while the less dense area has more space and that is why there are more visible light in the denser space and less visible light in the less dense space. The less dense area has more space and therefore it has less space in which light it can distribute light while the area where there is less space there is more light to distribute.

Light leaving the light denser area has more light that needs distributing because it has less space in which to distribute the light whereas the less dense area has more space to distribute the light. Again I wish to reaffirm that light will always run from a concentrated or overheating area to a less dense or cooler area. That also is a cosmic reality. Light that I visualise by seeing the light is coming to me and is coming from a

denser concentrated area holding more light in less space that my location in which I am.

The area from where the light is coming will light up because that area is giving me all the light it has, and for that reason it is bright. Then the very opposite must also apply where light moving away from me to a more spacious location where light is better distributed must be dark. Where light is shining the light is proof of an area overheating in relation to another area that seems to be dark while it is being cold and freezing in comparison. This is not true but in another book at another time I will delve into that… The area that musty be freezing is freezing because it expanded to the maximum and only the most expanded area is the hottest… and because it expanded maximum it holds maximum heat but as I said in another book and on another occasion more about that… An area being bright is releasing its heat by the heat expanding as light to an area that that area expanded to the maximum it can. Therefore the dark area is the hottest since it absorbs all the light and all and the light flows to another area which is dark because all the light (or heat) in the entire Universe can't get the area to expand more.

That dark area is as hot as it can ever get at this point and only time (the Big Bang) can heat it to expand more. Then that area that is the darkest is also the hottest since no additional heat can increase the heat levels in expanding that bit already contains. Hot particles expand and cold particles contract. Light coming towards me from an area makes that area seem bright and my area seems dark from such a point where the light leaves. The light moving away from me is dark because looking from the darkness my area will seem light. When I look at a bright spot I see light. That is because the light is coming towards me and not because there is something compared to the dark spot having nothing.

When I look at a dark spot it seems dark because my light is travelling towards that spot and that spot is taking my light because it is dark. I see the light moving away as the light expands and because the light moves away the light is black in contrast to the light coming towards me and looks white. It is not dark because it is nothing but it is dark because it holds more light than I have. It is even taking all the light I have to offer while I am taking the bright spotted area's light away leaving that spot as white. Therefore that area is dark in relation to me since it gives me all the light is has.

An Open Letter On Gravity Part 2   Page 663

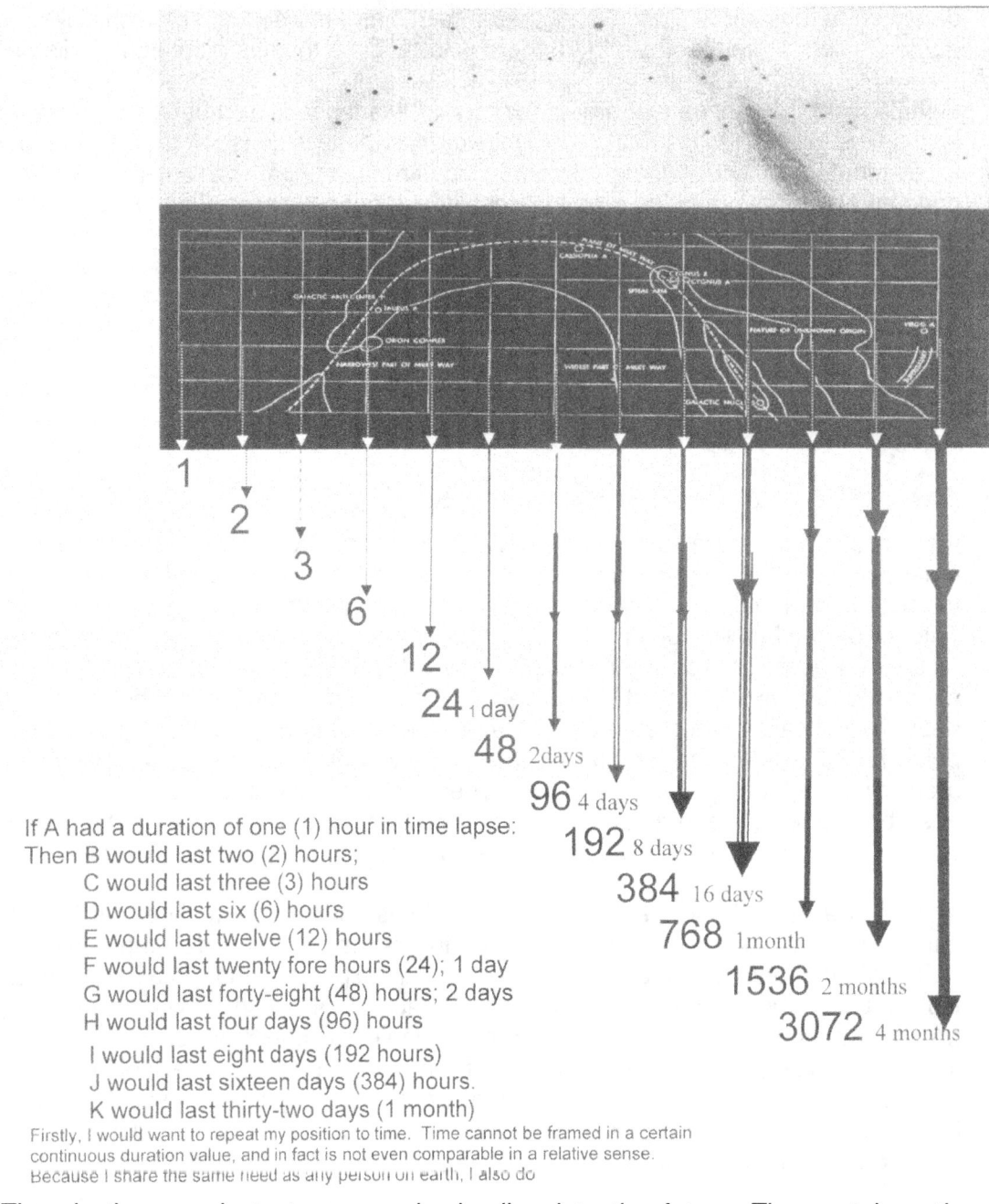

If A had a duration of one (1) hour in time lapse:
Then B would last two (2) hours;
    C would last three (3) hours
    D would last six (6) hours
    E would last twelve (12) hours
    F would last twenty fore hours (24); 1 day
    G would last forty-eight (48) hours; 2 days
    H would last four days (96) hours
    I would last eight days (192 hours)
    J would last sixteen days (384) hours.
    K would last thirty-two days (1 month)

Firstly, I would want to repeat my position to time. Time cannot be framed in a certain continuous duration value, and in fact is not even comparable in a relative sense. Because I share the same need as any person on earth, I also do

Time is the very instant you are in, leading into the future. The past is a three dimensional concept, with time moving away from the viewer in all directions simultaneously. As you view the past, the duration of time will progressively increase, therefore the further you look back in time (NOT SPACE), and the more the events will seem as an explosion. The events will be happening at a pace that seems to exceed time. However, the duration of the events was much slower than at present. On the inside, the duration will be at a pace, which would seem to be in slow motion, to the duration of time in which the viewer presently is.

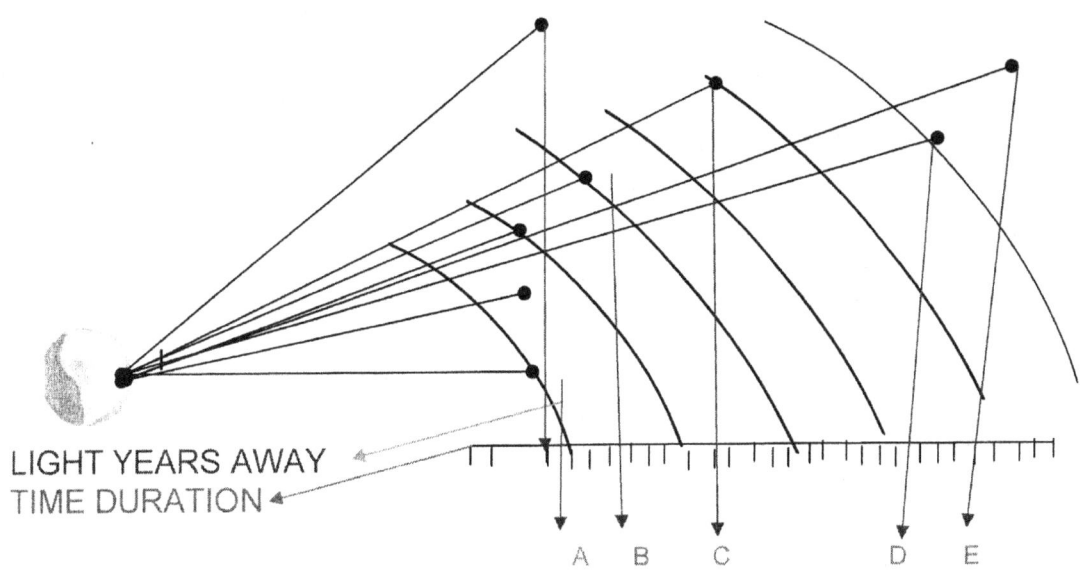

A =   300 seconds   = 5 minutes
B =   600 seconds   = 10 minutes
C = 1200 seconds   = 20 minutes
D = 2400 seconds   = 40 minutes
E = 4800 seconds   = 80 minutes

Ever since time formed our concepts we relied on determining distance in the form of space to survive. Our most basic need required that we depended on measuring distance because our time that the cosmos provided was fixed by nature and provided by the gravity that the Earth supplied. When we wished to hunt we had to judge the distance between us and the deer very carefully as to obtain the most accuracy. Determining the distance was most accurate and the Newtonian mind has not broke from this culture thus far. Space became vital.

Then as we shifted our values to include that stars we saw at night, stars were so many light years away and we hold onto this concept for deer life even today. It is still so strongly practised that even today Newtonians set out to convert this time we see as space into distance as if we are about to travel such a distance. The Big Bang proves that the distance is altering in the number by every instant that time develop and I mention the big Bang since Newtonians apparently have not yet in their hearts linked the Big Bang to the Hubble concept. To Newtonians the Hubble shifting is happening only at far off places and has no influence on us, notwithstanding the fact shown by test results it is proven beyond argument that the Earth and the Moon is rapidly departing. That can't be mentioned for any such thinking might reflect badly on Sir Isaac Newton and that will not be tolerated under any circumstances.

In **An Open Letter on Gravity's Recipe**, a book that should have preceded the two parts of An Open Letter On gravity, but due to some problems with a publisher by the name of Cork Hill Publishing the book is not yet published, I explain in much detail that when we look at the sky we see time and not space. Space is time delay or the history of time is written in space. We are not looking at space but what we see is time in history and we have to think of light travelling that is coming our way in terms of time and not distance. Time wrote down events as it happened in a signature we call space and the ink used to write in is light. But light is not only the ink as it is also the paper (space) the language (gravity) and every other name we can think of is summarised by one term, and that is light. It is hardly surprising the Bible only uses the one correct term to describe the cosmos and that is by naming light.

The beginning

In the beginning, there was time Zero to moment Alpha. There has never been a Big Bang, as such. Everything is a variation of time duration in space. During the period of time Zero to moment Alpha the value of 1 second was equal in duration to about 1 000 billion, billion, billion, billion years, measured in geodesic space-time values that currently applies. It could be even billions times this duration because the value of time then was measured far beyond the speed of light, since light did not yet exist. We have no way to calculate the duration of time that applied during this geodesic space-time era, except to put it down to eternal.

Life story of the universe

There is no point in calculating time, because the duration of time is relevant to the space in which it is contained. As there was no actual **moment of time Zero** and no actual **moment Alpha,** there will be no **moment Omega**. In this **very last era**, which **is moment Omega** time will have a duration in space where the **last 100 000 billion years** would have the value of about **one hour. The last 100 000 million years** would have a comparable duration of **one minute** when compared to the value of the time duration as it is found in the current geodesic space-time era. Afterwards the last billion years would have a comparable value of one second, then one moment. There will always remain the last billion years as time fades away becoming eternity.

The most important fact that one derives from this is that each galaxy has its own space-time value and even in any given galaxy, time would have a different value from space to space. This would then differ from the geodesic space-time value. If science does not recognize this fact, the determined value of the Hubble Constant (Ho) would have the inaccuracy and diversity that applies currently.

Light is bordering the margin of the Universe at the point where space runs into time. It will be rather obvious to look for borders in space-time limitation at such a point. Light will use singularity to compliment space-time and at that point one will find what drives the Universe. It is not coincidental that the Bible says what is started with the Command "Let there be light" because light shows what was as it was when it is. At the point where light finds form we find the principle holding the Universe erect and that is what the author of the Bible in Geneses claimed thousands of years ago. If only science was not visually incapacitated and mentally blind to find truth, they would have seen it as easy as I am seeing it. Being and atheist is the worst form of stupidity. It would be impossible for someone with a small and under developed brainpower, which promotes and is inspiring a mentality such as an atheist, has to realise that it is light that preserves time. It is light that forms space in which time can duplicate a Universe. Light travels in time while preserving space as the past and we don't see light, but we merely see a message of what the past was when light left. Light conveys the past of time in the present and only tells historical facts. The future is dark and for that there is a Biblical reason. When The Creator commanded the Universe to be, He commanded light to be, but also He commanded the Universe to be in light and in light it is. When He commanded light to be, He not only commanded light to be, but He ordered a past within which the Universe could be and from that point on, light brought on the past while the future is still covered in darkness. Looking at light we see the history of time preserved in space as space-time, while time moves through the dark into the future. Light conveys the past while preserving the Universe for the future. There are as much towards as there are behind and we will forever hold a position in the centre of the middle, not only because reality permits, but also because singularity dictates.

This was

AN OPEN LETTER ON GRAVITY PART 1
ISBN
Introducing
MATTER'S TIME IN SPACE: THE THESIS   ISBN 0-9584410-8-1
© KOSMOLOGIESE EN ASTRONOMIESE TEGNIKA

BEST WISHES,

PETRUS (PEET) S. J. SCHUTTE

Firstly: The four Cosmic Pillars
The Roche limit in a more practical sense.

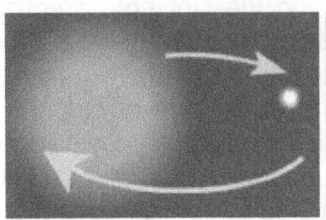  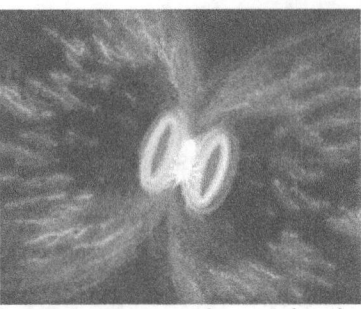

The formula $F = G(M_1.m_2)/r^2$ cannot explain the comic occurrence shown in the pictures above, but I can explain what is occurring in this instance and this occurrence connects directly to the Roche limit, as explained above. Not only does the Roche limit explain this phenomenon, but it ties directly to the Titius Bode principle, also inexplicable to the formula $F = G(M_1.m_2)/r^2$. According to the science formula of $F = G(M_1.m_2)/r^2$ the orbiting structures should collide with a bang, but instead they do the tango until one drops, but when dropping it still does not collide with the larger structure as would the formula $F = G(M_1.m_2)/r^2$ suggest.

The Roche limit is:
The region surrounding each star in a binary system, within which any material is gravitationally bound to that particular star. The boundary of the Roche lobes is an equipotential surface, and the lobes touch at the inner Lagrangian point, L_1, through which mass transfer may occur if one of the components expands to fill its lobe. It names after the French mathematician Edouard Albert Roche (1820-83).

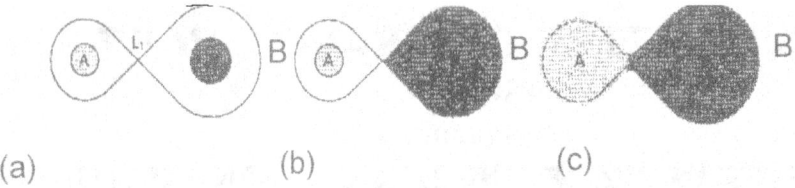

(a) (b) (c)

THE ROCHE LOBE: In a binary system, the Roche lobes of components A and B meet at the L_1 Lagrangian point. (a) In a detached system, neither star fills its Roche lobe. (b) In a semidetached system, one massive component, B, fills its Roche lobe. (c) In a contact binary, both components overfill their Roche lobes and share a common envelope.

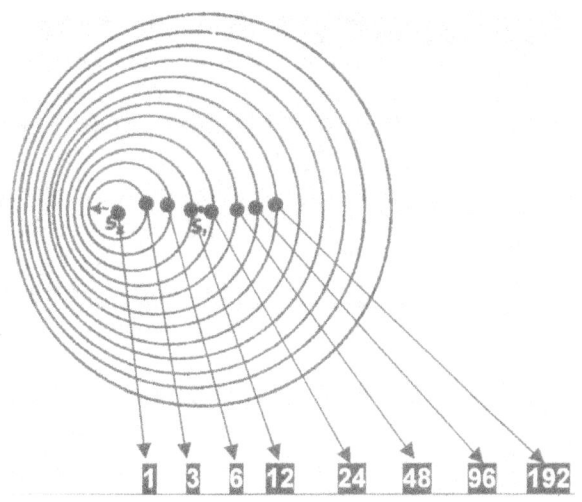

Planet	Mercury	Venus	Earth	Mars	Ceres	Jupiter	Saturn	Uranus	Neptune	Pluto
Bode's Law Distance	4	7	10	16	28	52	100	196	-	388
Actual Distance	3.9	7.2	10	15.2	28	52	95.4	191.8	300.7	394.6

Bode's Law:
A numerical sequence announced by J.E. Bode in 1772, which matches the distances from the Sun of the six planets then known. It is also known as the Titus-Bode law, as it was first pointed out by the German mathematician Johann Daniel Titius (1729-96) in 1766. It is formed from the sequence 0,3,6,12,24,48,96, and 192 by adding 4 to each number. The planets were seen to fit this sequence quite well – as did Uranus, discovered in 1781.

However, Neptune and Pluto do not conform to the 'law'. Bode's Law stimulated the search for a planet orbiting between Mars and Jupiter that led to the discovery of the first asteroids. It is often said that the law has no theoretical basis, but it does show how orbital resonance can lead to commensurability. The importance that becomes known is the sequence the Ties – Bode law saw in the number arrangement of 3; 6; 12; 24; 48; 96 etc. The incorrect application of the Titus Bode law lies in subtracting the figure of 3 from 10 leaving 7. The other way of reasoning is to add four each time to the first value of three starting with 3 and so on. The true significance of the Titus-Bode law is that it points directly to a circular growth of 7 stages. The 7 relating to 10 is a precise derogative of the Roche limit or the Roche limit is a precise derogative of the Titius Bode principle because the two systems interlink.

The formula $F = G(M_1.m_2)/r^2$ is unable to explain the principle discovered by Titius and later by Bode and it is not coincidental. From this one can arrive at the origins of the solar system.

LAGRANGIAN POINT: The Lagrangian points are five equilibrium points in the orbit of one body around another, such as a planet around the Sun.

LAGRANGE (-TOURNIER), JOSEPH LOUIS DE (1736-1813)

French mathematician, born in Italy. In celestial mechanics he studied perturbations and stability in the Solar System. He examined the three-body problem for the Earth, Moon and Sun (1764) and the motion of Jupiter's satellites (1766). In 1772 he found the particular solutions to the problem that give rise to the equilibrium positions now called Lagrangian points. Lagrange also studied the Moon's liberation. LAGRANGIAN POINT: One of five points at which small bodies can remain the orbital plane of two massive bodies; also known as liberation points. Three of the points lie on the line joining the two massive bodies: L_1 lies between them, while L_2 and L_3 have the two bodies between them. These three points are unstable, slight displacements of a body from then resulting in its rapid departure. the fourth and fifth points (L_4 and L_5) each form an equilateral triangle with the two massive bodies, 60° ahead of and behind the smaller body in its orbit around the larger one. A well-known example of bodies flying at the L_4 and L_5 Lagrangian points are the Trojan asteroids in Jupiter's orbit. Among Saturn's satellites, Telesto and Calypso lie at the L_4 and L_5 Lagrangian points in the orbit of the much larger Tethys. In similar fashion, tiny Helene precedes Saturn's satellite Dione, keeping 60° ahead of Dione. The Lagrangian points are named after the French mathematician J.L. de Lagrange, who first calculated their existence.

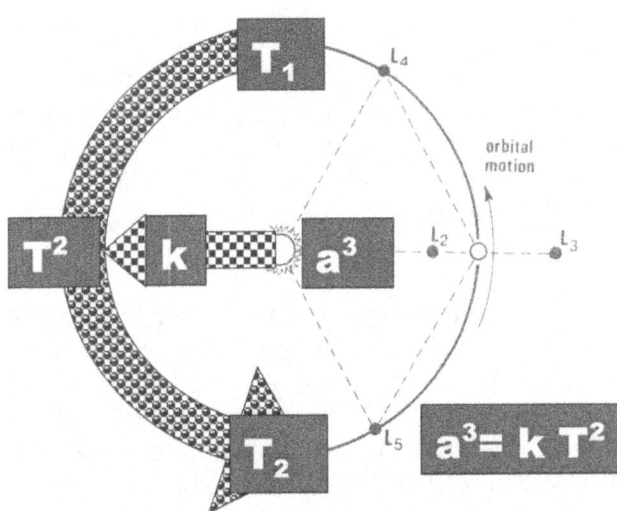

■■■■■■■■■:
The manner in which liquid relates to spinning solids by relative motion applying

Post Script
For your convenience and help to information you might be uncertain about

PLACED AS: L. I. T. F. B.

+........!......4......U......2......?

What the Christians find hard to admit, is that they destroyed thousands of years of knowledge through their insanity, and if it were not for the conservative efforts of Islam followers, like **AL-BATTÄNÏ, MUHANNAD IBN JÄBÏR** (858 – 929) an Arab astronomer, things would have been so much different today. He was born in modern Turkey and is known by the Latinized name Albetegnius. He was one of the first Arab astronomers to grasp the importance of accurate observations. He produced a set of tables, including a catalogue of star positions more accurate than in Ptolemy's Almagest that was to influence medieval European astronomers. Al-Battänï refined the values of the precession of the equinoxes, the obliquity of the ecliptic, and the length of the tropical year, and found that over the course of the year the Earth-Sun distance varies.

APOLLONIUS OF PERGE (c.262-c.190BC) who was a Greek mathematician was presumably born in the region, which now is the modern Turkey. He showed that the ellipse, parabola, and hyperbola are all curves formed by a plane intersecting a cone in different ways, i.e. that they are conic sections. The orbital path of an unperturbed body moving in a gravitational field follows one of these three curves, as would come to be appreciated by later astronomers such as E. Halley, who translated Apollonius's book Comics. Apollonius also originated the mathematical concept of motion based on epicycles and deferent's, later taken up by Hipparchus and Ptolemy to explain planetary motion.

Some sources say the **BIG BANG THEORY** is the most widely accepted theory of the origin and evolution of the Universe, but it seems as if this theory nowadays is the only accepted theory. According to the Big Bang theory, the Universe originated from an initial state of high temperature and density and has been expanding ever since. The theory of general relativity predicts the existence of a singularity at the very beginning, where the temperature and density were infinite. Most cosmologists interpret this singularity as meaning that general relativity breaks down at the Planck era under the extreme physical conditions of the very early Universe, and that the very beginning must be addressed using a theory of quantum cosmology. With our present knowledge of hinge-energy particle physics, we can run the clock back through the lepton era and hadron era to about a millionth of a second after the Big Bang, when the temperature was 10^{13}K. By adopting a more speculative theory, cosmologists have tried to push the model to within 10^{-35} of the singularity, when the temperature was 10^{28}K.

The Big Bang theory accounts for the expansion of the Universe; the existence of the cosmic background radiation; and the abundance of light nuclei such as helium, helium-3, deuterium, and lithium-7, which are predicted to have been formed about 1 second after the Big Bang when the temperature was 10^{10} K. The cosmic background radiation provides the most direct evidence that the Universe went through a hot, dense phase. In the Big Bang theory, the background radiation is accounted for by the fact that, for the first million years or so (i.e. before the decoupling of matter and radiation), the Universe was filled with plasma that was opaque to radiation and therefore in thermal equilibrium with it. This phase is usually called the primordial fireball. When the Universe expanded and cooled to about 3 00 K it became transparent to radiation. The discovery of the microwave background in 1965 resolved a long-standing battle between the Big Bang and its then rival, the steady-state theory, which cannot explain the blackbody form of the microwave background. Ironically, the term Big Bang was initially intended, to be derogatory and was coined by F. Hoyle, one of the strongest advocates of the steady state.

BIG BANG CHRONOLOGY		
ERA	TIME AFTER BIG BANG	TEMPERATURE
Planck era	0 to 10^{43}s	7 to 10^{34} K
Radiation era[a]	10^{-43}s to 30 000 years 10^{34}	10^4 K
Matter era[b]	30 000 years to present 10^4	3 K

[a] The time from about 10^{-6} or 10^{-5}s to about 1s or so is subdivided into the hydron and lepton eras.

[b] Includes the recombination epoch, which took place about 300 000 years after the Big Bang, at a temperature of about 3 000 K.

The BLUE SHIFT and the A Doppler shift of light towards the blue end of the spectrum, caused when the emitting source is approaching us.

The German mathematician and astronomer **BODE, JOHANN ELERT** (1747 – 1826), published a formula in 1772, now known as Bode's law, which yielded the approximate distances of the six known planets, from which he predicted the existence between Mars and Jupiter of an undiscovered planet. His major publication was Uranographia (1801), a comprehensive atlas of the entire sky showing over 17 000 stars and nebulae. For fifty years, he oversaw the publication of astronomical data in the Berlin Academy's yearbook. BODE'S LAW

A numerical sequence announced by **J.E. Bode in** 1772, which matches the distances from the Sun of the six planets then known. It is also known as the Titus-Bode law, as it was first pointed out by the German mathematician **Johann Daniel Titus** (1729-96) in 1766.

It is formed from the sequence 0, 3, 6, 12, 24, 48, 96, and 192 by adding 4 to each number. The planets were seen to fit this sequence quite well – as did Uranus, discovered in 1781. However, Neptune and Pluto do not conform to the 'law'. Bode's law stimulated the search for a planet orbiting between Mars and Jupiter that led to the discovery of the first asteroids. It is often said that the law has no theoretical basis, but it does show how orbital resonance can lead to commensurability.

Titius and later Bode took a series of numbers, beginning with zero and increasing geometrically in steps of 3: 0, 3, 6, 12, 24, 48, 96, 192, 384, 768. To each of these numbers he added 4, and then divided the result by 10. This gave him a series, which matched rather closely the distances from the Sun of the planets, which had always been known to man, when these distances were given in astronomical units.

Planets	Bode's Number	Add 4	Divide by 10	Distances in Astronomical Units
Mercury	0	4	0.4	0.39
Venus	3	7	0.7	0.72
Earth	6	10	1.0	1.00
Mars	12	16	1.6	1.52
Asteroids	24	28	2.8	2.65
Jupiter	48	52	5.2	5.20
Saturn	96	100	10.0	9.54
Uranus	192	196	19.6	19.19
Neptune	384	388	38.8	30.07
Pluto	768	772	77.2	39.52

In this one can trace the sphere (7) interacting with the 6 sides the cube has, but loses one side when coming into contact with one of the seven points forming the sphere. The five points hold both sides of the sphere thus forming ten in relation to seven In that we have space-time ($6 - 1 = 5 \times 2 = 10$) as well as the four adding time as a space-time developing factor. In this relation we find the foundation that brings about gravity as Π^2.

A close agreement with actual distance is shown here in the case of the nearer planets, particularly the asteroids, whose existence was not known until 1801. The legend is that Piazzi discovered Ceres, the first asteroid, when he was sweeping the skies at the proper distance between Mars and Jupiter, looking for something that might fit into Bode's law.

Uranus falls beautifully into place, and it was not found until 1781. Neptune is very much out of line, but Pluto fits much better into the Neptune slot than it does into its own place. Nothing is known to exist at the distance, which was assigned to Pluto in this strange "law." It is possible that Bode's law was nothing more than an exercise involving the distances of the planets then known, and that the agreement of Uranus and the asteroids is pure accident. Some recent theories of the formation of the solar system and of the distribution of its members, however, have tried to make use of Bode's figures.

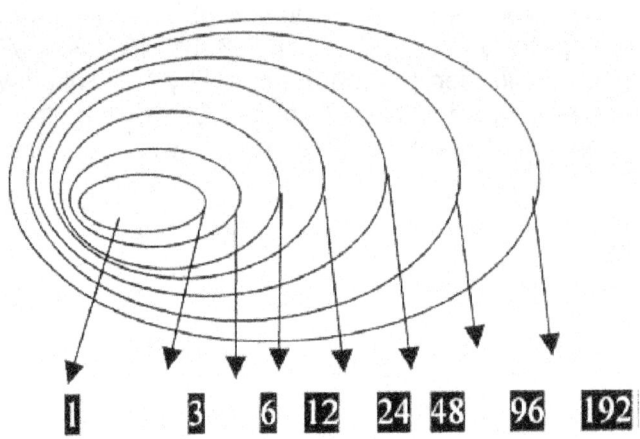

BODE'S LAW								
Planet	Mercury	Venus	Earth	Mars	Ceres	Jupiter	Saturn	Uranus
Bode's law distance	4	7	10	16	28	52	100	196
Actual distance (10^{-1} AU)	3.9	7.2	10	15.2	28	52	95	192

BOK, BARTHOLOMEUS ('BART') JAN (1906 – 83)
Dutch-American astronomer. His lifelong work was the study of the Milky Way, much of it in collaboration with his wife, Priscilla Book, née Fairfield (1896-1975). In particular, he investigated its structure, its distribution of stars, interstellar matter, and star-forming regions. In the 1930's, he discovered the objects now called Bok globules, and demonstrated that stellar associations are made up of young stars. In the early 1950s, with J.H. Oort and others, he pioneered the mapping of the Galaxy at radio wavelengths.

BOLTZMANN CONSTANT
(Symbol k or k_B) A constant that relates the kinetic energy of a particle in a gas to the temperature of that gas. It has the value $1.380\ 658 \times 10^{-23}$ joules per Kelvin. The particles can be molecules, atoms, ions, or electrons. The Boltzmann constant relates pressure, p, and temperature, T, by the equation $p = nkT$, where n is the number of particles per unit volume. In astrophysics, this equation is of importance in understanding the interiors and surface layers of stars, and the atmosphere of planets. The constant is named after the Austrian physicist Ludwig Edward Boltzmann (1844 – 1906).

BRAHE, TYCHO (1546 – 1601)
was the man who Johannes Kepler was associated with, in their joint study about the orbiting routes of the planets This Danish astronomer died before he could accomplish his life task. Nevertheless, Kepler completed the task. He was the most accomplished observer of the pre-telescope era, expert in

constructing instruments for making accurate naked-eye positional measurements. He first gained fame through his report (De nova stella, 1573) of the 1572 supernova in Cassiopeia. In 1576, he constructed Uraniborg, an observatory on the island of Hven in the Baltic (a second observatory, Stjerneborg, was built in about 1584). He calculated that the comet seen in 1577 had a highly elongated orbit, which would pass through several of the 'spheres' on which the planets were supposedly carried, and this led him to doubt the reality of Aristotle's planetary model. However, he rejected the heliocentric system proposed by Copernicus. In the Tychonic system, although the planets orbit the Sun, the Sun itself (and the Moon) revolves around a stationary Earth. Tycho made major contributions to the study of the Moon's orbit. In 1597 he moved to Prague, and employed J. Kepler as his assistant. Kepler later made use of Tycho's observations when deriving his laws of planetary motion.

BRANS-DICKE THEORY was the rival theory to Einstein's work; publish to serve as an alternative to Einstein's theory of general relativity. I will later in this article come to this, the most widely excepted theories of all, which attempts to incorporate Mach's principle. Among many other things, this theory predicts that the value of Newton's presumption of the so-called, gravitational constant, G, should change with time. The two persons responsible for introducing this theory were the American physicists Carl Henry Brans (1935 -) and R.H. Dicke.

BUTTERFLY DIAGRAM
A graph on which the latitudes of sunspots are plotted against time. It shows how spots migrate from higher latitudes (30 – 40° north or south) towards the equator (latitude 5° or so) throughout each sunspot cycle, in accordance with Spörer's law. The shape of the distributions, when plotted for both northern and southern hemispheres, resembles the wings of a butterfly.

CALLIPPUS (c. 370-c.300 BC.) was yet another Greek astronomer and mathematician who lived in the golden Greek Roman era. He modified the theory, which was held, by an earlier scientist by the name of Eudoxus. Adding to the existing Earth-centred spheres, extra ones for the Sun, Moon, and some of the planets changed Eudoxus's principle. This meant that there was a total to 34 different spheres. Aristotle further refined Callippus's model. Callipus calculated accurate lengths for the seasons, as measured between solstice and equinox.

CHANDRASEKHAR, SUBRAHMANYAN (1910 – 95)
Indian-American astrophysicist, born in modern Pakistan. He was the first to identify whit dwarf stars as end-products of stellar evolution, showing how a star collapses when there is no longer any radiation pressure to counteract its own gravity, producing degenerate matter. He calculated an upper limit (the Chandrasekhar limit) beyond which a star would enter a more dramatic final phase, presaging the existence of neutron stars. He studied how stars transfer energy by radiation in their atmospheres, publishing his findings in *Radiative Transfer* (1950). Chandrasekhar shared the 1983 Nobel Prize for physics with W.A. Fowler.

CHANDRASEKHAR LIMIT
The maximum possible mass of a degenerate star, above which it will be unable to support itself against the inward pull of its own gravity. For a star with no hydrogen content the limit is 1.44 solar masses, which is thus the maximum possible mass for a white dwarf. A degenerate star with a mass greater than this limit would collapse under gravity to become either a neutron star or a black hole. It is named after S. Chandrasekhar.

CHANDRASEKHAR-SCHÖNBERG LIMIT
The maximum mass of a star's helium core that can support the outer parts of the star against gravitational collapse, once the hydrogen as its centre has been exhausted. The limit is about 10 – 15% of the total mass of the star. If the mass of helium in the core exceeds this limit, the central parts collapse while the outer part expands rapidly to become a red giant. Calculations suggest that this happens only in massive stars. The limit is named after S. Chandrasekhar and the Brazilian astrophysicist Mario Shonberg (1916 -)

COPERNICUS, NICOLAUS (1473-1543) was according to the Anglo Americans a Polish churchman and astronomer although this is just another political inspired propaganda because his parents were both German (in Polish, Mikolaj Kopernigk). While he was completing his studies, he had realized that the Earth revolves around the Sun and not vice versa. Such a view was in that time, held to be heretical. As I pointed out in the first few articles, the Church regarded the geocentric world-view of Ptolemy as consistent with its doctrines. Copernicus set down his basic ideas around 1510 in the Commentariolus, which he circulated anonymously, because of the Islam link. In 1512-- 29 he conducted his study and concluded the observations that he needed to support his theory, while carrying out ecclesiastic and local administrative duties. In this time, he had to defend his mother in court on charges of witchcraft. In 1539, the Austrian astronomer and mathematician Georg Joachim von Lauchen (1514-74), known as Rheticus, became a pupil of Copernicus and began to spread his ideas. The published work was openly spread as the Copernican system, in spite of the life-threatening dangers connected with such a "crime", in 1543 in the book De revolutionibus orbium coelestium. However, the reality of a heliocentric Solar System was only commonly accepted, after the work of Galileo and J. Kepler.

CURVATURE OF SPACE-TIME
A property of space-time in which the familiar laws of geometry no longer apply in regions where gravitational fields are strong. In general, relativity the geometry of space-time is intimately connected with the distribution of matter. In a space of only two dimensions, such as a flat rubber sheet, Euclidean geometry applies so that the sum of the internal angles of a triangle on the sheet is 180°. If a massive object is placed on the rubber sheet, the sheet will distort and the paths of objects moving on the sheet will become curved. This is, in essence, what happens in general relativity.

In the simplest cosmological models, based on the Friedmann universe, the space-time curvature is simply related to the mean density of matter, and is described by a mathematical function called the Robertson-Walker metric. If a universe has a density greater than the critical density, it is said to have positive curvature, meaning that space-time is curved in on itself, like the surface of a sphere; the sum of the angles of a triangle drawn on a sphere is then greater than 180°. Such a universe has finite size and also finite lifetime; it is a closed universe. A surface of a saddle, on which the sum of the angles of a triangle is less than 180°. Such a universe would be infinite and would expand forever; it is an open universe. As Einstein-de Sitter universe has critical density and so is spatially flat (Euclidean) and infinite in both space and time.

CURVE OF GROWTH
A method for determining the temperature and chemical abundances of stellar atmospheres. From the different profiles of weak and strong absorption lines of a given element, a diagram can be constructed showing how the equivalent line widths increase (or 'grow') from weak to strong. The shape of this diagram is related to the total abundance of the element. Computer modelling of stellar atmospheres has now largely superseded the technique

DICKE, ROBERT HENRY (1916-97)
American physicist and astronomer. In 1961, he suggested that the gravitational constant varies with time. In 1964, with the Canadian-born American physicist Phillip James Edwin Peebles (1935 -) and others, he began to develop a hot Big Bang theory, independently of G. Gamow. The theory predicted the existence of the cosmic background radiation, discovered shortly after by A.A. Penzias and R.W. Wilson. He also invented the Dicke radiometer and Dicke switch, and in 1957 set out what has become known as the weak anthropic principle.

DE SITTER, WILLEM (1872-1934) a Dutch mathematician and astronomer, was an early supporter of the theory of relativity, assessing its implications for astronomy. From the relativity of time and space theory, he derived, what is now, the de Sitter universe, the first theoretical model of an expanding Universe. In other publications, he refined the orbits and masses of Jupiter's Galilean satellites, and showed the rotation of the Earth to be gradually slowing. The **DE SITTER UNIVERSE** is a model of an expanding universe in which there is no matter or radiation but a cosmological constant drives the expansion. W. De Sitter published this proposed theory in 1917 when he was strongly influence by the work and views of Albert Einstein. This model is physically unrealistic. The vantage this theory holds is that it introduces the idea that the real Universe is expanding at a certain precise rate. An expansion phase, which is very similar to that in the De Sitter model, also plays an important role in modern theories of the inflationary universe.

At the time the **DURAC COSMOLOGY** principal was firstly introduce it was receive with great scepticism by the scientific community of the day. Today more and more scientists are less sceptical about his views His is a cosmological theory built around the so-called large numbers hypothesis, which relates the fundamental constants of subatomic physics to large-scale properties of the Universe such as its age and mean density. It is due to the British mathematical physicist Paul Adrien Maurice Durac (1902-84). Durac's theory is not widely accepted, but it introduced ideas related to the anthropoid principle .My personal opinion is that Durac's ideas is dismissed rather on ground that does not suit the fashion trend of those in influence, rather than on grounds of in acceptance.

ALBERT EINSTEIN, (1879-1955) is worldwide viewed as the Mozart and Beethoven of the world of physics. This German-Swiss-American theoretical physicist's theories on the relativity still is widely not yet fully understood, by ALL members of science. They may think it helped to shape 20th century science but if implemented correctly, the profound implications that it will have on astronomy, would change all current views and wish full thinking. The special theory of relativity (published in 1905) arose out of the failure to detect the ether, and built on the work of the Dutch physicist Hendrik Antoon Lorenta (1853-1928) and the Irish physicist George Francis Fitzgerald (1851-1901). It yields the relation $E = mc^2$ between mass and energy, which was the key to understanding energy generation in stars. The general theory of relativity (announced 1915, published in expanded form 1916), which encompasses gravitation, assumes great importance in very large-scale systems, and rapidly had an impact on cosmology. Astronomy has furnished observational evidence to support these theories. Einstein produced no subsequent work of great significance, searching unsuccessfully for a theory that would link relativity with electromagnetic forces (a so-called grand unified theory).

Then the two wizards, mentioned above, joined forces and shared opinions to produce the **EINSTEIN-DE SITTER UNIVERSE**. They suggest a type of universe in which the mean density of matter matches precisely the critical density of matter through out the universe. Such a model will not actually collapse, but will expand forever with a continually decreasing expansion rate. This model lies on the dividing line between a closed Friedmann universe (which collapses) and an open Friedmann universe (which does not). This model has the mathematical virtue of simplicity, in that it is spatially flat. It names after A. Einstein and W. de Sitter.

ETHER
A hypothetical medium once thought to permeate all space, through which electromagnetic radiation supposedly travelled; formerly spelt aether. Based on this supposition, the Earth should move with respect to the ether, and it was predicted that the speed of light would vary when measured in different directions. Experiments in the 19th century (e.g. the Michelson-Morley experiment) failed to detect any such variation in speed. The ether is now regarded as unnecessary, since it is recognized that electromagnetic radiation can propagate through empty space.

EXPANDING UNIVERSE
Any model universe in which the space between widely separated objects is expanding. In the real Universe, neighbouring objects such as close pairs of galaxies do not move apart because their mutual gravitational attraction exceeds the effect of the cosmological expansion. However, the distance between two widely separated galaxies, or clusters of galaxies, will increase as the Universe expands.

Again, I as an outsider, get the distinct impression that in the **FRIEDMANN-UNIVERSE** there is such an enormous strive to pleas the opinions of as many persons as possible, whilst bringing it all under one umbrella. The, all inclusive, all pleasing model, portrays an expanding universe containing matter and radiation, but without a cosmological constant. Such a universe is both homogeneous and isotropic. There is, in fact, a family of such universes including those, which expand forever (open universe), those that eventually collapse (closed universe), and the particular example of the Einstein-de Sitter universe, which has a critical density of matter. The geometric of space-time in these universes is described by the Robertson-Walker metric and is, in the preceding examples, negatively curved, positively curved, and flat respectively. The Friedmann models, originated by the Russian mathematician Alexander Alexandrovich Friedmann (1888-1925), form the bases of the standard Big Bang theory.

The **GENERAL THEORY OF RELATIVITY**, is by far, the most widely excepted, best known and least understood universal theory of them all. I do realize that I have a very good chance of landing myself before the modern day version of the inquest for saying this, but I do sometimes wonder if Einstein himself understood his concept. At a certain place in this book, I mention the fact that I am the only one that grasps Einstein's theory. This I say, not because I am opinionated by myself, but to the contrary, I find it unacceptable that the modern day scientists go out of their way in ignoring the factuality of this theory. If I am wrong then why does scientist, go on and on about achieving the speed of light, when this fact contradicts the whole meaning of every thing Einstein ever tried to explain. In this, well known theory, A. Einstein introduced to the world in 1915 a first ever view that describes how the gravitational fields of matter affect space and time. The theory predicts that gravitational fields change the geometry of space and time, causing it to curve. This curvature is apparent in a number of ways. First, light is bent in a gravitational field, a prediction that was confirmed by photographic measurements of the positions of stars near the limb of the Sun made during a total solar eclipse in 1919. The same effect manifests itself in a delay in radio signals from distant space probes as the signals pass the limb of the Sun. The curvature of space near the Sun also causes the perihelion point of Mercury's orbit to move forward, by 43" per century more than predicted by Newton in his view on the orbit of gravity. In the orbits of pulsars in binary systems, the advance of perihelion can amount to several degrees per year.

Another effect predicted by general relativity is the red shift of light caused by gravity. This later proved correct in the demonstration of the red shift of lines, which is present in the spectra of the Sun and, more noticeably, white dwarfs. Other predictions of the general theory include the gravitational lens effect; gravitational waves; singularities; and the invariance of the universal gravitational constant, G. General relativity was developed from the principle of equivalence between gravitational and inertial forces.

The **GÖDEL UNIVERSE** is the most bizarre concept ever thought up by any one. It has Hollywood written all over it. This is a most outrageous and unusual cosmological model, which represents a rotating universe. This model possesses a number of strange mathematical features, including the fact that it allows time-travel to occur within it. It is due to the Austrian-American mathematician Kurt Gödel (1906 – 78).

GOULD, BENJAMIN APTHORP (1824 – 96) was an American astronomer, which, in 1849 he founded the Astronomical Journal. In 1870, he moved to Argentina, founding the National Observatory at Córdoba, where he initiated the Córdoba Durchmusterung, a southern equivalent of the catalogue produced by F.W.A. Argelander for the northern stars. The name Gould's Belt is in his honour.

GOULD'S BELT
A band of hot, bright stars (types O and B) forming a circle around the sky. It represents a local structure of young stars and interstellar material tilted at about 16° to the galactic plane. Among the most prominent components of the belt are the bright stars in Orion, Canis Major, Puppis, Carina, Centaurus, and Scorpius, including the Sco-Cen Association. The belt has a diameter of about 3 000 l.y. (about one-tenth the radius of the Galaxy), and the Sun lies within it. Viewed from Earth, Gould's Belt projects below the plane of our Galaxy from the lower edge of the Orion Arm, and above the plane in the opposite direction. The belt is estimated to be about 50 million years old, but its origin is unknown. It is named after B.A. Gould, who established its existence in 1879.

To curb the confusion that presented it in the accumulation of the various theories a new theory was officially accepted and named the **GRAND UNIFIED THEORY (GUT).** So many new, information, became apparent, because of the development in technology, an attempt to describe the weak and strong nuclear forces and electromagnetism in a single mathematical theory. The unification of the weak force with electromagnetism was achieved in the electro weak theory. Before about 10^{-12} seconds after the Big Bang, by which time the universe had cooled to about 10^{15} K, the electromagnetic and weak interactions acted as a single physical force; in the cooler temperatures since then, they have been distinct. Attempts to unify the electro weak force with the strong nuclear force have been only partially successful. It is thought that the temperature for their unification is of the order of 10^{27} K, which occurs only 10^{-36} s after the Big Bang. Particles surviving to the present day from this phase are possible candidates for non-baryonic dark matter. Unification of the GUT interaction with gravity may take place at higher energies still, but there is no satisfactory theory, which unifies all four physical forces. Such a theory would be called a theory of everything (TOE).

GRAVITATIONAL COLLAPSE
The collapse of a body that is unable to support itself against its own gravity. Gaseous bodies undergo such collapse if they are not hot enough for their gas pressure to balance gravity. This can happen in the early stages of star formation, or when nuclear burning ceases in a star's core. The time taken for such collapse decreases rapidly with increasing density, varying from about 100 000 years for the birth of a new star to less than a second for the formation of a neutron star. Star clusters may undergo a similar collapse if the random motions of their constituent stars are

insufficient to offset gravitational effects, either during their formation or at an advanced stage of their evolution.

GRAVITATIONAL REDSHIFT
The red shift of light or other electromagnetic radiation caused by a strong gravitational field; also known as the Einstein shift. It arises because radiation loses energy as it passes out of the gravitational field of the emitting body. Therefore, the frequency of the radiation decreases and its wavelength is shifted to the red end of the spectrum. The red shift at wavelength λ is given by $Gm\lambda/c^2r$, where m is the mass of the body, r is the distance of the emitting region from the centre of mass, c is the speed of light, and G is the universal gravitational constant. A gravitational red shift has been observed in the light from some white dwarfs, and would result in the rapid fading out of a black hole in the process of formation as seen from outside.

GRAVITATIONAL WAVE
A wave-like motion in a gravitational field, produced when a mass is accelerated or otherwise disturbed. Gravitational waves travel through space-time at the speed of light, and their amplitude is proportional to the rate of acceleration of the body producing them. The strongest sources are those with the strongest gravitational fields although the waves, like the force of gravity itself, would be very weak. Gravitational waves have not yet been observed directly. However, the decay in the orbital period of the binary pulsar PSR 1913 + 16 is attributed to loss of energy through gravitational waves.

GRAVITON
A hypothetical particle or quantum of gravitational energy, predicted by the general theory of relativity. Gravitons have not been observed but are predicted to travel at the speed of light and to have zero rest mass and charge. A graviton is the gravitational equivalent of a photon.

GRAVITY ASSIST
The technique of using the gravitational field and orbital velocity of a planet to alter a spacecraft's trajectory and velocity; also known as a gravitational slingshot. As the spacecraft makes a close fly-by of a planet, its direction of travel is altered and it picks up additional speed from the planet's orbital velocity. The technique was first used by Mariner 10, which flew past Venus on its way to Mercury in 1974. The two Voyager probes made fly-bys of Jupiter, considerably shortening the time they took to reach Saturn. Voyager 2 subsequently used gravity assists from Saturn and Uranus to take it to Neptune. Other probes to use gravity assists were Giotto, Galileo and Ulysses.

GRAVITY GRADIENT
The direction of a gravitational field at a point within the field. Near a massive body such as a star or planet, the gravity gradient points to the body's centre. An elongated object in orbit about the body will revolve with its long axis pointing towards the body's centre.

In example, the Moon's longest axis lies along a line towards the body's centre. Artificial satellites can be oriented in orbit by making use of the gravity gradient.

H_2O maser
A maser source in which the water (H_2O) molecule is excited to maser action. They are the most widely distributed of all the cosmic masers. There are many different H_2O maser lines. The first to be discovered, in 1969, was the powerful line at 22.2 GHz (13.5 mm) in the Kleinmann-Low Nebula in Orion. Other H_2O lines at higher frequencies are difficult to observe with ground-based radio telescopes because of strong absorption by water vapour in the Earth's atmosphere. Water masers are found in star-forming regions, circum stellar envelopes, comets, and in the nuclei of some active galaxies in the form of mega masers.

Another link in the long chain of the period that blossomed during early Roman, late Greek period was **HIPPARCHUS OF NICAEA** (c.190-c.120BC) an astronomer, geographer, and mathematician, born in modern Turkey. He put Greek astronomy on a more scientific footing, introducing arithmetic and early trigonometric methods Without any aid of lenses his many accurate astronomical observations resulted in, as far as I know, a catalogue of 850 stars, which included their coordinates and dividing them into six magnitude classes. Ptolemy incorporated the catalogue and other findings by Hipparchus in the Almagest. Hipparchus made surprisingly accurate measurements of the precession of the equinoxes, the length of the year, and (from observations of eclipses) the Moon's distance. He may have been the inventor of the astrolabe.

The **HOT BIG BANG** is a logic conclusion derived from the **BIG BANG THEORY** and shares the same principles. The **BIG BANG THEORY** was the brainchild of Father Lemaitre's a Catholic priest who saw the universe growing the same way as an egg does.

Father **LE MAÎTRE, GEORGE ÉDOUARD** (1894-1966) was a Belgian priest and cosmologist who was the first person to embrace the fact that the universe expanded from an infant stage. His model of an expanding Universe (1927) was superior to that of W. de Sitter in that it took into account mass, gravitation and the curvature of space. Similar models had been proposed in the early 1920s by the Russian mathematician Alexander Alexandrovich Friedmann (1888-1925). Lemaître argued further (1931) that the quantum theory supported an origin in the explosion of a 'primeval atom' or 'cosmic egg' into which was originally concentrated all mass and energy. As modified by A.S. Eddington, Lemaître's model provided the springboard for G. Gamow's Big Bang theory.

LE MAÎTRE'S UNIVERSE
A model of the universe containing a cosmological constant term, named after G.E. Lemaître. In this model, space has a positive curvature but expands forever. The Lemaître universe is both homogeneous and isotropic. The most interesting aspect of such a universe is that it undergoes a so-called coasting phase in which the cosmic scale factor is roughly constant with time.

This theory was an evolutionary process to act as an alternative term for the standard Big Bang theory. The word 'hot' was initially used to distinguish it from a rival theory, which had a cold initial phase. The existence of the cosmic background radiation requires that the Universe must have been hot in the past if the Big Bang picture is correct.

HOYLE, FRED (1915 -)
English astrophysicist and cosmologist. In 1948, with H. Bondi and T. Gold, he proposed the steady-state theory of the Universe in which matter is continuously created. Subsequently abandoned by most astronomers in favour of the Big Bang (so named from a dismissive remark by Hoyle), the steady-state theory nevertheless stimulated much important astrophysical research. Particularly significant was the work by Hoyle, W.A. Fowler, and G.R. and E.M. Bridge on nucleosynthesis in stars. As well as his suggesting for example those viruses and perhaps other life forms have been brought to Earth by comets. He has also been a popularizer of astronomy.

The man that (to my humble opinion) took cosmology into a new dimension was **HUBBLE, EDWIN POWELL** (1899-1953), the American astronomer. He first studied nebulae, concluding in 1917 that the spiral-shaped ones (which we now know as galaxies) were different in nature from diffuse nebulae, which he found to be gas clouds illuminated by stars. From 1923, using the 100-inch (2.5-m) telescope at Mount Wilson Observatory, he resolved the outer regions of the spiral nebulae M31 and M33 into star, identifying over 30 Cepheid variables in them. This proved that such 'nebulae' were truly independent star systems like our own – other galaxies. In 1925, he devised the so-called tuning-fork diagram of galaxies, dividing them into ellipticals, spirals, and barred spirals, which he believed to indicate an evolutionary sequence. By 1929, Hubble had good distance measurements for over twenty galaxies, including members of the Virgo Cluster. By comparing distances with their velocities, as revealed by the red shifts in their spectra, he concluded that galaxies were receding with speeds that increased with their distance, a relationship known as the Hubble law. This was powerful evidence that the Universe is expanding. The dynamics of his work was so far reaching everybody (including Einstein had to revise their theories to accommodate his findings. His findings are the most disputed, undisputed observations in all of history. The **HUBBLE CLASSIFICATION** is a widely used system for classifying galaxies according to their visual appearance, illustrated on the tuning-fork diagram. The sequence is based on three criteria: the relative sizes of the central bulge of stars and the flattened disk; the existence and character of spiral arms; and the resolution of the spiral arms and / or disk into stars and H II regions. The system was originated by E.P. Hubble.

The sequence starts with round elliptical galaxies (EO) showing no disks. Increasing flattening of a galaxy is indicated by a number which is calculated from $10(a-b)/a$, where, a, and b, are the major a minor axes as measured on the sky. No elliptical is known that is flatter than E7. Beyond this, a clear disk is apparent in the ventricular or SO galaxies. The classification then splits into two parallel sequences of disk galaxies showing spiral structure: ordinary spirals, S, and barred spirals, SB. The spiral types are subdivided into Sa, Sb, Sc, Sd (Sba, SBb, SBc, SBd for barred spirals). With each successive subdivision, the arms become less tightly wound (but more easily resolvable into stars and H II regions), and the central bulge becomes less dominant. Two types of irregular galaxy are defined. Irr I galaxies show rather amorphous, irregular structure with perhaps a hint of a spiral arm or bar, and can be placed at the far end of the spiral sequence. Irr II galaxies are sufficiently unusual to defy assignment to any of the other types, although this category encompasses only about 2% of bright or moderately bright galaxies in the nearby Universe. The original, erroneous idea that the sequence might be an evolutionary one led to the ellipticals refers to, as early-type galaxies, and the spirals and Irr I irregulars as late-type galaxies. Colour and amount of interstellar material vary systematically along the Hubble sequence: ellipticals are red and contain little interstellar gas or dust, whereas late spirals and Irr I galaxies are blue, with significant amounts of interstellar material. The relatively faint dwarf spheroidal galaxies were not recognized as a separate type in the Hubble classification. Some variants of the Hubble classification use plus and minus signs to subdivide classes, so that Sa^+ is later than Sa, but earlier than Sb^-. The importance of the **HUBBLE CONS**TANT is still to this day, underestimated. This "constant" is well explained, for the first time, I might add, in this book. The Symbol H_o is the figure that relates the speed of an object's recession in the expanding Universe to its distance in the Hubble law. It represents the current rate of expansion of the Universe. This important cosmological parameter is usually measured in units of kilometers per second per megaparsec. In the Big Bang theory, H_o varies with time and it is therefore more properly known as the Hubble parameter. Its value is not accurately known but is thought to lie between 50 and 100 km/s/Mpc, recent research tending to favour values towards the lower end of this range. In the **HUBBLE DIAGRAM,** a graph plots either the redshift, or velocity of recession of galaxies against their apparent magnitude or distance from us. The Hubble law appears in the form of a straight line on such a plot. The original diagram, presented by E.P. Hubble in 1929, was the first indication that the Universe is expanding. The Hubble diagram is now mainly used to test the geometry of the Universe, since at large distances any departures from the simple linear form of the Hubble law should show up as a curve. The **HUBBLE FLOW** is the general outward motion of galaxies resulting from the uniform expansion of the Universe. All motions lie in a radial direction from the observer, and the velocities are proportional to the distance of the galaxies. The real pattern of galaxy motions is not exactly of this form, particularly close to us, because of the mutual gravitational interaction between galaxies; some nearby galaxies are even moving towards the Milky Way. At large distances, however, the discrepancies are small compared with the Hubble flow. All these findings are incorporated in the **HUBBLE LAW**, which is the mathematical equation of the principle law that governs the expansion of the Universe. According to the law, the apparent recession velocity of galaxies is proportional to their distance from the observer. In mathematical terms, $v = H_o r$, where v is the velocity, r the distance, and H_o the Hubble constant. The law was put forward in 1929 by E. P. Hubble.

HUBBLE RADIUS
A distance defined as the ratio of the velocity of light, c, to the value of the Hubble constant, H_o, This gives the distance from the observer at which the recession velocity of a galaxy would equal the speed of light. Roughly speaking, the Hubble radius is the

radius of the observable Universe. Depending on the precise value of the Hubble constant, the Hubble radius lies between 9 and 18 billion l.y. This data is the basis on which the age of the universe depends and is the **HUBBLE TIME**. The time required for the Universe to expand to its present size, assuming that the Hubble constant has remained unchanged since the Big Bang. It is defined as the reciprocal of the Hubble constant, $1/H_o$. Depending on the precise value of the Hubble constant, the Hubble time is between 9 and 18 billion years. In the standard Big Bang theory, the actual age of the Universe is always less than the Hubble time, because the expansion was faster in the past.

KANT, IMMANUEL (1724 – 1804) was the German philosopher, which proposed a cosmogony, published in 1755, in which the Solar System forms, via a disk, which condensed out of primordial material. The Solar System was part of a larger system (what we would now call a galaxy), and many of the nebulae seen by astronomers were in fact other galaxies, which he termed island universes. Kant was influenced by I. Newton's theories, which he termed island universes. Kant was, as everyone ells up to now, influenced by I. Newton's theories and by the English philosopher Thomas Wright of Durham (1711-86).

The Scottish physicist, Lord **W. C KELVIN** (1824-1907) was born in Ireland. He originated the thermodynamic temperature scale, and considered the consequences of energy dissipation in the Universe. Kelvin made one of the first scientific attempts at estimating the Earth's age, based on known cooling rates of materials, although his result (20 – 400 million years) was far too low. He also calculated the solar constant. He produced the **KELVIN SCALE** as a temperature scale in which the zero point is defined to be equal to $-273.16°$ Celsius. This zero point is also known as absolute zero. The thermodynamic temperature is expressed in Kelvin, symbol K.

The German mathematician and astronomer **KEPLER, JOHANNES** (1571-1630) German mathematical and astronomer became Tycho Brahe's assistant in Prague in 1600 A. D. where he undertook to complete the tables of planetary motion Tycho had begun. Kepler first calculated the orbit of Mars. He spent much time trying to reconcile Tycho's accurate observations of the planet with a circular orbit, but concluded (in Astronomia nova, published in 1609) that Mars moved instead in an elliptical orbit. Thus, he established the first of his laws of planetary motion. A theory that the Sun controlled the planets by a magnetic force led him to the second and third of his laws, which were published as part of his treatise on theoretical astronomy, Epitome astronomiae Coernicanae (1618-21). The Rudolphine Tables (named after Tycho's patron, the Holy Roman Emperor Rudolph II) of planetary motion appeared in 1627 and were still in use in the 18th century. Kepler also wrote De Stella nova, on the supernova of 1604 and Diptirce on optics and the theory of the telescope. The overall view followed in this book **Matter's Time in Space** places the true significance of his work in true contents. In **KEPLER'S EQUATION** is the equation that relates the eccentric anomaly of a body in an elliptical orbit to its mean anomaly. The equation is $E - e \sin E = M$., where E is the eccentric anomaly, M the mean anomaly, and e the eccentricity of the orbit. It is important as one of the mathematical relations enabling the position of a planet about the Sun, or a satellite about is planet, to be calculated from the orbital elements for any time. However this only relates to the solar system, and **KEPLER'S LAWS** only apply in the contents of the solar system. The three laws governing the orbital motions of the planets, discovered by J. Kepler is as follows: The first law states that the orbit of a planet is an ellipse with the Sun at one focus of the ellipse. The second law states that the radius vector joining planet to Sun sweeps out equal areas in equal times. The third law states that the square of the orbital period of each planet in years is proportional to the cube of the semi major axis of the planet's orbit. The first law gives the shape of the planet's orbit; the second describes how the planet must continuously vary its speed as it follows its orbit, moving fastest at perihelion and slowest at aphelion. The third law gives the relationship between the planets' average distances from the Sun and their periods of revolution. Instead of placing, the true value to Kepler's laws I. Newton placed his own interpretation to Kepler's laws, and in doing this, he wilfully destroyed the principle working of the Creation. Through Newton's tunnel vision, he applied his own miss interpretations to the correct presumptions of Kepler. Newton reduced the implication that Kepler findings hold by introducing to the law of gravitation. he then went about and changed it to three laws of motion. I. Newton generalized Kepler's first law, verified the second law, and showed that the third law should be amended to the form; $4 \pi^2 a^3 / T^2 = G (m + m_p)$. In this, the value of T and a are the period of revolution and semi major axis of the orbit of a planet of mass m_p about the Sun of mass m, and G is the gravitational constant. The major aim of this book is to correct these misgivings of Newton.

KERR'S view of the **BLACK HOLE** is one of a rotating black hole, as distinct from a non-rotating Schwartzschild's black hole. Black holes are expected to rotate rapidly, since the stars that formed them would have been rotating; hence, they will be Kerr black holes. Several consequences arise from the addition of rotation to a black hole. First, the event horizon becomes elliptical, and its surface area, become less than that for a static black hole of the same mass. If the black hole were rotating sufficiently quickly, the area of the event horizon would reduce to zero, leaving the central singularity visible from outside (a naked singularity). Second, there is a region around a rotating black hole, the ergosphere, in which objects are forced to spin around the black hole. The outer edge of the ergosphere is the static limit. Third, a new, inner event horizon forms, and it becomes possible to travel through the black hole, and emerge into a new universe or perhaps another part of our own Universe, through this second event horizon. Rotating black holes are named after the New Zealand mathematician Roy Patrick Kerr (1934 -), who first described their properties in 1963. Rotating black holes with electric charge are called Kerr-Newman black holes, but in practice, black holes are unlikely to have any significant electric charge.

KINETIC ENERGY
The energy possessed by a body by virtue of its motion in space. It is equivalent to the work that would be done if the moving body were brought to rest. Kinetic energy is equal to $\frac{1}{2}mv^2$, where m is the mass of the body and v is its velocity. A rotating body has kinetic energy $\frac{1}{2}I\varpi^2$, where I is its momentum of inertia and ϖ is its angular velocity.

KIRCHOFF, GUSTAV ROBERT (1824-87)
German physicist. With the chemist Robert Wilhelm Bunsen (1811-99) he established the principles of spectral analysis. In 1859 he reasoned that the Fraunhofer lines in the solar spectrum indicated that light from the photosphere was being absorbed at those wavelengths by the Sun's atmosphere. Furthermore, he realized that the Fraunhofer D lines were produced by sodium in the Sun's atmosphere, and other Fraunhofer lines would therefore reveal which other elements were present in the Sun. From then on, astronomical spectroscopy developed rapidly in the hands of others such as P.A. Secchi in Italy and W. Huggins in England.

KIRCHOFF'S LAWS
Three laws concerning spectra, stated in 1859 by the German physicist G.R. Kirchoff:
1. A solid, liquid, or gas under high pressure, when heated to incandescence, produces a continuous spectrum.
2. A gas under low pressure, but at a sufficiently high temperature, produces a spectrum of bright emission lines.
3. A gas at low pressure (and low temperature) , lying between a hot continuum source and the observer, produces an absorption line spectrum, i.e. a number of dark lines superimposed on a continuous spectrum.

KUIPER, GERARD PETER (1905-73)
Dutch-American astronomer. In a search for new planetary satellites he discovered Miranda (in1948, orbiting Uranus) and Nereid (in 1949, orbiting Neptune). His spectroscopic studies revealed methane in the atmospheres of Uranus and Neptune. He also found methane bands in Titan's spectrum, demonstrating that the satellite has an atmosphere. He suggested the existence of the Kuiper Belt as the source of short-period comets. Kuiper was advisor on many American lunar and planetary missions, and proposed the idea of flying infrared telescopes on board high-altitude aircraft, which led to the Kuiper Airborne Observatory.

KUIPER BELT
A region of the outer Solar System containing an estimated 10^7-10^9 icy planetesimals, or comet nuclei. The Kuiper Belt is an inner, flattened extension of the Oort Cloud. It lies more or less in the same plane as the planets and extends outwards from around 30 AU (the orbit of Neptune) to perhaps 1 000 AU. Such a vast reservoir of comets beyond Neptune was first proposed in 1951 by G.P. Kuiper. In 1992, the British-born American astronomer David Clifford Jewitt (1958 -) and the Vietnamese-born American astronomer Jane Luu (1963 -) discovered the first Kuiper Belt object, 1992 QB_1. This has a diameter of about 200km, semi major axis 44.0 AU, orbital period about 296 years, perihelion 40.9 AU, aphelion 47.1 AU, and inclination 2°.2. Since than, dozens more have been found. The Kuiper Belt could be the source of most periodic comets. Unusual objects such as Chiron and Pholus may have originated in the Kuiper Belt.

MAGELLANIC CLOUDS
The two irregular galaxies that are satellites of our own Galaxy, easily seen with the naked eye in the southern hemisphere like detached portions of the Milky Way. They

are named after the Portuguese explorer Ferdinand Magellan (1480 –1521), who described them during his voyage around the world. Both Clouds are believed to orbit our Galaxy in a plane nearly perpendicular to its disk, and may eventually spiral into the Galaxy.

MATTER ERA
In the Big Bang theory, the era that started when the gravitational effect of matter began to dominate the effect of radiation pressure. Although radiation is massless, it has a gravitational effect, which increases with the intensity of the radiation. Moreover, at high energies, matter itself behaves like electromagnetic radiation because it is moving at a speed close to that of light. In the very early Universe, the expansion rate was dominated by the gravitational effect of radiation pressure but, as the Universe cooled, this effect became less important than the gravitational effect of matter. Matter is thought to have become predominant at a temperature of around 10^4K, roughly 30 000 years after the Big Bang. This marked the start of the matter era.

MEAN DENSITY OF MATTER
The density of material that would be obtained if all the matter contained in galaxies were smoothed out across the universe. Although stars and planets have densities greater than the density of water (about 1 g/cm^3), the cosmological mean density is extremely low (less than 10^{-29} g/cm^3), or 10^{-5} atoms/cm^3) because the Universe consists mostly of virtually empty space between galaxies. The mean density of matter determines whether the Universe will continue to expand.

NEWTON, ISAAC (1642-1727)
English physicist and mathematician. He developed his principal theories of gravitation, optics and mathematics in 1665 and 1666. In 1668 he made the first working reflecting telescope. Most of his work remained unpublished for long periods, partly because of criticisms by c. Huygens and the English scientist Robert Hooke (1635-1703) of his early work on the corpuscular theory of light. However, in 1684 E. Halley persuaded him to organize his work on the celestial mechanics of the Solar System, which was published as the Principia. Newton's other major work, Opticks, was not published until 1704. It contains his corpuscular theory of light, and the theory of the telescope. His greatest mathematical achievement was his invention of calculus, independently of the German mathematician Gottfried Wilhelm Leibniz (1646-1716). His profound influence on physics and astronomy is reflected in the phrase 'Newtonian revolution'.

NEWTON'S LAWS OF MOTION
Three laws published in 1687 by I. Newton concerning the motion of bodies.
1. A body continues in a state of uniform rest of motion unless acted upon by an external force.
2. The acceleration produced when a force acts is directly proportional to the force and takes place in the direction in which the force acts.
3. To every action there is an equal and opposite reaction.

OORT CLOUD

A roughly spherical halo of comet nuclei surrounding the Sun out to perhaps 100 000 AU (over one-third of the distance to the nearest star). Its existence was proposed in 1950 by J.H. Oort to account for the fact that new comets approach the Sun on highly elliptical orbits at all inclinations. The Oort Cloud remains a theoretical concept, since we cannot currently detect inert comets at such great distances. The cloud is estimated to contain some 10^{12} comets remaining from the formation of the Solar System. The most distant members are loosely bound by the Sun's gravity. There may be a greater concentration of comets relatively close to the ecliptic, at 10 000 – 20 000 AU from the Sun, extending inwards to join the Kuiper Belt. Oort Cloud comets are affected by the gravitational influence of passing stars, occasionally being perturbed into orbits, which take them through the inner Solar System.

OPEN UNIVERSE

A universe, which expands forever and has an infinite lifetime. A Friedmann universe with a density less than the critical density is an example. It is not yet known whether our Universe is of this type.

OPPENHEIMER-VOLKOFF LIMIT

The maximum mass that a neutron star can have without it being overwhelmed by its own gravity. Calculations put this between 1.6 and 2 solar masses, although the exact figure is uncertain. A neutron star with a mass greater than this is expected to collapse further into a black hole. The limit is named after the American physicist (Julius) Robert Oppenheimer (1904-67) and the Russian-born Canadian George Michael Volkoff (1914 -).

PARSEC

A measure of astronomical distance. A parsec is the distance at which the Earth and Sun would appear to be 1 second of arc apart. A star at this distance would therefore show a shift in position (Parallax) of 1 arc second in the sky as observed from opposite sides of the Earth's orbit. (Actually, no star is quite this close.) The origin of the term parsec – a contraction of *par*allax of one *sec*ond – is attributed to the English astronomer Herbert Hall Turner (1861 – 1930), an expert on measuring star positions. A thousand parsecs is termed a *kiloparsec*; a million parsecs is a *megaparsec*. This distance of a star in parsecs is the reciprocal (inverse) of its parallax in seconds of arc. One parsec is 3.2616 light-years, 206,265 astronomical unites, 19.174 trillion miles, or 30.857 trillion km.

PENROSE PROCESS

A process for extracting energy from a rotating black hole (i.e. a Kerr black hole). The process requires sending a mass into a trajectory in the ergosphere around the black hole, against the direction of rotation of the black hole. While inside the ergosphere the mass splits into two parts, one of which enters the black hole while the other escapes. Given a suitable trajectory, the emerging fragment may possess a total energy (i.e. rest mass plus kinetic energy) greater than the total energy of the mass that went in. The extra energy has been extracted from the rotational energy of the black hole, which must therefore slow down slightly. The process is named after the English mathematician Roger Penrose (1931 -), who discovered it in 1969.

PLANCK CONSTANT
(Symbol h) A constant that relates the energy of a photon to its frequency. It has the value 6.62076×10^{-34} Js. It is named after the German physicist Max Karl Ernst Ludwig Planck (1858 – 1947)

PLANCK ERA
In the Big Bang theory, the fleeting period between the Big Bang itself and the so-called Planck time when the Universe was 10^{-43} s old and the temperature was 10^{34}K. In this period, quantum gravitational effects are thought to have dominated. Theoretical understanding of this phase is virtually non-existent. It is named after Max Planck (1858-1947)

PLANCK'S LAW
A mathematical description of the energy radiated at different wavelengths by a black body: E = hf, where E is the energy of a photon and f its frequency. It was formulated in 1900 by Max Planck (1858-1947), who realized that energy is radiated in discrete packets, which he called quanta, and it formed the basis of quantum theory. The quantum of light is a photon, the energy of which depends on its wavelength.

PTOLEMAIC SYSTEM
The ancient Greek geocentric model of the Solar System, as described by Ptolemy . It may be traced back through the work of, for example, Hipparchus, Apollonius, Callippus and Eudoxus. The Earth is placed at the centre of the Universe, and around it revolve the Moon, Mercury, Venus, the Sun, Mars, Jupiter and Saturn. Beyond Saturn is the sphere of the fixed stars. In the basic model, each body moves along the circumference of a small circle, the epicycle, whose centre in turn follows the circumference of a larger circle, the deferent, centred on the Earth. In later refinements, Ptolemy introduced two points equally spaced on either side of the Earth: the eccentric and the equant. The centre of the epicycle revolved around the eccentric, not the Earth, and the orbiting body moved uniformly with respect to the equant. As a computational device the Ptolemaic system predicted planetary movements, including their retrograde motion, tolerably well, and survived with minor amendments until displaced by the Copernican system in the 16th century.

PTOLEMY (CLAUDIUS PTOLEMAEUS) (2nd century AD)
Egyptian astronomer and geographer. He produced the Almagest , a compendium of contemporary astronomical knowledge, drawing on writers, such as Plato and Hipparchus, whose works were kept in the great library at Alexandria. His Ptolemaic system was a geocentric model of the Universe. Highly contrived as it now appears, it successfully accounted for the observed apparent motions of the planets, and remained largely unquestioned until the 16th century, when it was challenged by N. Copernicus. Ptolemy's Geography enjoyed a similar period of dominance (it convinced Columbus that he could sail westwards to India); his Tetrabiblos was an astrological treatise.

ROBERTSON-WALKER METRIC
A mathematical function describing the geometry of space-time in a model which incorporates the cosmological principle. In general, a metric relates physical distances or intervals between events separated in space and / or time to the coordinates used to describe their position. General relativity deals with four-dimensional space-time in which the separation between space and time coordinates is not obvious. In a homogeneous and isotropic cosmology, however, it is possible to define a unique time coordinate, called cosmic time, and three spatial coordinates. The Robertson-Walker metric is the most general possible four-dimensional metric function compatible with homogeneity and isotropy. In general, it describes a curved space which is either expanding or contracting with cosmic time. It is named after the American mathematician and cosmologist Howard Percy Robertson (1903 – 61) and the English mathematician Arthur Geoffrey Walker (1909 -)

SCHWARTZSCHILD BLACK HOLE
A non-rotating black hole with no electrical charge. This is the simplest case of a black hole, predicted by K. Schwartzschild in 1916 but is unlikely to be found in reality. The most likely form for black holes is the rotating Kerr black hole.

SCHWARTZSCHILD KARL (1873 – 1916)
German astronomer. He established a method for determining stars' brightness from photographs, comparing their visual and photographic magnitudes to obtain their colour index. In 1905 he obtained ultraviolet photographs of a solar eclipse, and went on to study energy transfer in the Sun, deducing that its outer regions had a layered structure. In1916 he showed that, in the general theory of relativity, a sphere of material (approximating to a star) collapsing under its own gravitational field past its Schwartzschild radius would cease to radiate energy (i.e. it would become a black hole). His son, Martin Schwartzschild (1912 – 97), became a naturalized American and studied stellar evolution.

SCHWARTZSCHILD RADIUS
The radius of the event horizon of a black hole. At the Schwartzschild radius the escape velocity becomes equal to the speed of light. The more massive the black hole, the larger the Schwartzschild radius. For a body of mass M the Schwartzschild radius is $2GM?c^2$, where G is the gravitational constant and c is the speed of light. It was first calculated in 1916 by K. Schwartzschild.

SPECIAL THEORY OF RELATIVITY
A theory proposed by A. Einstein in 1905, based on the proposition that the speed of light in a vacuum is constant throughout the Universe, and is independent of the motion of the observer and the emitting body. A consequence of this proposition is that three things happen as an object's velocity approaches the speed of light: its mass goes up, its length shortens in the direction of motion, and time slows down. Hence, according to special relativity, no object can ever reach the speed of light because its mass would then become infinite, its length would become zero, and time would stand still. In addition, Einstein concluded that the mass of a body is a measure of its energy content, according to the famous equation $E = mc^2$, where c is the speed

of light. This equation describes the conversion of mass into energy in nuclear reactions within stars.

STEFAN-BOLTZMANN CONSTANT
(Symbol σ) A constant (appearing in the Stefan-Boltzmann law) that relates the luminosity of a black body to its thermodynamic temperature in Kelvin. It has the value 5.67051×10^{-8} W/m^2/K^4. Also called the Stefan constant.

STEFAN-BOLTZMANN LAW
A law relating the energy emitted by a black body (such as a star) to its temperature; also known as Stefan's law. According to the law, the total energy radiated in watts per square meter is proportional to the fourth power of its thermodynamic temperature in Kelvin; hence a doubling of temperature leads to a sixteen-fold increase in energy output. Expressed mathematically, $E = \sigma T^4$, where σ is the Stefan-Boltzmann constant. The total power per square meter can vary from 3 μW for the microwave background radiation, to 75 MW for the sun, and thousands of gigawatts for hot stars such as whit dwarfs. The law was discovered by Joseph Stefan (1853-93) and derived theoretically by Ludwig Edward Boltzmann (1844-1906).

THERMAL EQUILIBRIUM
1. A state in which two objects, or an object and its surroundings, have the same temperature so that there is no exchange of heat energy between them. For example, a telescope mirror should ideally be in thermal equilibrium with its supports and with the atmosphere to prevent distortion of the mirror or the creation of air currents within the telescope's tube.

2. A state in which the available energy of an object is distributed uniformly among all the possible forms of energy; also known as thermodynamic equilibrium. for example, deep inside a star the radiation field, the kinetic energy, the excitation energy, and the ionization levels will all have equal amounts of energy. Furthermore, all processes are in balance so that, for example, there will be as many ionizations of helium per second as there are combinations of free electrons and helium ions. The condition of local thermodynamic equilibrium is often taken as an approximation when modelling stellar atmospheres.

UNIVERSAL TIME (UT)
A world-wide standard time-scale, the same as Greenwich Mean Time. Universal Time is the mean solar time on the meridian of Greenwich. It is defined as the Greenwich hour angle of the mean sun plus 12 hours, so that the day begins at midnight rather than noon. It is closely linked to Greenwich Mean Sidereal Tim (GMST), since the mean sidereal day is a precisely known fraction of the mean solar day. In practice, UT is determined by a formula from GMST, which in turn is derived directly from such observations of the meridian transits of stars. The version of UT derived directly form such observations is designated UTO, which is slightly dependent on the observing site. When UTO is corrected for the variation in longitude due to the Chandler wobble, a version of Universal Time, UT1, is derived which has genuine world-wide application. When UT1 is compared with International Atomic Time (TAI), it

is found to be losing approximately a second a year against TAI. Broadcast time signals use the time-scale known as Coordinated Universal time (UTC). This is TAI with an offset of a whole number of seconds. The offset is adjusted when necessary by the introduction of a leap second, and UTC is always kept within 0.9 s of UT1.

UNIVERSE

Everything that exists, including space, time, and matter. The study of the Universe is known as cosmology. Cosmologists distinguish between the Universe with a capital 'U', meaning the cosmos and all its contents, and universe with a small 'u' which is usually a mathematical model derived from some physical theory. The real Universe consists mostly of apparently empty space, with matter concentrated into galaxies consisting of stars and gas. The Universe is expanding, so the space between galaxies is gradually stretching, causing a cosmological redshift in the light from distant objects. There is growing evidence that space may be filled with unseen dark matter that may have many times the total mass of the visible galaxies. The most favoured concept of the origin of the Universe is the Big Bang theory, according to which the Universe came into being in a hot, dense fireball about 10-20 billion years ago.

VIRIAL THEOREM

A way of estimating the total mass of an object such as a galaxy or a cluster of galaxies from the movement of its individual members. The theorem states that the average gravitational potential energy of the constituent objects is twice their average kinetic energy. Calculations with the virial theorem show that galaxies and clusters contain up to ten times as much mass as can be seen telescopically, providing strong evidence for this existence of large quantities of dark matter. A modified version of this theorem, called the cosmic virial theorem, applies on cosmological scales. It relates the statistics of galaxy motions an the correlation function (which describes the way galaxies cluster in space) to the average density of the Universe. Since the first two quantities are measurable, the density parameter can thus be estimated. The usual result obtained is around 0.2, indicating that there is dark matter on cosmological scales, but not enough to reach the critical density.

VOGT-RUSSEL THEOREM

The theorem, found valid except in rare circumstances that there is only one internal structure possible for a star of given mass and chemical composition. The calculation of that structure depends on knowing how quantities such as pressure, rate of energy production, and opacity depend on local gas properties such as temperature and chemical composition. The mass-radius and mass-luminosity relations in main sequence stars are among the theorem's consequences. It is named after the German astronomer Heinrich Vogt (1890-1968) and H.N. Russell.

WEIZÄCKER, CARL FRIEDRICH VON (1912 -)

German theoretical physicist and astrophysicist. In 1938, independently of H.A. Bethe, he proposed that stars generate their energy via the carbon-nitrogen cycle, converting hydrogen into helium by nuclear fusion. In 1944 he set out a modern version of the nebular hypothesis proposed in the 189th century, first by I. Kant and later by P.S. de Laplace, to account for the origin of the Solar System.

www.ingramcontent.com/pod-product-compliance
Lightning Source LLC
Chambersburg PA
CBHW080235180526
45167CB00006B/2279